Reinhard Haas · Lutz Mez ·
Amela Ajanovic
Editors

The Technological and Economic Future of Nuclear Power

Editors
Reinhard Haas
Technische Universität Wien
Wien, Austria

Lutz Mez
Freie Universität Berlin
Berlin, Germany

Amela Ajanovic
Technische Universität Wien
Wien, Austria

This open access publication was funded by BMNT Austria, BMVIT Austria, City of Vienna, Austria, Province of Upper Austria, Province of Vorarlberg.

https://doi.org/10.1007/978-3-658-25987-7

Preface

In November 1978 the Austrian Electorate, in a nationwide referendum, decided against the commissioning of the first Austrian Nuclear Power Plant in Zwentendorf. History proved this decision having been rather forward-looking.

Today, 40 years later, having experienced the severe accidents of Chernobyl in 1986 and of Fukushima in 2011, both with far ranging and long lasting severe impacts, quite a number of countries is phasing-out, or has decided to phase-out, nuclear power. Nevertheless, other countries are determined to continue the use of nuclear power, or have decided to embark on a nuclear power programme.

From our point of view, nuclear power cannot be considered compatible with the concept of sustainable development. Consequently, reliance on nuclear power must not be considered a viable option to combat climate change. Sustainable development, if fully applied to the energy sector, would require substantial increases in energy efficiency and in energy saving as well as switch to renewable sources of energy.

However, the global debate on the potential role of nuclear power in the decarbonisation of our energy systems is ongoing. This debate urgently needs to benefit from sound scientific evidence. This book highlights various important aspects of nuclear power from an up-to-date perspective. Thus, we are convinced that this book will serve as an important input to the debate and that it will have a lasting impact.

Elisabeth Köstinger (Minister of Sustainability and Tourism)
Norbert Hofer (Minister of Transport, Innovation and Technology)
Ulli Sima (Vice-mayor, City of Vienna)
Rudi Anschober (Regional Minister, Province of Upper Austria)
Christian Gantner (Regional Minister, Province of Vorarberg)

Acknowledgements

The authors are grateful to Veronika Kiesenhofer, Andreas Molin, Markus Niedermair, David Reinberger and Theo Zillner for their valuable support.

Contents

Legislation

Technical

Nuclear Waste, Proliferation

Major Accidents

Introduction: Why Discuss Nuclear Power Today?

Reinhard Haas, Lutz Mez, and Amela Ajanovic[1]

Introduction

Commercial nuclear power was once depicted as an infinite source of energy to meet growing energy demand. In spite of costs increases, several reactor accidents and remaining challenges regarding radioactive waste, nuclear reactors still play an important role in the energy policy of several countries. However, the future use of nuclear power is a disputed issue in the policy community. There is wide disagreement about the scale of the future use of nuclear power for electricity generation.

While the interest in new nuclear power plants (NPP) in the Western EU-countries and the USA is very modest with few or no plants under construction and planned, there are still talks about new-build in Eastern Europe (Poland, Hungary, Czech Republic, Slovakia, Belorussia, Ukraine), in the Middle East, in Russia, in China and in South-Eastern Asia.

According to the International Atomic Energy Agency (IAEA) 454 NPP are "operational" in 31 countries (as of 11 December 2018). Moreover, 54 reactors were under construction end-2018, with the largest number of new-build in China. The major arguments brought forward in favor of nuclear power are:

---

1 Reinhard Haas, Technische Universität Wien, Austria, haas@eeg.tuwien.ac.at; Lutz Mez, Freie Universität Berlin, Germany, lutz.mez@fu-berlin.de; Amela Ajanovic, Technische Universität Wien, Austria, ajanovic@eeg.tuwien.ac.at

R. Haas et al. (Eds.), *The Technological and Economic Future of Nuclear Power*, Energiepolitik und Klimaschutz. Energy Policy and Climate Protection, https://doi.org/10.1007/978-3-658-25987-7_1

1. There are urgent and huge needs for energy especially for electricity;
2. Nuclear energy is necessary to provide base load
3. Nuclear power is indispensable for combatting global warming.

However, an unofficial argument in some countries is to retain (or build up) their military nuclear applications, particularly weapons but also nuclear submarines. Other non-outspoken drivers can be geo-political interests and mere status symbols, as belonging to the rather exclusive club of three dozen countries—15% of United Nation member states—and mastering what is still considered supreme science remains a goal for some countries or rather regimes.

To understand current and future role of NPP in different countries it is important to learn from the history of nuclear power in electricity generation from a technological and economic point of view. Already in the 1950s, when the economies of many countries (e.g. USA, Europe, Japan, Russia) started to grow fast, and electricity demand grew with the economic development. At that time a desperate need for additional sources for electricity generation beyond limited fossil fuels was seen as essential for continued economic growth. The promises of the nuclear dream emerging in the 1950s were to have cheap electricity in abundance.

In the USA after the early military applications, there was the announcement of the nuclear program "Atoms for Peace" by Eisenhower in 1953. As a consequence, the USA exported reactors to a number of key countries at low prices, launching nuclear ordering. Countries included Germany, France, Spain, Japan, India. The first nuclear plants started operation in the period 1954–1956 in the UK, the Soviet Union and the USA. In the USA "Atoms for Peace" was accompanied by the slogans "too cheap to meter" and the "all-electric home". The idea was that most efforts in households, agriculture and other sectors could be managed by electricity and that electricity could be produced at such low marginal cost that metering would not be justified.

However, as described in this book – given a broad range of subsidies revealed – it is more than doubtful whether such cheap nuclear power has ever existed and prospects of the future competitiveness of nuclear energy appears jeopardized by competition from the surprisingly rapid expansion of renewable energies.

Yet, up to the mid-1970s, nuclear power was booming in many countries and plant by plant was added. Hopes were strong that costs would start coming down with increasing experience.

In the history of nuclear power, there have been three major shocks to the system: The first was the accident in Three Mile Island in the USA in 1979. While reduced economic support slowed construction already before, this accident stopped new nuclear power plants constructions in the USA for a long time. The catastrophe beyond the maximum credible accident in Chernobyl in 1986 was a major set-back for the nuclear power industry in Europe. However, the list of large orders has ended already before Chernobyl. The accident in Fukushima in 2011 accelerated Germany's program of phasing-out completely all NPPs by 2022, and profoundly impacted other countries, e.g. Belgium, China, South Korea, Switzerland, Taiwan. Many other accidents with radioactive contamination have happened in the past, however without getting the same attention, e.g. the Kyshtym disaster, and the Windscale fire both in 1957.

Yet the issue of safety is only one amongst a series. Aside from the issue of safety, technical risks, security issues and the unsolved problem of nuclear waste disposal and the economic performance are currently the major barriers for acceptance of NPP. Costs, especially the construction costs of NPP have been increasing since the 1970s and in recent decades they have skyrocketed especially in European countries and North America. Construction time has also become even more subject to overruns. At the same time the costs of renewable energy technologies such as photovoltaics and wind turbines have significantly decreased.

In light of these still falling costs the economic performance of nuclear power in comparison to these renewable technologies is also getting worse. It already has become much harder for nuclear investors to recover money in competitive electricity markets The "base-load" concept has lost its meaning in these markets when low cost renewable sources outperform any thermal power plant. Even the pure operation & maintenance costs are difficult to recover in today's electricity markets leading to more unfavorable prospects for future nuclear competitiveness.

The increasing construction costs and durations have exacerbated the major problems faced by nuclear power plant vendors. The French vendor AREVA went bankrupt, was bailed out and broken up by the government in 2017. In the same year, the US-based Westinghouse (then owned by Toshiba) declared bankruptcy and, apart from Russian and Chinese vendors, none of the historic vendors has a healthy order book.

Around the turn of the century there were forecasts of a 'nuclear renaissance' or at least of 'rising expectations' based on reactor designs evolved from existing ones, so-called Generation III+. This temporarily revived interest in nuclear power but this new interest has generally not been translated into firm orders. This was sometimes supported by vision of new "cheap and safe" radical new technologies of a so-called generation IV reactors that have been discussed since the early 2000s. "Small Modular Reactors" (SMRs) are the most recent flavour of the year, but they are still technologically underdeveloped, and financially out of reach.

The core objective of this book is to provide a comprehensive appraisal of the technical and economic aspects of nuclear power in the next decades. It discusses whether the claimed renaissance of nuclear is really on the horizon or whether the eroding economics as well as technical, legal and industry/vendor issues will continue to close the curtain for this technology. It is organized as follows.

In the next chapter a historical review of the Nuclear Dream is conducted by Rosaria di Nucci from military to early civilian applications. In the following a comprehensive analysis of recent developments and the current state of the world nuclear industry is provided by Mycle Schneider. The collision of atomic and flow renewable power in decarbonization of electricity supply is analyzed by Aviel Verbruggen and Yuliya Yurchenko.

One of the most heavily discussed current issues – costs and economics – are analysed by Reinhard Haas, Steve Thomas and Amela Ajanovic. The major focus of this chapter is on the question, why the real investment costs of NPPs have increased at such a high rate. The specific aspects of one of the most prominent disputed new planned NPP – Hinkley Point C – and its economics compared to renewable energy technologies are investigated by Gustav Resch and Demet Suna in the chapter following.

Another economic issue, which is often ignored, is the cost of nuclear decommissioning. It is treated in two chapters – Wolfgang Irrek analyzes the problem of "financing nuclear decommissioning", and Christian von Hirschhausen, Jan Paul Seidel, and Ben Wealer investigate the decommissioning of NPP, storage of nuclear waste and provide the experiences from Germany, France, and the UK.

Dörte Fouquet reflects on the legacy around the setup and foundation of the EURATOM treaty, the clashing economic nuclear interests of France and the United States on civil use of nuclear power and the respective technologies, the limits and shortcomings of this treaty and its strong support angle at odds with the liberalization of the European energy market.

Tomas Kåberger analyzes the legislation of the economic liability for accidents and back-end costs. David Reinberger, Amela Ajanovic and Reinhard Haas provide a historical overview on nuclear technologies with special focus on the intended new concepts of Generation IV.

The problem of nuclear waste and its long-term disposal is investigated by Gordon MacKerron. As China is currently the most important country world-wide for the development of nuclear power in this book it is treated by M.V. Ramana.

Regarding the large accidents in Chernobyl and Fukushima, three chapters are dedicated to these negative highlights in nuclear history. Nikolaus Müllner analyses the technical versus the human causes about three decades after Chernobyl. The reality after Fukushima in Japan with focus on actual damage to local people is investigated by Tadahiro Katsuta. Eri Kanamori and Tomas Kåberger analyze Japan's experience the distribution of the costs of nuclear core melts about six years after Fukushima.

An outlook on the prospects of a future democratic and sustainable electricity generation system is provided in the last chapter by Reinhard Haas and Hans Auer.

From Military to Early Civilian Applications

An Appraisal of the Initial Success of the Light Water Reactor Technology

Maria Rosaria Di Nucci[1]

Abstract

The article presents a historical overview of the development path of nuclear technology, from its military application to its civilian start up (Atoms for Peace) and early commercialisation of nuclear power plants. The chief aim is to demonstrate and analyse the commercial exploitation of nuclear energy and the beginnings of the nuclear industry by means of the link with the military research and development and production and its success based on tight industry-government integration. We describe the gradual growth of national nuclear frameworks and of the industry as a result of a combination of exogenous and endogenous factors originating with the military spill-over effects. These factors evolved during the subsequent phases of the technology development following the transfer of technological know-how from military establishments to civilian atomic agencies and the creation of a private industry.

1 Maria Rosaria Di Nucci, Freie Universität Berlin, Germany, dinucci@zedat.fu-berlin.de

R. Haas et al. (Eds.), *The Technological and Economic Future of Nuclear Power*, Energiepolitik und Klimaschutz. Energy Policy and Climate Protection, https://doi.org/10.1007/978-3-658-25987-7_2

1 Introduction

This chapter presents a historical overview of the development path of nuclear technology, from its military application to its civilian start up, under the aegis of *Atoms for Peace* and early commercialisation of nuclear power plants. The chief aim is to demonstrate and analyse the commercial exploitation of nuclear energy and the beginnings of the nuclear industry by means of the link with the military research and development and its success based on tight industry-government integration. Military nuclear technology and production issues were decisive not only in the USA, but also in the UK and in France, where fissionable material was indispensable for nuclear weapons.

The gradual growth of national nuclear frameworks and of the industry are described as a result of a combination of exogenous and endogenous factors originating with the military spill-over effects. These factors evolved during the subsequent phases of the technology development, following the transfer of technological know-how from military establishments to civilian atomic agencies (occurring first in the US in 1946, and later in France and the UK in 1954), and the creation of an independent private industry.

The key development stages of military and civilian applications will be analysed chronologically. The primary concern will be the US, British and French experiences, due to their early start and close military ties. In addition to technology development, we briefly describe the agreements, policies and regulations of the US government and the International Atomic Energy Agency (IAEA) which affected the domestic and international marketing of nuclear power plants. Looking at the international framework will allow us to consider the role of the international organisations fostering the development of nuclear power such as the IAEA, the European Atomic Energy Community (Euratom) and the schemes that acted as vehicles for the international sale of nuclear power plants. Afterwards, we look briefly at two national frameworks. Britain receives more attention because initially it was the only country competing with the US in size and government backing in the international market.

The discussion of early military and civilian deployment of nuclear energy will be roughly subdivided into the following stages:

1. Initial military applications characterised by US leadership of the allies who took part in the Manhattan project leading to the A-bomb (Britain, Canada, France and the US), roughly corresponding to 1940–47;
2. Co-existence between military and civilian applications preceding commercial development, characterised by the creation of a national public civilian

Atomic Agency. Knowledge accumulated under the military was transferred to the Atomic Agency, but tight military control was maintained in the choice, implementation, and deployment of the technology. During this stage, basic and applied research leading to reactor prototypes was conducted and several different technologies were concurrently tested with the primary goal of obtaining plutonium for military purposes;
3. Exclusively civilian applications. Here the focus is on the evolution of a specific industrial autonomy and the establishment of the market duopoly by Westinghouse and General Electric (GE). Their marketing policy was articulated through pre-existing and newly established licensing agreements with European countries and Japan.

In analysing the success of Light Water Reactor (LWR) – the American technology par excellence – and its establishment in the world market, we argue that the success of LWR can be explained by examining the interplay of various factors at work in the US market, along with certain exogenous influences in both domestic and export arenas.

The early technological and economic success of LWR technology can be attributed to a rather heterogeneous set of interdependent factors. In this analysis, however, we find it useful to separate these factors and determine which had the greatest impact at any given time. Some of them, like the military spill-over effect, can be considered exogenous to the industry or as a socialisation of its costs. Others, like the industrialisation nurtured under the US Atomic Energy Commission (AEC), can be viewed as the result of a dynamic interrelation between organisational, industrial and institutional factors. Although at these stages, the industry's role was limited to receiving government contracts and procurements and/or working under strict military control, this period marks the beginning of a close interdependence between the military establishment (and later the atomic agencies) and industry that enhanced growth of internal capabilities and nurtured what can be called the military-industrial complex.

2 Early military and civilian applications in the USA[2]

Early military applications

With the US entry into war in 1942, and amid widespread fears that Nazi Germany was progressing in atomic research for military ends, the American authorities stepped up research on the military application of nuclear physics. In September 1942, the US Army formed an organisation known as the "Manhattan Engineering District" under General Grove. The District's task was to reorganise work formerly undertaken by the National Defence Research Committee, which was established in 1940, and subsequently part of the office for Scientific Research and Development. The "Manhattan Project" pooled together the best scientists of North America and Western Europe. The programme of the Manhattan District was initially managed by the Army Corp of Engineers, before responsibility was entrusted entirely to the US Army in 1943.

The discovery of Plutonium 239 in December 1940 and its devastating explosive potential catalysed great efforts to build nuclear reactors whose main purpose was the production of plutonium. It should be noted that the military-scientific establishment encountered severe problems in obtaining self-sustaining nuclear chain reactions. In particular, the production of sufficient quantities of uranium and graphite in the state of purity required by laboratory and industrial operations was well beyond that currently available. By 1942, the first chain reaction took place at Fermi´s Chicago Pile-1 (CP-1) using natural uranium as fuel and graphite as a moderator, with a power of 200 Watts.[3] This experiment demonstrated the feasibility of using atoms as an energy source, although applications for civilian use were to be a long time in the making.

In fact, a period of strict secrecy was ushered in, during which the atomic project had top military priority which led to a pooling of scientific and financial resources. The construction of additional, larger reactors quickly followed, such as the reactor at Oak Ridge and a second "Chicago Pile". A third offspring of Fermi´s pile used enriched uranium and heavy water (in place of graphite as a moderator) to produce 300 kW of power. To put it with Wellersten (2017) "virtually overnight, the University of Chicago had become a major wartime contractor".

Subsequently, several additional plutonium reactors of the so-called "Hanford" type were built under the control of the Manhattan District and under management of the DuPont Company. Natural uranium was deployed in the form of

2 This section draws on my dissertation (Di Nucci 1986), on Sanger and Wollner (1995) and De Wolf Smyth (wd).

3 http://www.atomicarchive.com/History/firstpile/firstpile_01.shtml

rods encased in aluminium, with graphite as a moderator and water as a coolant, while uranium concentrate was obtained from a diffusion plant at Oak Ridge. This work was backed up by scientific research performed at the University of Chicago (Argonne National Laboratories), Ames Laboratories and other universities. The chemistry of plutonium was studied at the University of California in Los Angeles (UCLA) while Columbia University handled the experimental nuclear data central to the uranium technology. Reactor construction and the entire nuclear fuel cycle were characterised by a close interaction and cooperation between scientific and university establishments and supported by industrial groups responsible for the production of key equipment (bound by military secrecy).[4]

This spectacular pooling of resources led to a rapid advance in pure and applied scientific knowledge and in engineering skills as well as innovative production techniques. Increased confidence in the programme's potential paved the way for a special group based out of UCLA to take charge of the design, construction and testing of the atomic bomb. This project, known as the Los Alamos Laboratory, began operation in 1943. At its disposal were minimal amounts of fissionable materials, before more could become available from the first working reactor.

By 1944, Los Alamos had its own reactor for use in research on the A-Bomb. Uranium was present as a solution in the form of uranyl sulphate, rather than in metallic form. The reactor was referred to as the "water boiler". Its explosive properties were rapidly exploited and brought under control, culminating in the successful nuclear test in New Mexico in July 1945. In August 1945, the first uranium 235 bomb fell on Hiroshima, to be followed by a plutonium bomb on Nagasaki three days later.

The transition period

With the end of World War II, it became clear that nuclear energy would open a new era. Under President Truman, the US decided to keep the development of nuclear energy under secrecy and control in the name of national and international security. To ensure this, a domestic regulatory framework was required. It was provided by the May-Johnston Atomic Energy bill of 1945, which advocated military control over atomic technology and information. The bill further underlined the military use of the new technology and confined industrial application within highly restricted boundaries. Against the background of the Soviet hydrogen bomb tests in the 1950s, nuclear weapons remained a highly delicate and divisive political issue. Conservatives feared the loss of US military supremacy while liberals feared nuclear war. The scientific community was similarly divided.

4 See news releases about Argonne's nuclear science and technology legacy, http://www.ne.anl.gov/About/hn/

Meanwhile, the countries that had contributed to the Manhattan Project were denied access to the technology, due to the American fear of espionage. This put an end to US-UK nuclear cooperation in 1946, except in the field of uranium ore procurement. While there was no evidence to suggest that the other countries involved in the Manhattan Project could not independently handle the theoretical aspects of nuclear energy, the predominant view was that the US would retain its leadership through its greater experience and sheer weight of human capital, infrastructure, overall organisation, industrial planning and supply. However, the scientific community, including many scientists and engineers who collaborated on the project leading to the bomb, saw the military supervision and control as an unjustifiable interference. The critique and subsequent protest against the May-Johnston Bill led to a compromise in the form of a committee composed of scientists, engineers, industrialists and politicians chaired by Senator McMahon. After many hearings, the committee concluded that nuclear energy was essential for both national defence and industrial growth. It recommended that this development be entrusted to a civilian commission with access to the knowledge obtained through the A-bomb research. The resulting bill went through various amendments before being passed in August 1946 as the "Atomic Energy Act".

The United States Atomic Energy Commission (AEC) was established to nurture and control the development of nuclear science and technology and its civilian applications. The McMahon Atomic Energy Act, signed on August 1, 1946, transferred the control of nuclear matters from military to civilian hands. The AEC was entrusted with the control of the plants, equipment and laboratories which were involved in the production of the atomic bomb. The transfer of the atomic establishment into civilian hands hardly diminished the conflicts between the military and working scientists, a considerable number of which resigned from the laboratories and plants.[5] The transfer of knowledge, plants and equipment was to be overseen by five commissioners who were approved by the military and the president. Yet, disagreements were rife and the commission sometimes refused to approve the requests made by army officials for exemptions from transfer. When General Eisenhower became president in 1953, the Joint Chiefs of Staff of the armed forces recommended that the president emphasise the superiority of the US stockpile of nuclear weapons, while science advisors recommended a programme called "Operation Candor" to communicate the dangers of nuclear war (Etzkowitz. 1984: 419). The concept of a nuclear industry was then proposed.

5 President Truman signed an executive order to transfer the Manhattan Engineering District on December 31, 1946.

The start of the civilian applications

A civilian nuclear industry was established in part to legitimatise the development of nuclear weapons. Eisenhower's "Atoms for Peace" speech to the United Nations on December 8, 1953, marked the beginning of large-scale US government funding to develop civilian nuclear power plants (Camilleri 1977).

Most applications were developed through the government's national laboratory system, in which Argonne played a key role. The AEC launched the "Power Reactor Demonstration Program" (PRDP), and within this framework a number of various demonstration reactors (light water, gas cooled and breeder reactors) were built.

At the time the AEC was born, the US assets in nuclear energy were constituted by:

1. The Oak Ridge and Clinton Laboratories, working on gaseous diffusion and producing enriched uranium;
2. The Chicago Group, which continued to carry out work on the Hanford reactor but also handled research on the fast breeder reactor (Argonne Laboratories), with General Electric taking over from DuPont. In Chicago, there was also a project using natural uranium and heavy water as a moderator, in addition to fabrication techniques for producing and testing alternative moderators such as beryllium;
3. The Manhattan District, which had also begun studies on various combinations of fuel-moderator-coolant for different types of reactors, in particular the gas-cooled natural uranium reactor deemed feasible for civilian use;
4. The Knoll Atomic Power Laboratory at Schenectady, which had been organised in 1949 under contract for civilian development in the years to come; and
5. GE, to promote nuclear power for ship propulsion, under direct control of the Navy[6].

Ahead of Eisenhower's "Atoms for Peace" speech, the AEC started considering civilian nuclear power in 1952. By 1953, it was making plans to build a power reactor in collaboration with industry. Plans to develop nuclear-powered surface ships were abandoned in order to provide funds for the civilian project (Cowan 1990: 561).

In 1961, the AEC Development Program was upgraded to include the subsidisation of large scale reactors (400 MWe and larger), in an effort to realise economies of scale and reduce capital costs. In 1962, the Joint Committee on Atomic Energy responded to growing pressures from industry and utilities by specifically allotting

6 A similar project was undertaken at Oak Ridge under the Air Force, the Manhattan District and the Fairchild Aircraft and Engine Corporation's control. Finally shelved in 1961, the project was known as NEPA: Nuclear Energy Propulsion for Aircraft.

$20 million in subsidies from the AEC to the design, construction and operation of large scale LWRs (Burness et al. 1980: 189).

3 Military spill-over: LWR as by-product of submarine research

The development of LWR technology is intimately connected with military applications. Companies involved in its development are positioned to profit from the economic advantages and scientific know-how nurtured within the military-industrial establishment (Di Nucci 1986, Cowan 1990). Following World War II, Navy submarine's reactors and their operating procedures became the prototype for the most widespread commercial nuclear power plants.[7]

The spill-over effect by the US Navy significantly influenced technological development in the years to come, and military officials played a key role in this new development pathway. Admiral Rickover, known as the "Father of the Nuclear Navy", was put in charge of the US naval propulsion programme in 1946. In May 1946, Rickover, who originally had been assigned to work with General Electric (GE) at Schenectady to develop a nuclear propulsion plant for destroyers, started pushing the idea of nuclear marine propulsion. Subsequently, Rickover became chief of a new section in the Bureau of Ships, the Nuclear Power Division, and began work with Weinberg, the Oak Ridge director of research, both to establish the Oak Ridge School of Reactor Technology and to begin the design of the pressurised water reactor for submarine propulsion (The Economist 2012).

While the earliest studies were performed at Oak Ridge, work was transferred to Argonne in 1948 and development was jointly taken over by Argonne and Westinghouse.[8] That same year, Argonne's Naval Reactor Division was established. Whereas Argonne scientists and engineers performed much of the early research, design and feasibility studies, Westinghouse improved and implemented the designs, first in the S1W prototype at NRTS, and then in the Nautilus submarine reactor. The first test reactor plant, a prototype referred to as S1W, began operations in 1953 at the

7 The difference between a submarine reactor and a PWR for electricity generation is that they have a high power density in a small volume and run either on low-enriched uranium (as do some French and Chinese submarines) or on highly enriched uranium (>20% U-235). U.S. submarines use fuel enriched to at least 93%. (Wiki)

8 Argonne is a direct successor of the University of Chicago's Metallurgical Laboratory, where Fermi supervised construction and testing of the Chicago Pile 1.

Naval Reactors Facility in Idaho. Bettis Laboratory and the Naval Reactors Facility were in charge of the reactor operation and were managed by Westinghouse.

According to official documents (Argonne, w.d.), researchers faced difficulties in designing a high-efficiency nuclear reactor small enough to fit in a submarine hull and still produce enough energy to drive the vessel. They used high-pressure water to cool the reactor core, a breakthrough in reactor technology. The first prototype, Submarine Thermal Reactor Mark I, was completed in 1953 by Westinghouse. STR Mark II was installed in 1954 in the USS Nautilus, the world's first atomic-powered submarine.

A second type of reactor was installed on the submarine USS Seawolf (SSN-575). It was initially powered by a sodium-cooled S2G reactor and supported by the land-based S1G reactor at the Kesselring site, under Knolls Atomic Power Laboratory and operated by GE. An additional S2G was also built, but never used. USS Seawolf was plagued by super-heater problems and thus the higher-performing USS Nautilus was selected as the standard US naval reactor type. Even though GE's technology was not a success, the corporation gained the experience necessary to enter the civilian market with the LWR technology through its participation in the Navy programme. (Di Nucci 1986; Cowan 1990). The efficient safety and control methods mandated by the Naval Reactor Program were transferred to the civilian market: another notable spill-over from military research.

By 1962, the US Navy had 26 nuclear submarines in operation and 30 under construction. Further development of LWR technology was based on the experience from the naval nuclear programme and on a series of experiments performed using the nine Argonne research reactors (Argonne, w.d.).

The early choice of LWR for the US Navy Program provided for substantial learning about this technology at a very early stage. Thus, by the time the civilian programmes started in the early 1960s, the LWR technology was "well advanced along its learning curve while the other technologies were late entrants which failed to catch up" (Cowan 1990: 545). LWR submarine technology was shared with the United Kingdom, while technological development in France, China and the Soviet Union proceeded independently.[9]

9 Rolls Royce built similar units for Royal Navy submarines and then developed the design further to the PWR-2. The Soviet Union concentrated also on PWR in submarines and never engaged in the development of Boiling Water Reactors (BWR).

4 The establishment of the international atomic framework[10]

At the time of the first Geneva Conference in 1955, there were roughly 100 different kinds of reactors under scrutiny (Cowan 1990). Approximately 70% of the development programmes for nuclear reactors were in military hands. However, by the second Geneva Conference, only 12 reactor types were being seriously considered (Mullenbach 1963: 38–39).

At the second conference, negotiations began for the establishment of an International Atomic Energy Agency under the auspices of the UN. Also under way were negotiations for a set of bilateral agreements between the US and other nations for cooperation on the development of civilian use of atomic energy. The US set aside 200 kg of Uranium 235 to assist in international R&D under the aegis of the "Atoms for Peace" programme.

By the end of 1961, 37 bilateral agreements were in effect between the US and other countries and 24 grants were available for the construction of research and experimental reactors as well as laboratory equipment. Though the 1955 Geneva conference was hailed as a breakthrough for the peaceful exploitation of the atom, its outcome was rather limited in terms of transfer of technology and international cooperation. What it did achieve was the legitimation of nuclear power, inducing an optimism in many parts of the world that a plentiful supply of cheap energy was around the corner. Difficulties with the new technology, especially on the production side, were downplayed or overlooked. As a result of the strict secrecy of the previous military development stage, the industry was not yet in a position to master the technicalities associated with scaling up from the prototype stage to commercial sized plants (Cohn 1997, Di Nucci 1986).

In the years following the Geneva Conferences, numerous international organisations were established with the aim of fostering cooperation between countries. Following the New York Conference on Atomic Energy, the International Atomic Agency (IAEA) was created in 1956. Based in Vienna, its goal was to enhance and increase the contribution of nuclear energy for peaceful purposes. Of the initiatives involving Western Europe, many can be considered as extensions of the Marshal Plan and of the Organisation for European Economic Cooperation (OEEC). OEEC countries gave life to a European Nuclear Energy Agency in 1957, unifying their legal restrictions and nuclear regulations.

10 This section draws heavily on Di Nucci (1986).

Alongside this development, the six countries that formed the Coal and Steel Community set up the European Atomic Energy Community (Euratom) in 1958.[11] Though the IAEA was not to substitute for cooperative agreements between countries, this agency was expected to establish safeguards, foster the growth of nuclear energy and the exchange of information and serve as a supply agent for materials, services and equipment. Despite its ambitious aims, the agency achieved limited results. It drew up regulatory procedures for nuclear reactors, but never had the power to implement them, nor did it achieve concrete results in connection with third party liability. Euratom launched an ambitious nuclear programme envisaging 15 GWe by 1967, but this was revised in 1960 to 10 GWe of installed capacity by 1970. In addition to the establishment of four joint research centres, a number of initiatives were adopted within the Euratom framework; the most important of which was the joint programme with the US.[12] This established the terms of cooperation between the US and Euratom member states in nuclear R&D and reactor construction. Notably, US manufacturers willing to participate were then obliged to supply design and cost specifications to Euratom and to set up licensing agreements or subsidiaries within Euratom countries. The latter took place between 1957 and 1960. Although minor in terms of American penetration in the European market (the programme resulted in the sale of only three nuclear power plants, partly backed by the Export-Import Bank), it was of great significance for the transfer of LWR technology to Europe.

The first invitation for proposals on nuclear plant construction was jointly made by Euratom and the US AEC in 1959. This joint programme was decisive for the US industry because it required that proposals for nuclear plant construction include a reactor type on which R&D had already been carried out to an advanced stage in the US. For projects to qualify and be approved, one or more US manufacturers and one or more Euratom member countries had to pay a determinant role in the construction of the nuclear plants. The selected and approved projects were eligible for loans from the Export-Import Bank at preferential rates. The fuel could be purchased by Euratom from the US AEC on a deferred payment basis, while the US AEC provided fuel burn-up guarantees. The Euratom Supply Agency entered into long term contracts with the reactor operators.

11 The signing of the Treaty was preceded by the release of the report "A Target for Euratom" which recommended the cooperation between Euratom and the U.S. nuclear reactor programme. This cooperation can be seen as another vehicle for the penetration of LWR technology, ousting the French and British gas-cooled technology (Cowan 1990, Di Nucci 1986)

12 For details, see Lucas (1977) and the inquiry known as "The three wise men report" leading to the establishment of Euratom.

Given the availability of technical information, the "joint programme" went further than previous initiatives, for it worked exclusively through licensing agreements and joint ventures which allowed European firms to achieve a gradual build-up of nuclear capabilities and to carry out subsequent autonomous nuclear R&D.

5 Other national frameworks

In the late 1950s, the United Kingdom and France were independently developing gas graphite reactors – drawing knowledge and experience from the Manhattan Project – and Canada was working on heavy water reactors. In both European countries, the technological option was also influenced by military concerns. The US' near monopoly on uranium enrichment technology left France and the UK no other choice but to develop natural uranium technologies for their civilian power programmes. In the following, we analyse the political and technical contexts in these two countries.

Significant emphasis has been laid on certain factors which are the success of the US nuclear framework and technological choice, as compared with other national paths such as the British one. In this context, the work of Burn (1967, 1978) provides a detailed and complete analysis which represented a reference for those advocating the benefits of free market forces for technological development. Burn attributes successes to policies inspired by a mixture of free market criteria and industrial promotion, but hardly considers that the US reactor vendors strongly profited from the R&D and subsidy support within AEC "infant industry" development strategy (Cohn 1997: 75).

In the USA, nuclear fuel and uranium enrichment remained a government preserve and was highly subsidised (Mullenbach 1963), a fact which gave the US industry a kind of artificial competitiveness vis-à-vis other countries that at that time could not hope to match. A questionable feature of Burn´s analysis is his account of the handing-over of the US AEC´s knowledge to the market and the "industrialisation" of the experience matured under the US AEC. Burn (1967) shares the same criticism as Mullenbach (1963) that the technological choice was functional to the possibilities of the national industry and that the US nuclear policy was in some sense subordinate to the interests prevailing in the industry. Our claim, rather, is that while it is true that the American industry displayed internal economies, it is also the case that this was achieved on the heels of a phase in which military targets received absolute priority. Only later was it possible to pursue the needs for a civilian development of nuclear power. The high priority assigned to military targets in the

USA remains therefore one of the chief factors explaining the development. Britain, France and Canada were countries which – like the USA – had a military nuclear experience, but one that began later, was more limited in the extent of its operations and thus lagged behind the USA in terms of its developments and spill-over effects.

Another factor distinguishing the European from the US development – one which Burn (1978) emphasises – is the incubation period provided for nuclear technology. While the US authorities allowed for the parallel development of the most promising projects before the most feasible technologies – in the economic, military and technical sense – emerged in the market, France and Britain ventured into a sort of technological wager which led to the pursuit of single projects to an advanced stage, beyond the point at which they could be easily be shelved in the case of limited commercial and technical success.

The United Kingdom[13]

As in the US, nuclear power in the UK began with the military. In 1941, a team of British scientists established the general feasibility of a bomb design and reactor construction. By 1942, it had become clear that research cooperation with the US would significantly benefit both nations. Once the Manhattan Project fell under a cloud of strict secrecy, however, Britain had difficulties accessing US laboratories and was forced to rely on independent R&D. The US Atomic Energy Act of 1946, with its tight security restrictions, thus effectively deprived Britain of access to the technology following World War II.

A British Atomic Energy Act became effective in November 1946 and allowed for nuclear development under largely similar conditions to those in the US, though without the same harsh penalties for security violations. Responsibility for production, use and disposal of nuclear material was assigned to the Ministry of Supply (MoS). Government authorities established a strictly centralised control over all the activities connected with the development of nuclear energy. In 1948, the Radioactive Substance Act entrusted to the MoS complete authority over the control and use of all radioactive substances, including the regulation of their import and export.

Nuclear power development was pursued by the Atomic Energy Production Division of the Atomic Energy Department of the MoS at Risley. The first scale reactors were built at the Atomic Energy Research Establishment at Harwell in Oxfordshire, which was the established in 1946 under the MoS. The reactors were planned as prototypes exclusively to produce plutonium for military purposes. They used natural uranium and graphite as a moderator; they were first cooled

13 This section draws on Burn (1978; 1967) and on Williams (1980).

by air, then by gas. The original plan to have water cooling, as in the US Hanford reactors, was dropped for safety reasons and for the lack of suitable sites (Gowing 1964; Williams 1980). By the end of the 1940s, British efforts had resulted in a plant for the manufacture of uranium, located at Springfield, and facilities for uranium enrichment and the production and separation of plutonium.

In 1948, Britain's Harwell Group initiated a project to study the steam aspects of dual reactors, with the aim of awarding a construction contract to the most successful firm. However, conflicts with the Risley group impeded the project. The main objections were based on the need for plutonium for military applications and the fact that Lord Hinton – who would a decade later become the first head of the newly created Central Electricity Generating Board – and the Risley group mistrusted the industry's reliability (Gowing 1964: 185–190). All the same, data was accumulated at Harwell from feasibility studies that lasted from 1951 to 1953. This resulted in the design of a dual-purpose reactor to be used jointly for the production of electricity and plutonium. With the growing demand for plutonium, the government approved a proposal to build the Harwell-designed reactor in February 1953, a decision that led to the first Calder Hall reactor, known first as Pippa and later as Magnox. It used natural uranium, graphite as a moderator and carbon-dioxide as a coolant. According to Gowing (1964), while there was plenty of support for the LWR option and for later development of Heavy Water Reactors (HWR) and High Temperature Reactors (HTR), the LWR solution was dropped because its potential for being scaled up was not recognised. Thus, what Burn (1967) calls the "Lord Hinton approach" set in, namely a concentration of efforts on the gas-graphite technology with an eye towards fast breeder reactor technology, and, importantly, fissile material production for the UK's developing a nuclear warhead programme. The Calder Hall reactor began producing electricity in July 1956. The work itself was the Harwell and the Risley Groups, since all development of the technology had been fully centralised in government establishments.

In 1954, and in parallel to the change in the US Atomic Act, a new legal framework was established in Britain. The change was less radical than that in the US since it amounted to amending and extending some points of the existing acts in order to establish a civilian atomic energy authority and assign suitable power and liabilities. Unlike the US, there were no special provisions for declassification of information; the 1946 Act had already laid guidelines for an automatic process of declassification for matters of no strict significance for defence purposes. The UK Atomic Energy Authority (UKAEA) was created to embrace both civilian and military atomic activities and to act as a consulting agency on all nuclear affairs. The new agency was entrusted with R&D, including fuel elements, the prototype stage and the phase leading up to the construction of the first commercial plant.

Only at that point was it envisioned that industrial firms would enter in the field and realise projects based on UKAEA R&D results. With the Calder Hall plant in its early stage of development, it was nevertheless decided that this would constitute the grounds for a large scale nuclear power programme. The provisional nuclear power programme was announced in a government White Paper in February 1955 and anticipated the construction of 12 plants with a total capacity of 1,500-2,000 MW by the end of 1965. The first plant was to be of the Magnox type.

The Harwell Group's experience in design, research and preliminary work pointed to the advantages of an integrated system in which the civilian, mechanical and nuclear parts of the plant would be jointly designed. Difficulties arose, however, in developing the industry and in establishing a coordinated approach for design and construction. The industry was handicapped by its virtual ignorance of almost nine years of nuclear growth, and further limited by the lack of a precise framework for collaboration with the UK AEA.[14]

The UK AEA retained for itself most of the reactor development and prototype construction; industry was allowed to undertake further design and development in connection with its role in full scale plant construction. Thus, when contracts were awarded to the chosen consortia, the UK AEA acted as a consultant on a turnkey basis and the industrial consortia undertook most of the R&D required to improve the Calder Hall technology. The UK AEA began R&D on other technologies like the advanced gas cooled reactor (AGR), but it simultaneously encouraged industry to take on its own R&D with the aim of contracting successful technologies in the future. Following the Suez crisis, the atomic energy plan was tripled and the power of projected plants was raised to 5,000-6,000 MW by 1966, with each plant to be as large as technically feasible with the expectation of reducing costs (Burns 1978). The five plants using Calder Hall Magnox technology were built by five different groups and involved design changes such that each could be regarded as a prototype. The first "commercially" operated Magnox plant was technically modified, at the instruction of the Ministry of Supply to optimise plutonium production for future UK military use, and later export to the U.S military nuclear programme run by the U.S AEC (Lowry 1989). At the end of 1957, the introduction of a new AGR technology was announced. It was based on the Magnox design but used enriched uranium. With this, a second nuclear programme began in 1965 with the contracting of the Dungeness B plant. While the choice of AGR received some criticism, it was preferred to LWR for technical and political reasons. However, the

14 According to Burn (1978: 277–78) some of the consortia were seeking a greater degree of freedom and eventually to take up licensing arrangements with US companies.

emphasis on this technology over all others brought the UK AEA to technological and commercial disaster.

According to Burn (1967, 1978), the commercial failure of the British technology can be attributed to both design and project management. The AGR reflected three fundamental mistakes in decisions made between 1955 and 1957. They were:

1. to have started a nuclear plan exclusively based on the Magnox technology;
2. to have tripled it by 1957;
3. to have limited the subsequent R&D to gas-graphite reactors.

The results of the implementation of AGR technology was to concentrate the industry on the production of systems without a secure future. The public monopoly over R&D was responsible for the excessive rigidity of the structure, which according to the plans, should have been highly dynamic.[15]

Although the UK AEA assisted potential buyer countries in obtaining credits extended for five years from the commissioning of the nuclear plant, the only successful bilateral agreements were the two signed contracts with Italy and Japan. These resulted in the sale of a 200MW and a 150MW Magnox plant, respectively. Thereafter, Britain failed to capture any orders on the international market. Officials often justified this failure by alleging unfair competition from the US government, citing the US' preferential loans for construction, artificially low fuel prices, exceptionally good terms for reprocessing the fuel and guarantees backed by the federal government.

France[16]

Like Britain and the US, France had a military start to atomic energy development. The early French nuclear reactors were designed and built to produce plutonium. Unlike the US, which attempted to separate civil and military uses of nuclear power, France has never separated the organisation of nuclear energy and nuclear weapons. "This has remained the underlying rationale until today" (Schneider 2008:8).

Following the end of WW2, the Commissariat à l'énergie atomique (CEA) was established as a highly efficient agency regulating the use of nuclear energy. However, the general orientation of the agency, with its strong military bias in R&D, led to the retirement of Juliot-Curie who had been among the pioneers of

15 For a detailed analysis of the policy aspects and the role of the government in the development of AGR technology, see Williams (1980).

16 This section draws on Scheinman (1965), Di Nucci (1986) and Schneider (2008).

the nuclear discovery, but whose political convictions were incompatible with the military emphasis of the CEA.

In 1952, the first 5-year plan for atomic energy was launched by the Secretary of State. It was based on the production of plutonium for military ends by dual purpose reactors. Around the same time, a Commission for the production of electricity of nuclear origin (PEON) was established to liaise with the CEA, the nationalised electric utility Électricité de France (EdF) and the industry engaged in the production of nuclear components. In response to General de Gaulle's ambitious pursuit of *grandeur*, the atomic agency objectives became to further the country as a military power by means of atomic weapons and, given its limited internal resources, to free it from dependence on foreign supply and technology. Due to the excessive cost of uranium enrichment plants and to the US ban on the export of enriched uranium, France faced similar limitations as Britain in its choice of moderator, fuel and coolant.[17]

Cost considerations ruled out the heavy water option and with it, the need for fissionable material for military purposes. France, like Britain, settled on gas-graphite reactors. The first was built at Marcoule and the plutonium it produced charged the first French atomic bomb, exploded in 1960. And just as in the UK, the first large scale gas-graphite reactors, as Chinon, were presented publicly as civilian, and named EDF-1 and EDF-2 (Davis 1988). Major challenges for the establishment of a French commercial nuclear system resulted from the weakness of the power generation equipment industry. Thus, it was only with the cooperation between CEA and EdF that the industry could participate in the construction of three GCR s similar to Marcoule (38 MW, completed at the end of 1959). Prior to 1968, all plants were of the gas-cooled reactor (GCR) type, with exception of a 10% participation by EdF in the planned SELNI project in Italy. This project was to use the PWR commercial technology from Westinghouse. Framatome entered into licensing agreements with Westinghouse in 1958.[18]

While France was developing gas-graphite reactors as a long term strategic option, other technologies were being experimented with, including heavy water-moderated reactors (HWR) and the light water technology, pressurised water reactors (PWR). Following President's de Gaulle death in 1969, the gas-graphite technology was replaced by the LWR, mostly developed under licence agreements with the US duopoly. Akin to the US experience, this development had its origins

17 Uranium supply was guaranteed through Niger and Gabon, at that time still French colonies.

18 Framatome (Societe Franco-American de Construction Economique) was established in 1958 by seven companies of the Empain Schneider Group. Framatome terminated its licence in 1981 and negotiated a new agreement.

in research on nuclear submarine reactors. The establishment of the French civilian nuclear framework was entrusted to two main actors, the CEA and EdF. They served as the executive arm of the Ministry of Industry, which was responsible for energy policy. The CEA was responsible for the entire nuclear fuel cycle as well as research in nuclear physics. Its "Direction des Applications Militaires" (DAM) was responsible for bomb testing at Mouroa. CEA also built the plutonium production plants at Marcoule and La Hague (Schneider 2008).

On the whole, the French civil nuclear programme has largely profited from the military programme and vice-versa; the link with the military has remained strong. For instance, La Hague reprocessing plant was financed in equal shares by the civil and military budgets of the CEA. Schneider (2008: 8) describes this as military cross-subsidisation, which he considers a leading benefactor throughout the entire French nuclear programme.

From national technological options to LWR technologies under licensing agreement

Following the so-called "bandwagon market" of 1966–67 and massive investment in nuclear projects, the US industry compromised the European efforts to develop alternative technologies to LWR and nuclear energy was marketed as "too cheap to be metered" (Cohn 1997).[19] Between 1962 and 1976, the installed nuclear capacity approximately doubled every two years, with a growth rate of over 40% per year. Burness et al. (1990) consider this the fastest sustained growth rate for a US industry in the history of the country.

Because of the technical and commercial success of US LWR technology, European policymakers were torn between resisting the American marketing attack and simply taking up licensing agreements with leading US companies. Economic wisdom led the majority of European companies to strengthen their existing ties with Westinghouse and GE, with the aim of refining their internal capabilities by first gaining access to technical knowledge and then internalising the licence and solving technical problems on their own. Governments similarly made efforts to carve out and direct growth paths using a set of intermediate targets and instruments to foster technological autonomy as a final objective.

Naturally, the relative position of each country with regard to licence assimilation varied and was influenced by economic and political factors. With hindsight, however, one can say that licences brought about advantages when the recipient industries were able to pursue technological improvements on the "product" under

19 In such fixed price contracts, the reactor vendors had responsibility for design, construction and testing of a reactor, including regulatory guidelines.

licence. Given Euratom's failure to promote a European technology, it had become apparent that any nationalistic grounds for a country's autonomous technological path would have been inadequate in the face of the intrinsic fragmentation of the European market and the American oligopoly in the international arena.[20]

Once the LWR option had become the most widely chosen worldwide, both PWR and BWR (boiling water reactor) technology coexisted in national markets for more than a decade. France was the first country to abandon the BWR technology path and concentrate solely on PWRs in 1975.

6 Determinant factors of success for LWR market penetration[21]

The role of enriched uranium

The availability of enrichment technology and of enriched uranium has often been underestimated as a critical factor contributing to the success of the US technological path. Therefore, here we stress the importance of this element and the way in which pricing policy by the American authorities assisted the US industry's expansion throughout the 1970s and its imposition of LWR technology on the world market.

The availability of enriched uranium can be considered another spill-over from the military activities, and an example of the socialisation of costs as significant as that deriving from the knowledge and experience nurtured under the US AEC and Navy programme applications. The availability of enrichment facilities had a direct influence on the choice of LWR technology. Its impact was immediate, since the employment of enriched uranium permitted some degree of freedom in the alternative nuclear technologies. It also allowed for a certain latitude in the choice of materials and in reactor design, which prevented the "high construction costs and poor material economy" observed in the British case (Burn 1978).

One reason why little interest has been shown in uranium as a leading success factor may be the difficulty of ascertaining the start-up costs of uranium enrichment programmes exclusively for civilian reactor development projects. However, one may reasonably conjecture that whatever the hypothetical cost of such a programme, the expense would have been such that no economic or technical considerations would

20 Commercial strategies and paths differed in the case of experimental reactors and FBR. This field represents an atypical situation, especially for the high degree of European cooperation in R&D and as these projects were carried out independently of the USA.

21 This section draws heavily on Di Nucci (1986).

have justified developing a technology like LWR. On these grounds, countries like Canada, Britain and France, which used different moderators and coolants had no choice for fuel: it had to be natural uranium.

The crucial role played by enriched uranium is underlined by the fact that attempts to privatise the industrial phase of enrichment were unsuccessful, despite early promises by the US AEC that it was prepared to do so, until 1963–64. Moreover, the public monopoly over enriched uranium did not constitute a bottle-neck for the American industry abroad, even with the constraints on the international sale and re-purchasing of uranium. On the contrary, it allowed the industry to transfer costs to the taxpayer for a highly expensive operation. These incentive prices for uranium ore and plutonium were criticised by contemporary observers, but at the same time this move was justified with the need to accelerate the development of a civilian nuclear industry in order to support the uranium production industry (Mullenbach 1963:122).

The US remained the only Western nation where the home industry could benefit from strong military-linked government support. In Europe, the absence of a massive military programme left little hope for the autonomous and parallel development of European technologies. It might be objected that Britain and France also had this support at a later stage uranium enrichment plants, but the Capenhurst and Pierrelatte facilities were designed for military purposes. Their production was modest and insufficiently influenced a change in the preferred technology. While a mix of economic and military considerations (such as the costs of fuel moderators like heavy water and the need for plutonium) had motivated European countries to adopt gas-graphite technology, this proved to be a technical and economic flop. Subsequently, European nations entered into licence agreements with the US LWR vendors.

The search for the optimisation of the whole nuclear system

An additional success factor in the selection of technologies and the industrialisation of military nuclear assets was the active involvement of the US private industry in the fuel cycle. Reactor suppliers could profit from an efficient fuel industry because of their participation in the US AEC promotion programmes; this integration contributed to the optimisation of the system.

Compared to other technologies, nuclear power involved a greater degree of consideration in terms of creating an industry with high organisational and technological standards that were intricately linked to the political and institutional structures. In this respect, the growth of the industry and the progress of LWR technology in the US is an exemplary case of the development of an intimate relationship between industry and the institutional framework, and of the key role that public and private actors played in all aspects of the industry's development.

As a consequence of the many synergies created, the US nuclear system was able to reach a broad turn-key capacity guaranteeing plant construction, fuel rods, and further supplies of uranium as well as reprocessing. Being able to offer such a package from the outset meant that the US industry had a clear comparative advantage for its LWR technology. Our claim is that success did not depend – at least not exclusively – on the characteristics of the reactor offered for sale, but on the system of which the reactor was a part of. The case of Britain clearly shows that the lack of commercial success was mainly due to a nuclear framework which was self-sufficient and closed around its reactor, which ultimately confined national nuclear technology to a single domestic scene.

The British failures demonstrate that a pluralistic approach to technological development in which several alternative strategies are simultaneously pursued may be less costly, in terms of research outcomes, than a monistic approach concentrating on a single project. In the case of the US, the decisive factor in success was not the selection of LWR technology, but the compatibility of that option, of the many explored, with the industrial system that has to accommodate it.

The pragmatic approach by US authorities, what Burn (1978) calls the "selection principle" played an important role. But, unlike Burn, we argue that the choice itself was, in a certain sense, piloted rather than the result of market forces. The choice mechanism is evident from both the Five Year Programme and the various rounds of the Power Demonstration Reactor Programme. On the other hand, the best experimental results were obtained with prototypes that were later abandoned. In fact, the Joint Committee on Atomic Energy concluded in 1954 that of the five different reactor technologies developed for civilian use, the PWR appeared to be the least promising due to its conservative design (Cowan 1990). How then, was it able to emerge as the dominant technology? To address this question, the first thing to consider is that the LWRs, though less advanced than other technologies, were chosen for their commercial viability. Unlike other technologies, they presented fewer obstacles to being scaled up from the prototype stage. LWR was thus the only design ready for full-scale construction. Moreover, its deployment was necessary as a demonstration of the potentials of commercial nuclear power and to promote the "Atoms for Peace" Programme. With Cowan (1990: 566), we maintain that the first-comer technology which can advance along its learning curve will dominate the market.

This, of course, is not the whole story. An additional and more fundamental explanation is that unlike many national nuclear programmes, where the failure to export reactors was largely due to the inability to internationalise their productive structure, the US' technology was the only one to offer continuity and a greater flexibility. Any strategy can be adopted for the commercialisation of a product, but when an entire technological system is to be exported, the strategy which pays off is that which best fits what has been named the principle of technological-industrial continuity (Di Nucci and Pearce 1989). Solutions and systems that are too far from current technological frontiers are unlikely to succeed, since they would require the greatest amount of technical and industrial adjustment and transformation. The strength of the US nuclear system was its ability to be exported as a reactor-and-service- package, satisfying the technological-industrial continuity criterion.

The pull of the market

The success factors outlined thus far would have been of little avail if steps had not been taken to turn potential demand into orders. Such policies in the US and Europe differed not only in their manner of creating internal demand, but also in the paths pursued to reach this target. European strategies, at least initially, were inspired less by export and commercial criteria than by the urge for technological and energy autonomy. In contrast, the US AEC tried to stimulate demand from electric utilities by offering advantageous conditions and incentives, such as subsidies and a pricing policy for enriched uranium. With this approach, the size of the US domestic and export market were directly influenced by government policy until the mid-1960s.

A major turning point occurred in 1964, at the time of the third Geneva Conference. GE had established itself domestically and internationally as a reactor vendor. The company had moved quickly in improving the original design and scale-up of prototypes and offered turnkey contracts for large scale reactors at fixed prices. Thirteen reactors were ordered on a turnkey basis by electric utilities (Burness et al. 1980). Thus, when GE published its price list for BWRs in 1964, the price quoted for the Oyster Creek plant had already set new cost targets that neither national nor European competitors could ignore. This enabled the company to present themselves as having the most feasible and economic design (Cohn 1997).

GE's price list had an enormous impact. It represented a nuclear power plant as an "autonomous" commercial good and placed potential clients in a position to refer to a definite product with a definite price, much lower than those of its competitor, Westinghouse. Of course, the venture resulted in corporate losses; however, it also indicated that GE was not only likely to reap the benefits of cost socialisation of the early development phases, but that it was also prepared to take risks in commercial ventures.

Another important success factor is that in 1964, an amendment of the Atomic Energy Act granted the US AEC permission to lease nuclear fuel directly to market actors. The timing of the GE price campaign coincided with the readiness of the US utilities to begin adjusting to previous under-capacity. The combination of demand and of the low, seemingly competitive prices for power generation, plus the kind of contracts for which the vendor guaranteed a fixed turnkey supply, triggered a boom in US plant orders. Seven units were ordered in 1965 and eight in the first half of 1966, to be followed by 13 in the second half of 1966 and 31 in 1967 (Burness et al. 1980:190). Though turnkey projects were costly investments for vendors, the completion of the turnkey units stimulated demand for new reactors and subsequent sales. The time period which followed was characterised by a tremendous flow of new orders, so much so that this era has been referred to as the 'Great Bandwagon Market.' Whereas 78 reactors had been ordered over the 12-year period between 1955 and 1967, 166 reactors were ordered for projects across the USA between 1968 and 1973, with 38 units ordered in 1972 alone (Bernd and Aldrich 2015).

A key observation is that in the decade 1963–1973, the US domestic demand alone offset the aggregate demand of the world market. The size of the internal market enabled the industry to pass the minimal threshold in physical and investment terms necessary for an autonomous take off, and to speculate on the promise of possible economies of scale including learning effects, which was hoped to trigger success on the export market. Two aspects of this cumulative effect are illustrated by the widespread network of licence agreements that US companies started in Europe and Japan. The third Geneva Conference therefore marked the establishment of the US commercial and technological supremacy. It simultaneously dealt a blow to the commercial aspirations of many autonomous national nuclear technologies in Europe. Licences became the major vehicle in US export policy.

The scale of the reactors increased dramatically and most of the plants within the 'Great Bandwagon Market' were considerably larger than older reactors (400MWe or greater). By the end of 1970, the entire nuclear industry had only accumulated 11 years of operating experience on units of this size. As the demand for electricity in the US decreased dramatically in 1973, the first signs of the industry's problems emerged. Most reactors faced construction delays and massive cost overruns; orders for nuclear power plants started being cancelled, with 12 projects called off before the end of 1973 (Bernd and Aldrich 2015). In 1975, only four reactors had been ordered, and just nine more were ordered in the three years that followed. The last order for a new nuclear power plant came in 1978. A year later, in 1979, the Three Mile Island's accident occurred and the collapse of industry began.

7 Conclusions

Our analysis has pointed out exogenous and endogenous factors affecting the long-term development and diffusion of the LWR technology: the military spill-over effect; the use of enriched uranium and its restrictions; the subsidised price of enriched uranium; the choice of a commercial reactor based on the industry's capacity to accommodate the technology and internationalise the whole system; and, last but not least, the scale of the nuclear programmes. The causal interactions at work here were unidirectional, but their influences have been mutual. New nuclear technology was not simply tested and then integrated within a system; its development was embedded in interactions with this system. Although there are similarities between national experiences, very distinct national stories have emerged.

The US: The first-comer in enhancing the leading LWR technology was an innovator that benefited from early government infrastructure and support to commercialise the technology and benefited from military spill-over effects.

The UK: An example of the failure of a nuclear system based on a domestic autonomous technological pathway. The centralised framework and the idea that technical progress could occur despite negative signals from the international market, combined with institutional inertia, had a strong adverse effect on the industry. Ultimately, it could not rely on any suitable instruments to compensate for an international market that opted for LWRs.

France: A latecomer in the development of LWR technology. Its experience indicates the timeliness of giving up a national technology without a commercial future and taking on the risk of starting practically from scratch under licence. The extremely integrated decision-making framework, the nurturing attitude of the French government and the national electric utility bestowed a steadily growing market and learning economies.

The initial nuclear development in these leading countries was characterised by a plurality of base technologies, following the experience gained through strategic and military activities. National differences in these experiences led to correspondingly different national civilian frameworks and technological choices. In the initial development phase, the US enjoyed a virtual monopoly on uranium enrichment, backed by military financial support. It could opt for LWR technology accordingly.

The availability of enriched uranium is arguably the most crucial factor in the US' comparative technological success and market advantage. This can therefore be regarded as the cornerstone of the US export policy and success. In contrast, the absence of enrichment facilities in Europe, along with the need for plutonium for military purposes, guided France and the UK to pursue gas-graphite and Canada to select heavy water as a moderator.

Although the US was not alone in early nuclear development, its emphatic leadership came about through a series of crucial factors. The US AEC's financing of almost all nuclear R&D, along with its direct role in funding and subsidising the majority of the early plants, proved to have a decisive and positive influence on the industry's development. Domestic contracts became a key vehicle for eventual export sales, even though government regulations and administrative controls initially limited the scope of power plant export. The marketing of nuclear reactors occurred within a framework determined by international organisations, treaties, bilateral agreements and national laws, and was also influenced by economic and political relations with the recipient countries. The transfer of nuclear technology, mainly to Europe and Japan, was assisted by a certain liberality on the part of the licensors, but also by restricting expertise on the nuclear fuel cycle for military purposes. In this process, Britain was to suffer most and France was also adversely affected. Both countries eventually opted for LWR technology in 1979 in the former case, and a decade earlier, in 1969, in the latter.

The inherent characteristics of the nuclear plant as a saleable good, and the resulting implications for its development required a special role that only national governments could fill. At the same time, the industry needed specific government support, whether for selling abroad (as in the case of GE and Westinghouse), or for its engagement in the manufacturing of components imported from technology leaders (in the case of Europe). Either way, backing and promotion by governments was essential for various reasons, military development and spill-over and the nationalist incentive that made the industry desirable even before economic competitiveness had been achieved.

We have tried to separate the progressive attack by the US oligopoly on the national and international market into distinct temporal phases. The marketing of the US power plants followed two strategies: one amounting to the sale and later export of the nuclear reactor as a product; the other as the export of the entire nuclear productive structure. The former strategy applies to the period 1963–72, and in particular between 1962 and mid-1966. During this period, GE and Westinghouse sold turnkey plants to US public utilities and the contracts were available under fixed price terms (Burness et al. 1980; Bernd and Aldrich 2015).

GE ended its turnkey contracts sale offensive to US electric utilities in June 1966. According to Burness et al. (1980), GE and Westinghouse took combined losses on the contracts upwards of $1 billion. However, the financial losses they suffered during the turnkey era can be considered an investment "in obtaining information through "learning by doing" in an effort to capture rents from the second generation of reactors" (Burness et al. 1990: 189). Finally, the US companies ceased offering turnkey contracts because of the cost risks.

The second stage occurred through the transfer of knowledge via the sale of patent rights and licences, without direct industrial investment abroad, and by establishing subsidiaries and internationalising industrial capital via direct investment of risk capital in a foreign country.

As we have explored in this chapter, there are good grounds for claiming that the early commercial success of LWR and the establishment of the US oligopoly (first domestically and then in the global market) is a rich example of the combined effects of a technology push (via the US AEC activities and military spill-over effects), market pull (bandwagon market) and market push (via competition among utilities to improve technical standards through innovation). A civilian nuclear industry was created in part to legitimatise the continued development of atomic weaponry. In the haste to develop nuclear power plants, economic and technical considerations were often secondary. A military model was selected for civilian use because it initially provided a dual purpose fissile material production capacity, rather than simply because it was immediately available (Bupp and Derian 1978).

Alleging the importance of a production system which is common all over the world, Cowan (1990: 552) claims that "it is occasionally suggested that network externalities are also important in nuclear power. The network in this case has to do with information. Information about operating performance, appropriate accident response, and safety regulations can be passed among users of the same technology. This was seen (at least in retrospect) as a key factor in the explanation of the Belgian and Swedish decisions to adopt light water". In the US, the choice of technology to be pursued in commercialisation encouraged autonomous technological advances for the options which could be developed while maintaining a certain continuity with the pre-existing industrial structure (Di Nucci 1986). Other, more "innovative" paths were to stay at the experimental level and be undertaken under governmental support, not directly by the industry – and they have remained "experimental" until today.

References

Argonne National Laboratory, w.d., History of Argonne Reactor Operations. http://www.ne.anl.gov/About/reactors/History-of-Argonne-Reactor-Operations.pdf

Berndt, E., Aldrich D.P., 2016. Power to the people or regulatory ratcheting? Explaining the success (or failure) of attempts to site commercial US nuclear power plants: 1954–1996.

Bupp, I., Derian, J.C., 1968. Light Water: How the Nuclear Dream Dissolved, Basic Books, New York.

Burness, H.S., Montgomery, W.D., Quirk J.P., 1980. The Turnkey Era in Nuclear Power, Land Economics.Vol. 56, No. 2 (May, 1980), 188–202.
Burn, D., 1978. Nuclear Power and the energy crisis. London, MacMillan.
Burn, D., 1967. The political economy of nuclear energy. London, Institute of Economic Affairs.
Camilleri, J.A., 1977. The myth of the peaceful atom. Journal of International Studies 6 (autumn),111-127.
Cantelon, Ph. L., Hewett, R.G., Williams, R.C., (Eds), 1984. The American Atom: A Documentary History of Nuclear Policies from the Discovery of Fission to the Present – 1939–1984. University of Pennsylvania Press.
Clark, T., Woodley, R., De Halas, D., 1962. Gas-Graphite Systems, in "Nuclear Graphite" R. Nightingale, Editor. Academic Press, New York, p. 387.
Cohn, S.M., 1997. Too Cheap to Meter: An Economic and Philosophical Analysis of the Nuclear Dream. SUNY Press: Albany.
Cowan, R., 1990. Nuclear Power Reactors: A Study in Technological Lock-in. The Journal of Economic History, Vol. 50, No. 3 (Sep., 1990), pp. 541-567.
Davis, M. D., 1988. The Military-Civilian Nuclear. Link: A Guide to the French Nuclear Industry. London: Westview Press.
De Wolf Smyth, H. Atomic Energy for Military Purposes (The Smyth Report). The Official Report on the Development of the Atomic Bomb under the Auspices of the United States Government. http://www.atomicarchive.com/Docs/SmythReport/, retrieved 10.05.2016.
Di Nucci, M.R., 1986. Technology, Competition and State Intervention. Development paths and public policies in the promotion and commercialization of Light Water Reactors, PhD thesis, Science Policy Research Unit, The University of Sussex.
Di Nucci, M.R., Pearce, D.A., 1989. Economics and Technological Change: Some Conceptual and Methodological Issues, Erkenntnis 30 (1989) 101–127.
The Economist, 2012. A brief history. From squash court to submarine. http://www.economist.com/node/21549101, retrieved 12.05.2016.
Etzkowitz, H., 1984. Solar versus nuclear energy: autonomous or dependent technology? Social Problems, Vol. 31, No. 4. 417–434.
Gowing, M., 1964. Independence and deterrence: Britain and Atomic Energy 1945- 1952, London: MacMillan.
Lowry, D., 1989. Nuclear Weapons and Nuclear Power: Bias and Mythology in the Making of the British Magnox Nuclear Reactor Program. In M. Blunden and O. Greene, eds. Science and Mythology in the Making of Defense Policy. London: Brassey's Defense Publishers, 1989, 127–166.
Lucas, N.J., 1977. Energy and the European Community, London. Europa Publications.
Mullenbach, P., 1963. Civilian Nuclear Power: Economic Issues and Policy Formation. New York. The Twentieth Century Fund.
Nehert, L.C., 1966. International marketing of Nuclear Power plants, Bloomington, Indiana University Press.
Sanger, S.L., Wollner, C., 1995. Working on the Bomb: An Oral History of WWII Hanford. Portland State University.
Scheinman, L., 1965. Atomic Energy Policy in France under the Fourth Republic. Princeton University Press.
Schneider, M., 2008. Nuclear Power in France. Beyond the Myth, Study commissioned by the Greens-EFA Group in the European Parliament. http://www.nirs.org/international/westerne/258614beyondmythfr.pdf, retrieved 15.05.2016.

Walker, W., Lonnroth, M., 1983. Nuclear Power Struggles: Industrial Competition and Proliferation Control. London: George Allen and Unwin.
Wellerstein, A., 2017. Remembering the Chicago Pile, the World's First Nuclear Reactor Seventy-five years ago. The New Yorker, December 2.
Williams, R., 1980. The Nuclear Power Decision, London: Croom Helm.

Websites

https://www.iaea.org/
http://www.ne.anl.gov/About/reactors/lwr3.shtml
http://www.ne.anl.gov/About/hn/
http://www.atomicarchive.com/Docs/SmythReport/
http://www.ewp.rpi.edu/hartford/~er esto/F2010/EP2/Materials4Students/Misiaszek/NuclearMarinePropulsion.pdf

The Current Status of the World Nuclear Industry

Mycle Schneider, and Antony Froggatt[1]

Abstract

The following chapter is based on the World Nuclear Industry Status Report 2018 (WNISR2018). The annual WNISR is a comprehensive assessment of the status and trends of the global nuclear power industry.

1 Introduction

Heat. The planetary, record-breaking heatwave in 2017 gave a daunting hint on what the future on earth will almost certainly look like.

Water. The food system is the most sensitive to lack of water. As of early August 2017, it is already clear that the draught will severely impact harvests in many parts of the world.

1 Mycle Schneider, International Analyst on Energy and Nuclear Policy, Paris, France, mycle@sfr.fr; Antony Froggatt, Chatham House, London, United Kingdom, afroggatt@chathamhouse.org.uk

R. Haas et al. (Eds.), *The Technological and Economic Future of Nuclear Power*, Energiepolitik und Klimaschutz. Energy Policy and Climate Protection, https://doi.org/10.1007/978-3-658-25987-7_3

Heat, water and nuclear power. Thermal power plants need vast amounts of cooling water. It is estimated that in France 51 percent of freshwater takeout or about 10 percent of precipitation is absorbed in thermal power plants, with roughly three-quarters of its electricity generated by nuclear power over the years. No other electricity generating source needs more water than atomic fission energy. David Lochbaum, Director of the Nuclear Safety Project at the Union of the Concerned Scientists (UCS), who has produced a fact sheet on "Nuclear Power and Water"[2], stated: "We'll have to solve global warming if we want to keep using nuclear power".[3]

The European Pressurized Water Reactor (EPR) under construction at Flamanville on the coast of Normandy will have its own desalination plant to cope with freshwater needs. Four in-land reactor sites along French rivers with no cooling towers—Bugey (2 units), Fessenheim (2 units), St. Alban (2 units), Tricastin (4 units)—take out about 70 percent of all thermal power plant cooling water in the country. The two oldest French reactors at Fessenheim alone take up about 18 percent of *all* 17 billion cubic meters of France's annual freshwater takeouts.[4] While these sites consume a large portion of the nation's surface freshwater, they return about 90 percent back to the environment, but significantly heated up.

And that is a problem. In order to make sure reactors can be appropriately cooled, the uptake water temperature is limited for safety reasons, and to avoid excessive heating of the rivers, the operating licenses impose limits to downstream water temperatures. Consequently, as of 1 August 2018, operators in several countries, including Finland, France, Germany, Sweden and Switzerland, had put operational restrictions on some of their nuclear power plants. While in most cases, regulations required to lower the output of the reactors by 10 percent or so, some reactors were shut down, including at least four reactors in France, to deal with the problem.

The heat symptom occurred just after the first EPR (European Pressurized Water Reactor) and the first AP1000 had started up within 24 hours interval—both in China—end of June 2018. A shift towards better times for the global nuclear industry? By no means. On every piece of positive development follows an avalanche of bad news. For now, the heat wave is only a secondary problem for the industry.

---

2 UCS, "Nuclear Power and Water", 2011, see https://www.ucsusa.org/sites/default /files/l egacy/assets/documents/nuclear_power/fact-sheet-water-use.pdf accessed 2 August 2018.

3 Commons, "Amid climate concerns, nuclear plants feel the heat of warming water", Energy News Network, see https://energynews.us/2016/09/09/midwest/nuclear-plants-feel-the-heat-of-warming-water/ accessed 1 August 2018.

4 CGDD, "Les prélèvements d'eau par usage et par ressource", 21 June 2017, see http://www.statistiques.developpement-durable.gouv.fr/lessentiel/ar/234/1108/prelevements-deau-usage-ressource.html accessed 1 August 2018.

The general malaise about the uncertain future of the industry remains deep and disconcerting.

While China proudly presents the prowess of its construction industry with the completion of the first Generation-III reactors—designed by western companies, the EPR by Framatome-Siemens and the AP1000 by Westinghouse, the now-bankrupt worldwide largest historic builder—the rest of the world wonders at what rhythm the country will continue to expand its nuclear program. No new commercial reactor construction was launched in China since December 2016.

In France, the sub-standard pressure vessel of the Flamanville EPR was declared fit to operate by the safety authority, but the vessel head will have to be replaced after only six years of operation. Startup was delayed again by several months after numerous faulty welds were identified in the main steam supply system. After the technical bankruptcy, subsequent government bailout, breakup and name-change of AREVA to Orano, the new company renews with the old pattern and has been losing money again in 2017.

In Japan, the utilities managed to increase the number of operating reactors from zero in 2014 to nine by mid-2018. But this remains a very limited success with the plants contributing just 3.6 percent of the national electricity generation and 26 reactors remaining in Long-Term Outage (LTO, see definition below). Local populations and the general public remain overwhelmingly opposed to the restart of reactors. The attempts of the Japanese government to declare certain Fukushima evacuation zones as "decontaminated" and suitable for return did not convince many evacuees and most of them will likely never go back.

In the United Kingdom (U.K.), the Hinkley Point C project is underway but strangely still not officially under construction. After having spent at least €3 billion and thousands of workers on-site, apparently, the base-mat of the reactor building has still not being concreted—that marks the official construction start. Latest news on new-build in the U.K. is that Toshiba—former owner of Westinghouse—has stripped Korea Electric Power Company (KEPCO) of the preferred bidder status to acquire 100 percent of the company NuGen set up to build a nuclear power plant at the Moorside site in Cumbria.[5] KEPCO had been seen as the most promising candidate for the takeover, after other potent potential investors like the French Engie or Spanish Iberdrola left the U.K. new-build playing field. Toshiba got severely burnt in the Westinghouse bankruptcy and will not build any reactors any more. Prof. John Loughhead, Chief Scientist at the Business, Energy and Industrial Strategy

5 WNN, "Kepco loses preferred bidder status for NuGen", 1 August 2018, see http://www.world-nuclear-news.org/Articles/Kepco-loses-preferred-bidder-status-for-NuGen, 2 August 2018.

Ministry (BEIS), stated at a conference at the UK Royal Society on "Decarbonising UK energy": "There are clear issues with nuclear technology at present. The nuclear industry has created a product so expensive that no one can afford to buy it."[6]

In the United States (U.S.)., many reactors remain threatened to shut down long before their licenses expire because they cannot compete in the market. The nuclear industry and its supporters are clearly now focusing on efforts to come up with innovative subsidizing schemes, in particular on state level, to help avoiding "early closures" of uneconomic reactors. *Science Daily* titles a research paper[7]: "The vanishing nuclear industry" and is asking: "Could nuclear power make a significant contribution to decarbonizing the US energy system over the next three or four decades?", only to provide the answer: "Probably not." In May 2018, William Von Hoene, Senior Vice President and Chief Strategy Officer with Exelon, the largest nuclear operator in the U.S., had this to say: "I don't think we're building any more nuclear plants in the United States. I don't think it's ever going to happen... They are too expensive to construct, relative to the world in which we now live."[8] The recent revelation by the *Wall Street Journal* is therefore barely surprising: "A major donor to President Trump agreed to pay US$10 million to the president's then-personal attorney if he successfully helped obtain funding for a nuclear-power project, including a $5 billion loan from the U.S. government..."[9] The project in question is the Bellefonte plant in Tennessee, where the construction of two reactors was launched in the 1970s and abandoned in the 1980s—two of 42 nuclear construction sites abandoned in various stages of advancement in the U.S. alone. In 2016, the site was purchased by a private company for US$111 million with the stated-intention to invest up to US$13 billion to complete construction. Obviously, the project needed government support, as everywhere else, thus the willingness to pay President Trumps long-time fixer the extraordinary amount of US$10 million to help obtain a government loan.

Nuclear new-build is simply not competitive under ordinary market economy rules anywhere. Worse, like in the U.S., similar economic constraints continue to

---

6 David Lowry, personal communication, 4 October 2017.

7 Science Daily, "The vanishing nuclear industry", ScienceDaily, 2 July 2018, see https://www.sciencedaily.com/releases/2018/07/180702154736.htm, accessed 7 July 2018.

8 With 23 operational reactors, Exelon is the US' largest nuclear operator. S&P Global Platts, "No new nuclear units will be built in US due to high cost: Exelon official", 18 April 2018, see https://www.platts.com/latest-news/electric-power/washington/no-new-nuclear-units-will-be-built-in-us-due-26938511, accessed 22 May 2018.

9 WSJ, "Top Trump Donor Agreed to Pay Michael Cohen $10 Million for Nuclear Project Push", 2 August 2018, see https://www.wsj.com/articles/top-trump-donor-agreed-to-pay-michael-cohen-10-million-for-nuclear-project-push-sources-say-1533245330, accessed 3 August 2018.

press owners of currently operating, amortized reactors around the world, leading to an increasing number of units being closed permanently earlier than anticipated.

Finally, maybe the largest barrier to nuclear power development or its mere survival is still the time factor. The German electrical and electronics giant Siemens has just raised the stakes to an unprecedented level. In June 2018, Siemens connected 14.4 GW of turnkey natural gas combined cycle power capacity to the grid in Egypt 27.5 months aft construction start, three years after contract signature, boosting the national electricity generating capacity by 45 percent. An intermediate step of 4.8 GW, the first of the three giant plants, started up after only 18 months. With over 60 percent efficiency, these combined-cycle gas plants are almost twice as efficient as nuclear reactors. The next step is the implementation of up to 600 wind turbines with a total capacity of up to 2 GW, part of the goal of 7.2 GW wind power capacity spinning by 2020.[10]

2 General overview worldwide

The role of nuclear power

As of mid-2018, 31 countries were operating nuclear power reactors. That number has remained stable since Iran started up its first reactor in 2011.

The world nuclear fleet generated 2,503 net terawatt-hours (TWh or billion kilowatt-hours) of electricity in 2017[11], a one percent increase, but still less than in 2001 and four percent below the historic peak nuclear generation in 2006 (see Figure 1). Without China—which increased nuclear output by 35 TWh (+18 percent), more than the worldwide increase of 26 TWh—global nuclear power generation would have slightly decreased again in 2017. This is the third year in a row that China alone made up for the global decrease outside the country. In fact, in the past decade, only three years would have seen a global increase without China, 2010, 2013 and 2014, the year before 3/11 triggered the Fukushima disaster, and the two years after the 284 TWh (11 percent) production slump in 2011–2012.

10 Siemens, "Completion of world's largest combined cycle power plants in record time", 24 July 2018, see https://www.siemens.com/press/en/feature/2015/corporate/2015-06-egypt.php accessed 2 August 2018.

11 If not otherwise noted, all nuclear capacity and electricity generation figures based on International Atomic Energy Agency (IAEA), Power Reactor Information System (PRIS) online database, see http://www.iaea.org/programmes/a2/index.html. Production figures are net of the plant's own consumption unless otherwise noted.

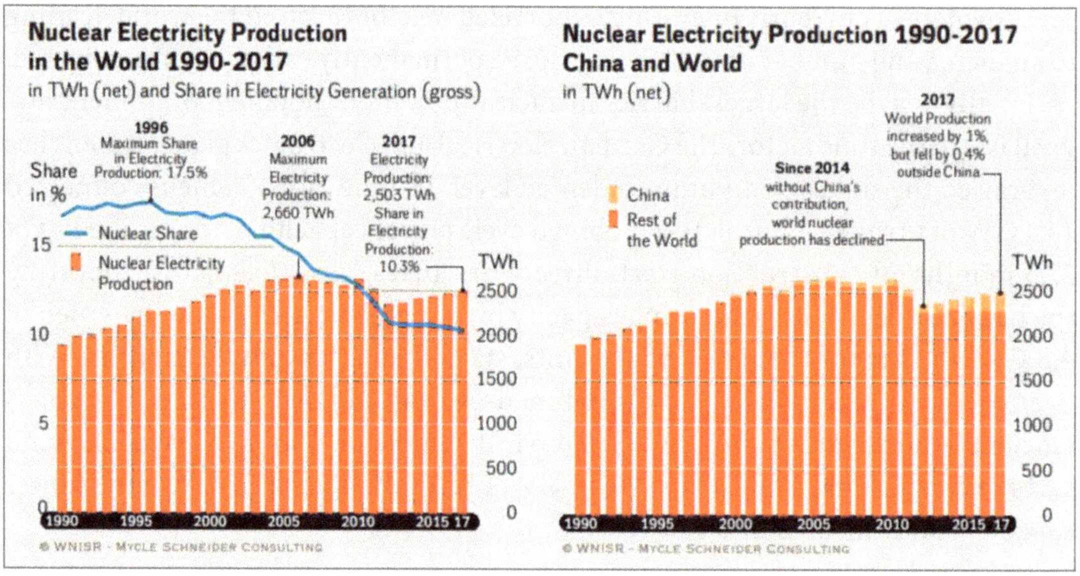

Fig. 1 Nuclear Electricity Generation in the World... and China
Sources: WNISR, with IAEA-PRIS, 2018, p. 28

Close to half of the world's nuclear power countries are located in the European Union (EU), and, in 2017, they accounted for 31.5 percent of the world's gross nuclear production, with half of the EU generation in France.

Nuclear energy's share of global commercial gross electricity generation remained almost stable over the past five years (–0.5 percent over the period), after dropping below 11 percent in 2012, for the first time in over three decades. The nuclear share declined slowly but steadily from a peak of about 17.5 percent in 1996 to 10.3 percent in 2017. Nuclear's primary energy remained rather stable after hitting a 30-year low at 4.4 percent in 2014.[12]

In 2017, nuclear generation increased in 13 countries, declined in 11, and remained stable in seven.[13] Five countries (China, Hungary, Iran, Pakistan, Russia) achieved their greatest lifetime nuclear production in 2017. Of these, China and Pakistan connected new reactors to the grid. China started up three units, and Chinese companies built the one that was commissioned in Pakistan.

As in previous years, in 2017, the "big five" nuclear generating countries—by rank, the United States, France, China, Russia and South Korea—generated 70 percent

12 BP, "Statistical Review of World Energy 2017", June 2018, see https://www.bp.com/content/dam/bp/en/corporate/pdf/energy-economics/statistical-review/bp-stats-review-2018-full-report.pdf accessed 28 July 2018.

13 Less than 1 percentage point variation from the previous year.

of all nuclear electricity in the world (see Figure 2, left side). In 2002, China held position 15, in 2007 it was tenth, before reaching third place in 2016. Two countries, the U.S. and France, accounted for 47.5 percent of global nuclear production in 2017.

Seven countries' nuclear power generation peaked in the 1990s, among them Belgium, Canada, Japan, and the U.K. A further eleven countries' nuclear generation peaked between 2001 and 2010 including France, Germany, Spain, and Sweden. Fourteen countries generated their maximum amount of nuclear power in the past seven years, five of which peaked in 2017.

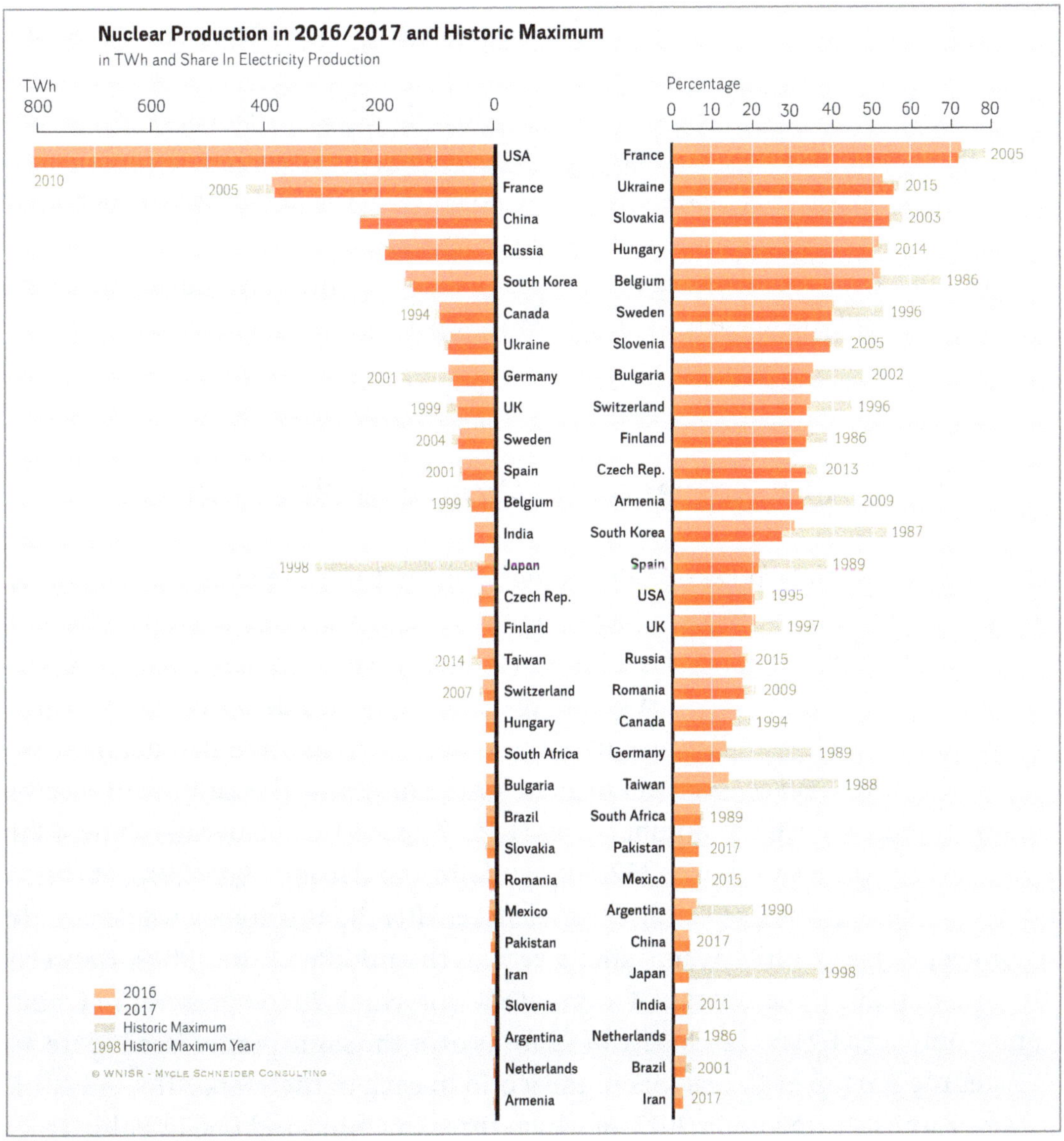

Fig. 2 Nuclear Electricity Generation and Share in Global Power Generation
Sources: WNISR, with IAEA-PRIS, 2018, p. 29

In many cases, even where nuclear power generation augmented, the development is not keeping pace with overall increases in electricity production, leading to a nuclear share below the respective historic maximum (see Figure 2, right side). It is therefore remarkable that in 2017, there were 19 countries that maintained their nuclear share at a constant level (change of less than 1 percentage-point), six countries increased and six decreased the relative share of their nuclear portion.

There were three exceptions in 2017, where countries peaked their respective nuclear share in power generation:

- Starting up three new reactors throughout the year, augmenting production by 18 percent, China increased the 2016 maximum of 3.6 percent, to reach a new record 3.9 percent nuclear share.
- Iran's only commercial reactor started up in 2011 after 33 years of construction but it took another five years to reach a reasonable grid-connection time and load factor in 2016. Increasing production by 7.6 percent, the nuclear share increased again slightly from 2.1 percent to 2.2 percent in 2017.
- Pakistan has been increasing production of existing units and started up a new one, which led the nuclear to jump from 4.4 percent to 6.2 percent.

3 Operation, power generation, age distribution

Since the first nuclear power reactor was connected to the Soviet power grid at Obninsk on 27 June 1954, there have been two major waves of startups. The first peaked in 1974, with 26 grid connections in that year. The second reached a historic maximum in 1984 and 1985, just before the Chernobyl accident, reaching 33 grid connections in each year. By the end of the 1980s, the uninterrupted net increase of operating units had ceased, and in 1990 for the first time the number of reactor shutdowns outweighed the number of startups. The 1991–2000 decade showed far more startups than shutdowns (52/30), while in the decade 2001–2010, startups did not match shutdowns (32/35). Furthermore, after 2000, it took a whole decade to connect as many units as in a single year in the middle of the 1980s. Between 2011 and mid-2018, the startup of 48 reactors—of which 29 (60 percent) in China alone—outpaced by six the closure of 42 units over the same period (see Figure 3).

After the startup of 10 reactors in the world in each of the years 2015 and 2016, only four units started up in 2017, of which three in China and one in Pakistan.

Three reactors were closed in 2017, respectively the oldest unit in Germany (Gundremmingen-B, 33.5 years), South Korea (Kori-1, 40 years) and Sweden (Oskarshamn-1, 46 years).[14]

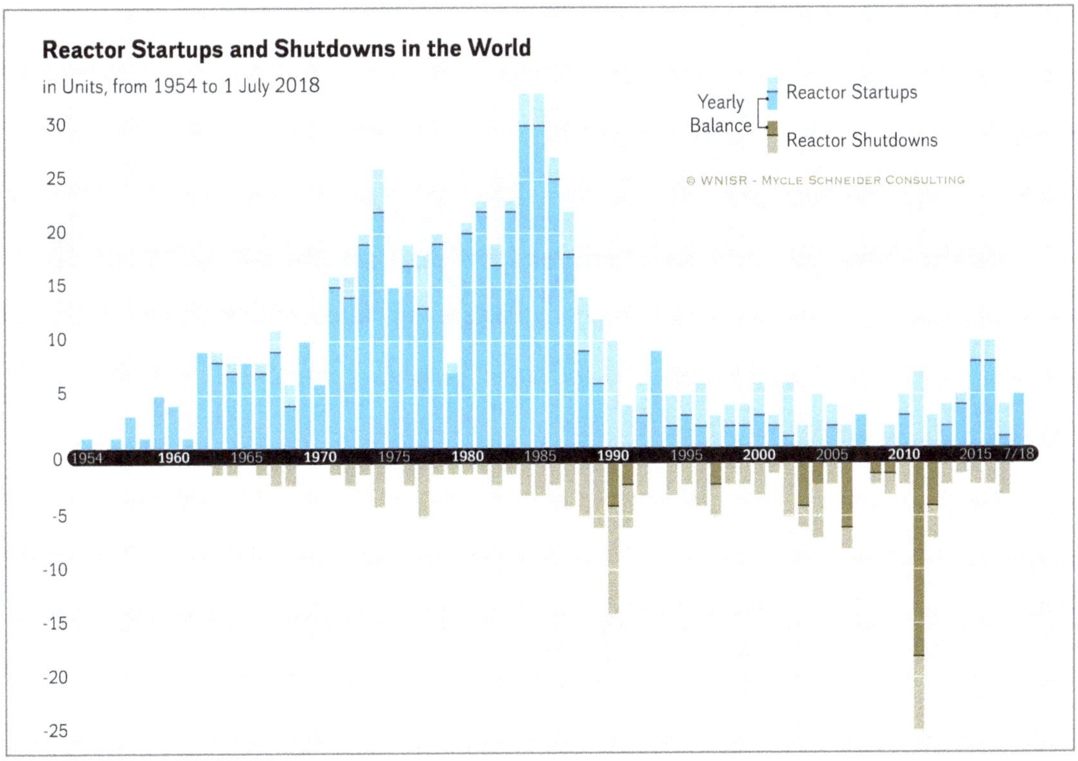

Fig. 3 Nuclear Power Reactor Grid Connections and Shutdowns
Sources: WNISR, with IAEA-PRIS, 2018, p. 30

In the first half of 2018, five reactors started up in the world, more than in the entire year of 2017, while none has been closed. Three units were connected to the grid in China, including the first EPR (Taishan-1) and the first AP1000 (Sanmen-1) to start up in the world (see Figure 4). And two reactors started up in Russia, Leningrad 2-1, and Rostov-4 that began construction 35 years ago.[15]

14 WNISR considers shutdowns from the moment of grid disconnection—and not from the moment of the industrial, political or economic decision—and as the units have not generated power for several years, in WNISR statistics, they are closed in the year of the latest power generation.

15 see https://www.worldnuclearreport.org/35-Years-After-Construction-Start-Rostov-4-Reactor-Connected-to-Russian.html.

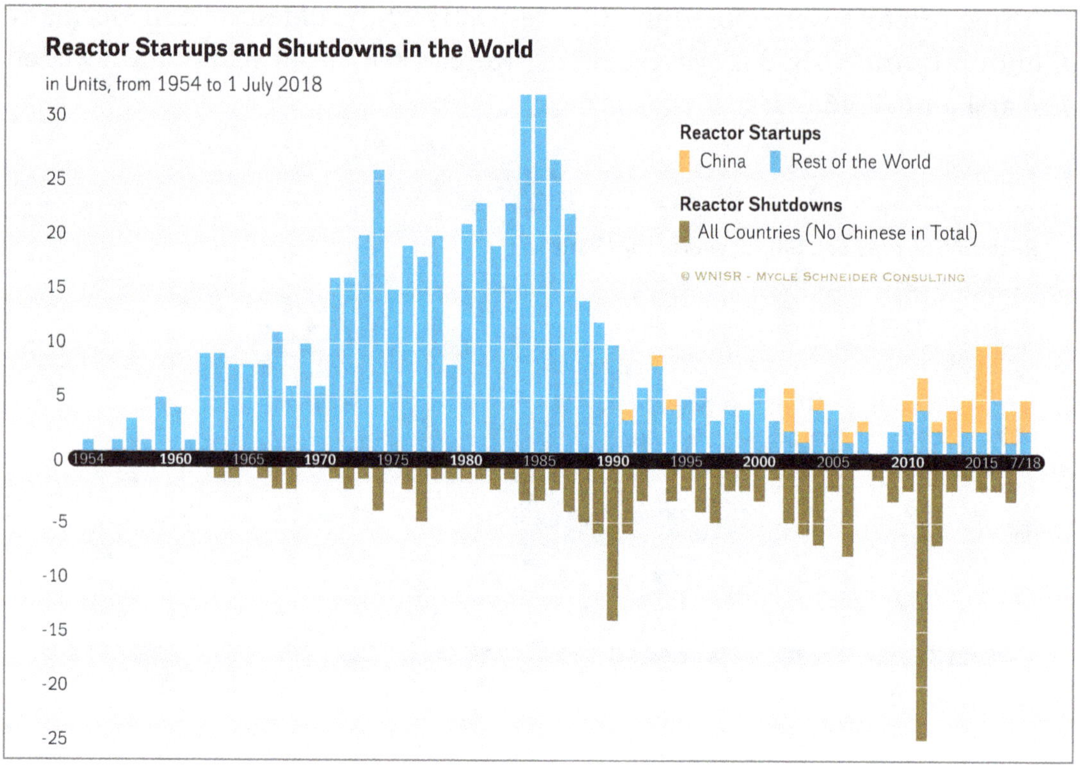

Fig. 4 Nuclear Power Reactor Grid Connections and Shutdowns – The China Effect
Sources: WNISR, with IAEA-PRIS, 2018, p. 31

The IAEA continues to count 42 units in Japan in its total number of 453 reactors "in operation" in the world[16]; yet no nuclear electricity was generated in Japan between September 2013 and August 2015, and as of 1 July 2018, only nine reactors were operating. Nuclear plants provided only 3.6 percent of the electricity in Japan in 2017.

The WNISR keeps calling for an appropriate reflection in world nuclear statistics of the unique situation in Japan. The attitude taken by the IAEA, the Japanese government, utilities, industry and many research bodies as well as other governments and organizations to continue considering the entire stranded reactor fleet in the country as "in operation" or "operational" remains a misleading distortion of facts. Steve Kidd, long-time industry strategist, agreed in a WNISR2016 review in Nuclear Engineering International:

16 IAEA, "Power Reactor Information System", see https://www.iaea.org /pris/ CountryStatistics/CountryDetails.aspx?current=JP, accessed 28 July 2018.

> Including reactors as "operable" along with those definitely in service, when they have not generated power for many years (and don't even have a licence to do so) is clearly ridiculous.[17]

Maybe as a result of such criticism, the World Nuclear Association (WNA), in its second "World Nuclear Performance Report", has distinguished between "generating" and "not generating" nuclear generating capacity.[18]. The IAEA actually does have a reactor-status category called "Long-term Shutdown" or LTS.[19] Under the IAEA's definition, a reactor is considered in LTS, if it has been shut down for an "extended period (usually more than one year)", and in early period of shutdown either restart is not being "aggressively pursued" or "no firm restart date or recovery schedule has been established". The IAEA lists zero reactors in Japan in the LTS category.

The IAEA criteria are vague and hence subject to arbitrary interpretation. What exactly are *extended* periods? What is *aggressively* pursuing? What is a *firm* restart date or recovery schedule? Faced with this dilemma, the WNISR team in 2014 decided to create a new category with a simple definition, based on empirical fact, without room for speculation: "Long-term Outage" or LTO. Its definition:

> A nuclear reactor is considered in Long-term Outage or LTO if it has not generated any electricity in the previous calendar year and in the first half of the current calendar year. It is withdrawn from operational status retroactively from the day it has been disconnected from the grid.

When subsequently the decision is taken to permanently close a reactor, the shutdown status starts with the day of the last electricity generation, and the WNISR statistics are modified retroactively accordingly.

Applying this definition to the world nuclear reactor fleet, as of 1 July 2018, leads to considering 26 Japanese units in LTO. Over the past year, four additional reactors were restarted and three more were officially closed. WNISR considers all ten Fukushima reactors shut down permanently—while the operator Tokyo Electric

---

17 *NEI*, "Nuclear power in the world – pessimism or optimism?", 13 October 2016, see http://www.neimagazine.com/opinion/opinionnuclear-power-in-the-world-pessimism-or-optimism-5031270/, /, accessed 13 August 2017.

18 The World Nuclear Performance Report was launched by WNA in 2016, "perhaps as a reaction to the success of successive WNISRs". In fact, in its September 2015 "Update for Members", WNA reported that its Fuel Report Working Group "discussed the merits of producing an annual nuclear capacity scenario update. Such an update would be a useful communications tool and a counter to the industry-critical World Nuclear Industry Status Report".

19 See IAEA Glossary, at www.iaea.org/pris/Glossary.aspx, accessed 1 July 2016.

Power Company (TEPCO) has written off the six Daiichi units, it keeps the four Daini reactors in the list of operational facilities. However, it is expected that the Daini plant will shortly be officially released for decommissioning.

As of 1 July 2018, besides the 26 Japanese reactors, two reactors in India (Kakrapar-1 and -2), and one each in China (CEFR), France (Paluel-2)[20] and Taiwan (Chinshan-1) met the LTO criterion. Besides the restarts in Japan, one reactor each in France (Bugey-5) and Switzerland (Beznau-1), that were categorized as being in LTO status in WNISR2017, were reconnected to the grid, and thus moved back to operational status. The total number of nuclear reactors in LTO as of 1 July 2018 is therefore 32; yet all are considered by the IAEA as "in operation".

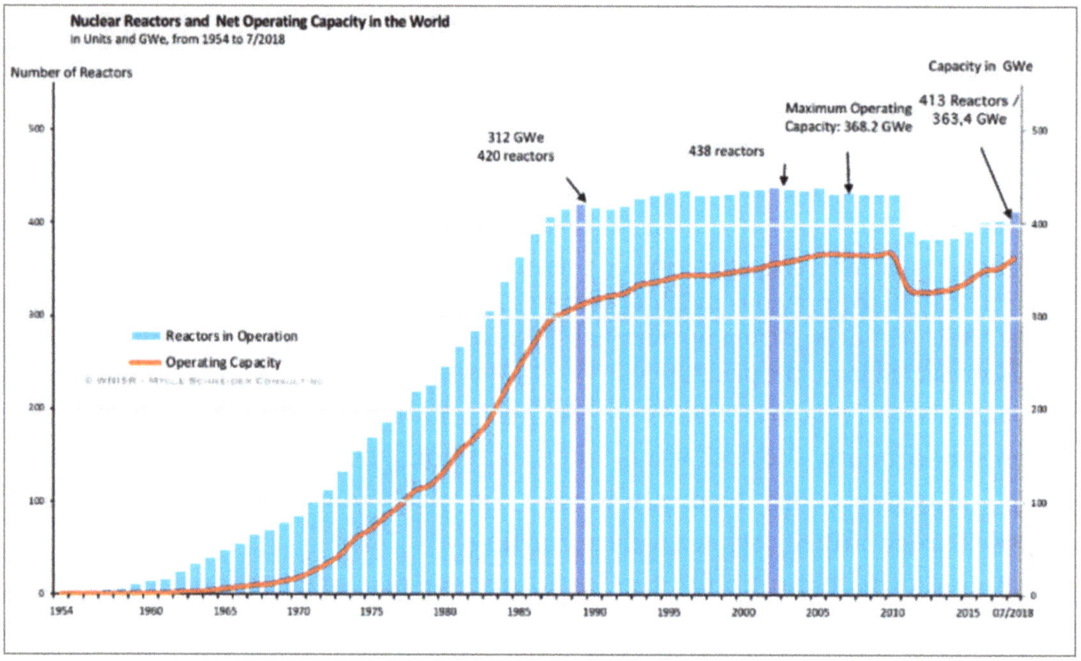

Fig. 5 World Nuclear Reactor Fleet, 1954–2018
Sources: WNISR, with IAEA-PRIS, 2018, p. 33

As of 1 July 2018, a total of 413 nuclear reactors are operating in the world. The current fleet has a total nominal electric net capacity of 363.4 gigawatts (GW or thousand megawatts), see Figure 5.

For many years, the net installed capacity has continued to increase more than the net increase of numbers of operating reactors. This is a result of the combined

20 After many delays, Paluel-2 was reconnected to the grid on 23 July 2018.

effects of larger units replacing smaller ones and, mainly, technical alterations at existing plants, a process known as uprating.[21] In the United States alone, the Nuclear Regulatory Commission (NRC) has approved 164 uprates since 1977. The cumulative approved uprates in the United States total 7.9 GW, the equivalent of eight large reactors.[22]

A similar trend of uprates and major overhauls in view of lifetime extensions of existing reactors has been seen in Europe. The main incentive for lifetime extensions is economic but this argument is being increasingly challenged as alternatives become cheaper.

4 Overview of current new-build

As of 1 July 2018, 50 reactors are considered here as under construction, the lowest number in a decade, three fewer than a year earlier, and 18 less than in 2013 (five of these projects have already been abandoned since). Four in five reactors are built in Asia and Eastern Europe, and China alone has nearly a third of all reactors under construction (16 out of 50). In total, 15 countries are building nuclear plants (see Table 1).

Five building projects were launched in 2017, two of which in India, and one each in Bangladesh, China (a non-commercial, demonstration fast breeder) and South Korea. As of 1 July 2018, there were two construction starts in the world so far in 2018, one in Russia, (Kursk-2-1) and Turkey (by a Russian company). Just prior to the official construction start in Turkey, the Turkish investors pulled out. It is remarkable that China has not launched a single new construction site for a commercial nuclear plant since December 2016.

Fifty is a relatively small number compared to a peak of 234 units listed as under construction—totalling more than 200 GW—in 1979. However, many of those projects (48) were never finished (see Figure 6). The year 2005, with 26 units under construction, marked a record low since the early nuclear age in the 1950s.

21 Increasing the capacity of nuclear reactors by equipment upgrades e.g. more powerful steam generators or turbines.

22 U.S. Nuclear Regulatory Commission (NRC), "Approved Applications for Power Uprates", Updated 4 May 2018, see http://www.nrc.gov/reactors/operating/licensing/power-uprates/status-power-apps/approved-applications.html, accessed 28 July 2018.

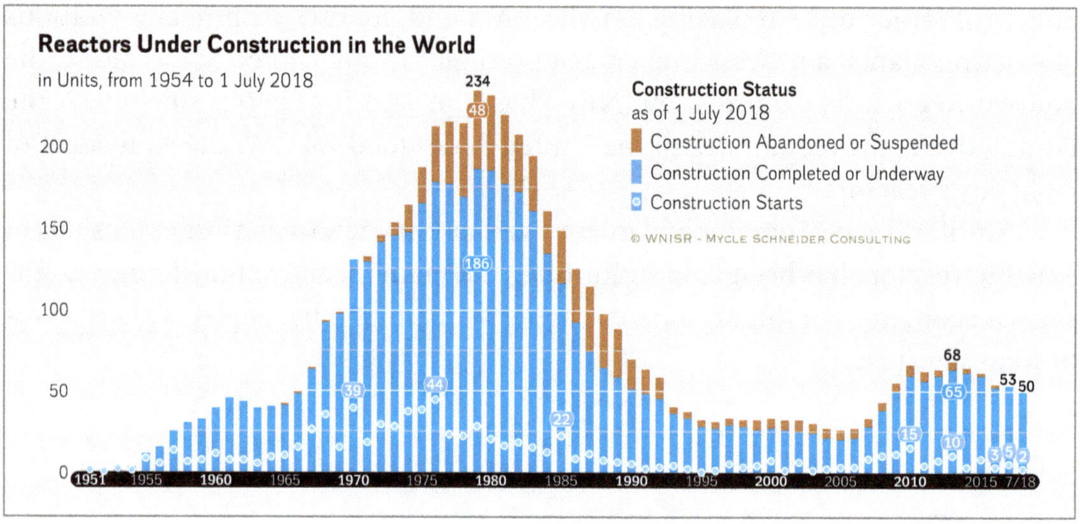

Fig. 6 Nuclear Reactors Under Construction
Sources: WNISR, with IAEA-PRIS, 2018, p. 34

Tab. 1 Nuclear Reactors "Under Construction" (as of 1 July 2018)

Country	Units	Capacity MW net	Construction Starts	Grid Connection	Behind Schedule
China	16	15 450	2009 – 2017	2018 – 2023	8-9?
India	7	4 824	2004 – 2017	2018 – 2023	5
Russia	5	3 378	2007 – 2018	2019- 2022	4
South Korea	4	5 360	2009 – 2017	2018 – 2022	4
UAE	4	5 380	2012 – 2015	2020 – 2021?	3-4?
Belarus	2	2 218	2013 – 2014	2019 – 2020	1-2?
Pakistan	2	2 028	2015 – 2016	2020 – 2021	-
Slovakia	2	880	1985 – 1985	2018 – 2019	2
USA	2	2 234	2013 – 2013	2021 – 2022	2
Argentina	1	25	2014 – 2014	2020	1
Bangladesh	1	1 080	2017 – 2017	2023	-
Finland	1	1 600	2005 – 2005	2019	1
France	1	1 600	2007 – 2007	2020	1
Japan	1	1 325	2007 – 2007	?	1
Turkey	1	1 114	2018 – 2018	2023	-
World	50	48 496	1985- 2018	2018 – 2023	33-36

Note: This table does not contain suspended or abandoned constructions.
Sources: Compiled by WNISR, 2018, p. 35

5 Construction times

Construction times of reactors currently under construction

A closer look at projects listed as "under construction" illustrates the level of uncertainty and problems associated with many of these projects, especially given that most constructors initially assume a five-year construction period:

- As of 1 July 2018, the 50 reactors being built have been under construction for an average of 6.5 years, many still far from completion.
- All reactors under construction in at least seven (possibly nine) out of a total of 15 countries have experienced mostly year-long delays. At least two thirds (33-36) of all building projects are delayed. Most of the remaining units under construction in the world, were begun within the past three years or have not yet reached projected startup dates, making it difficult to assess, whether or not they are on schedule.
- Of 33 reactors behind schedule, at least 14 have reported *increased* delays over the past year since WNISR2017.
- At the beginning of the year, 16 reactors were scheduled for startup in 2017, only four made it. Four reactors were scheduled to start up in the second half of 2017, only one did do so, the other three were connected to the grid in the first half of 2018.
- As of mid-2017, a total of 19 reactors were scheduled for startup in 2018. As of mid-2018, only three of these reactors were connected to the grid (one of which was already connected in 2017) and seven have been officially delayed until at least 2019.
- Two projects have been listed as "under construction" for more than 30 years, Mochovce-3 and -4 in Slovakia. Rostov-4 in Russia was finally connected to the grid in February 2018, 35 years after construction start.
- Four reactors have been listed as "under construction" for a decade or more, the Prototype Fast Breeder Reactor (PFBR) in India, the Olkiluoto-3 reactor project in Finland, Shimane-3 in Japan and the French Flamanville-3 unit.

It should be stressed that the actual lead time for nuclear plant projects includes not only the construction itself but also lengthy licensing procedures in most countries, complex financing negotiations, site preparation and other infrastructure development. As the U.K.'s Hinkley Point C illustrates, a significant share of investment and work can be carried out before even entering the official construction phase.

Construction times of past and currently operating reactors

There has been a clear global trend towards increasing construction times. National building programs were faster in the early years of nuclear power. As Figure 7 illustrates, construction times of reactors completed in the 1970s and 1980s were quite homogenous, while in the past three decades they have varied widely. The four units completed in 2017 by the Chinese nuclear industry in the homeland (3 units) and in Pakistan (1 unit) average an excellent 4.9 years construction time. This is only the second time since 2005 that world average construction time dropped below 5-year average. However, the five units that started up in the first half of 2018 show a much lower performance with an average of 13.4 years construction.

The longer-term perspective illustrates that short construction times remain the exceptions. Nine countries completed 55 reactors over the past decade after an average construction time of just over ten years (see Table 2). While the average has hardly moved since 2014, the range increased from 3.8–36.3 years to 4.1–43.5 years (the Watts Bar-2 in the U.S. record, which will remain the upper limit for some time to come).

Tab. 2 Reactor Construction Times 2008–2018

Construction Times of 55 Units Started-up 2008–7/2018				
Country	Units	Construction Time (in Years)		
		Mean Time	Minimum	Maximum
China	31	6	4.1	11.2
Russia	7	24.0	8.1	35.1
India	5	9.8	7.2	14.2
South Korea	5	5.3	4.1	7.2
Pakistan	3	5.4	5.2	5.6
Argentina	1	33.0		
Iran	1	36.3		
Japan	1	5.1		
USA	1	43.5		
World	55	10.1	4.1	43.5

Sources: WNISR, with IAEA-PRIS, 2018, p. 37

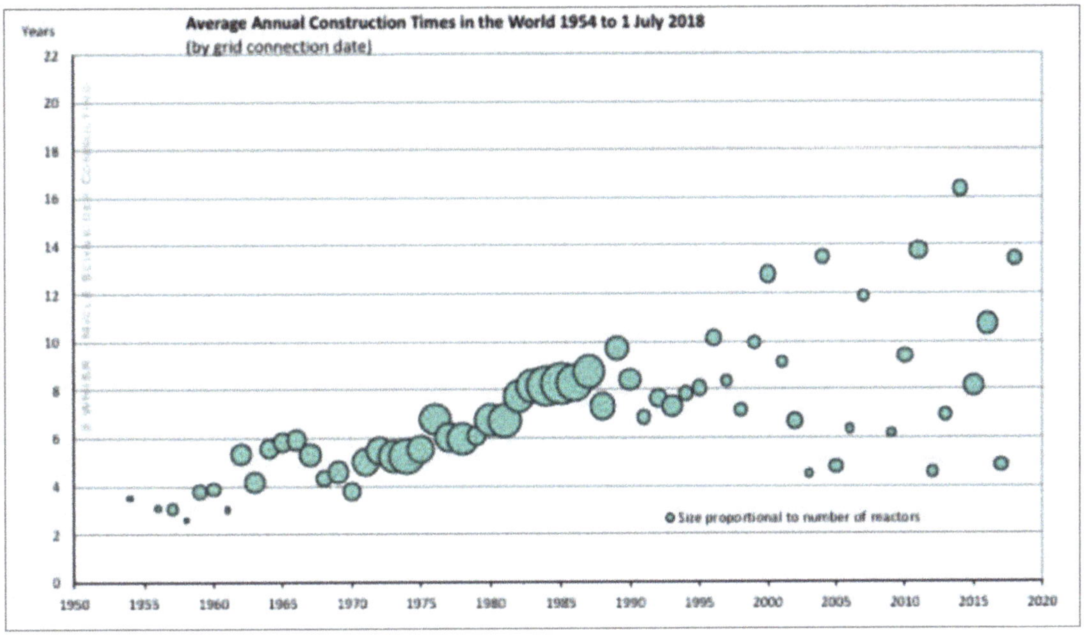

Fig. 7 Average Annual Construction Times in the World
Sources: WNISR, with IAEA-PRIS, 2018, p. 37

6 Construction starts and cancellations

The number of annual construction-starts[23] in the world peaked in 1976 at 44, of which 12 projects were later abandoned. In 2010, there were 15 construction starts—including 10 in China alone—the highest level since 1985 (see Figure 8). That number dropped to 10 in 2013, eight in 2015, five in 2017 and two in 2018 as of mid-year.

Seriously affected by the Fukushima events, China did not start any new building site in 2011 and 2014. While Chinese utilities began constructing six more units in 2015, the number shrank to two in 2016, only a demonstration fast reactor in 2017 and none in 2018 as of mid-year (see Figure 9). In other words, since December 2016, China has not started building a new commercial reactor.

23 Generally, a reactor is considered under construction, when the base slab of the reactor building is being concreted. Site preparation work, excavation and other infrastructure developments are not included.

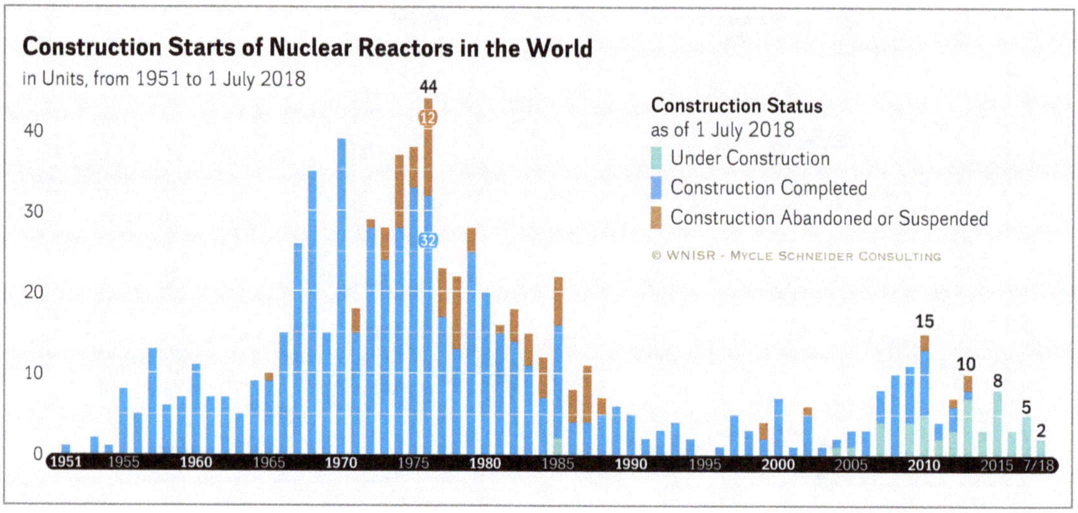

Fig. 8 Construction Starts in the World
Sources: WNISR, with IAEA-PRIS, 2018, p. 38

Over the decade 2008–2017, construction began on 76 reactors in the world (of which five have been later cancelled), that is twice the rate of the decade 1998–2007, when work started on 38 units (of which three have been abandoned). However, close to half (51) of these 114 building starts over the past two decades were in China alone (see Figure 9).

In addition, past experience shows that simply having an order for a reactor, or even having a nuclear plant at an advanced stage of construction, is no guarantee of ultimate grid connection and power production. The abandonment of the two V.C. Summer units at the end of July 2017 after four years of construction and a multi-billion-dollar investment is only the latest example in a long list of failed nuclear power plant projects.

French Atomic Energy Commission (CEA) statistics through 2002 indicate 253 "cancelled orders" in 31 countries, many of them at an advanced construction stage (see also Figure 10). The United States alone accounted for 138 of these order cancellations.[24]

24 French Atomic Energy Commission (CEA), "Elecnuc—Nuclear Power Plants in the World", 2002. The section "cancelled orders" has disappeared after the 2002 edition.

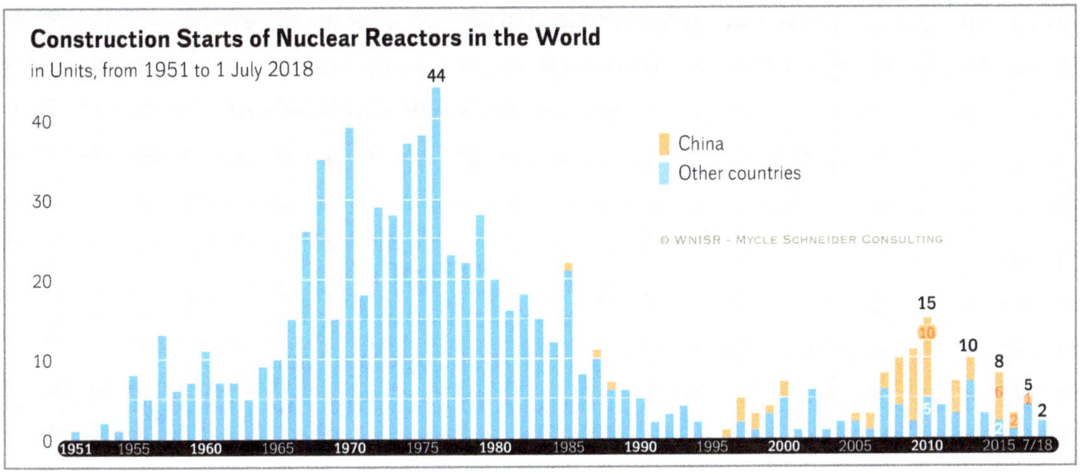

Fig. 9 Construction Starts in the World/China
Sources: WNISR, with IAEA-PRIS, 2018, p. 39

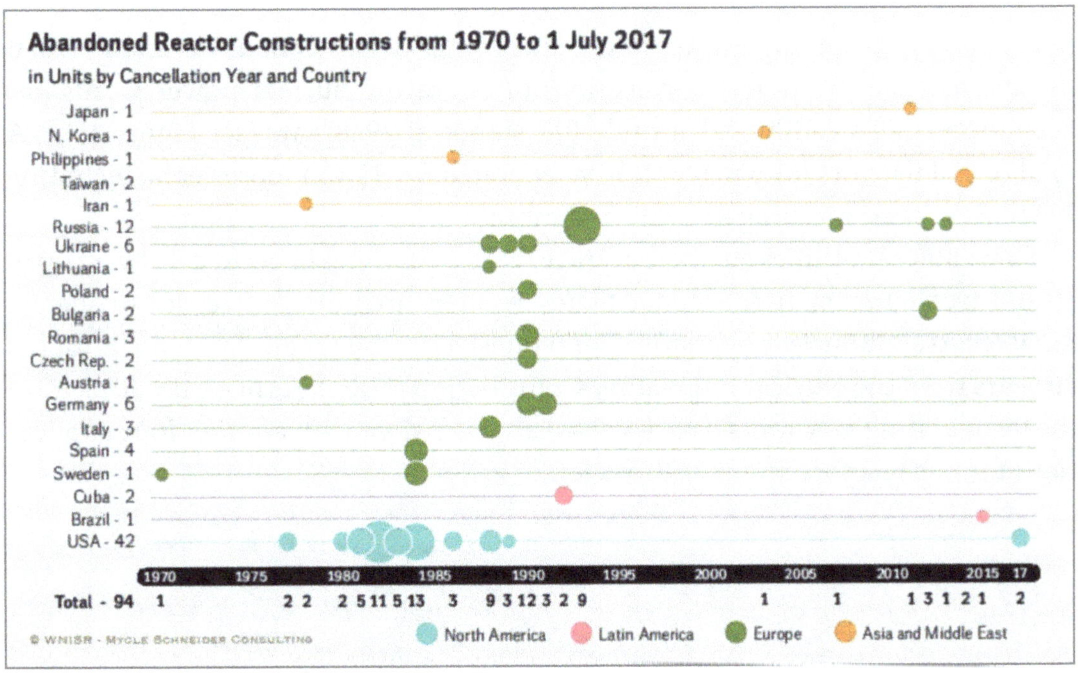

Fig. 10 Cancelled or Suspended Reactor Constructions (Note: This graph only includes constructions that had already officially started).
Sources: WNISR, with IAEA-PRIS, 2018, p. 39

Of the 762 reactor constructions launched since 1951, at least 94 units (12 percent) in 20 countries had been abandoned as of 1 July 2018. The past decade shows about half of the historic abandoning rate of one in eight constructions, as five in 76 building sites officially started during that period were later given up at various stages of advancement.

Close to three quarters (66 units) of all cancelled projects were in four countries alone—the U.S. (42), Russia (12), Germany and Ukraine (six each). Some units were actually 100 percent completed—including Kalkar in Germany and Zwentendorf in Austria—before the decision was taken not to operate them.

There is no thorough analysis of the cumulated economic loss of these failed investments.

7 Operating age

In the absence of any significant new-build *and* grid connection over many years, the average age (from grid connection) of operating nuclear power plants has been increasing steadily and at mid-2017 stands at 29.9 years (see Figure 11).[25] A total of 254 reactors (61.5 percent) have operated for 31 and more years, including 77 (18.5 percent) reaching 41 years and more.

Some nuclear utilities envisage average reactor lifetimes of beyond 40 years up to 60 and even 80 years. In the United States, reactors are initially licensed to operate for 40 years, but nuclear operators can request a license renewal for an additional 20 years from the Nuclear Regulatory Commission (NRC). As of 4 May 2018, 87 of the 99 operating U.S. units have received an extension, with another four applications for five reactors under NRC review.[26]

25 WNISR calculates reactor age from grid connection to final disconnection from the grid and "startup" is synonymous with grid connection and "shutdown" with withdrawal from the grid..

26 NRC, "Status of License Renewal Applications and Industry Activities", Updated 4 May 2018, see http://www.nrc.gov/reactors/operating/licensing/renewal/applications.html, accessed 29 July 2018.

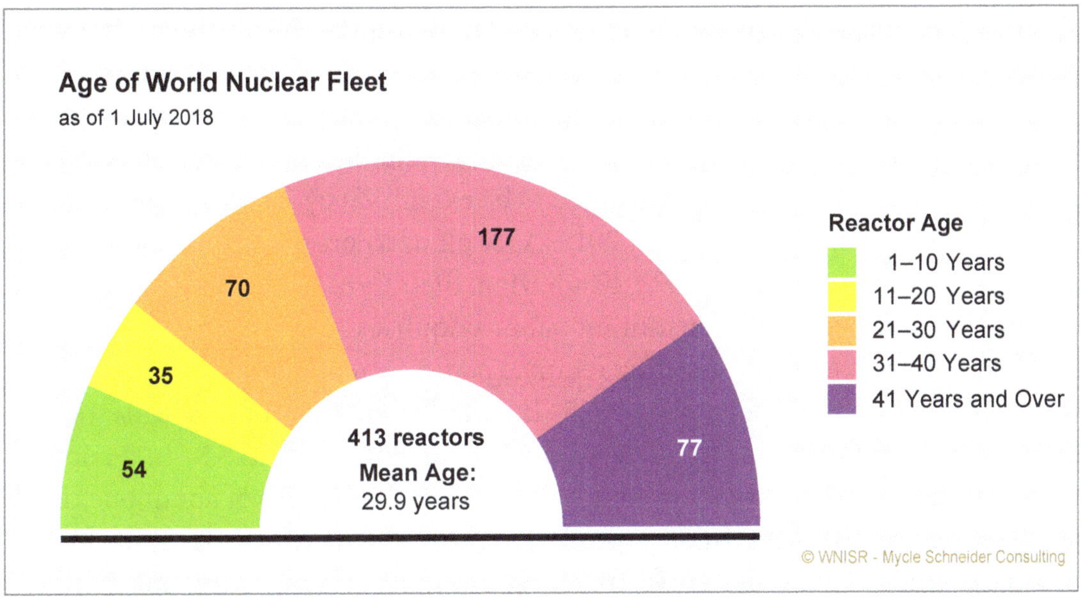

Fig. 11 Age Distribution of Operating Reactors in the World
Sources: WNISR, with IAEA-PRIS, 2018, p. 41

In the U.S., only two of the 34 units that have been shut down had reached 40 years on the grid—Vermont Yankee, closed in December 2014, at the age of 42, and Fort Calhoun, shut down in October 2016, after 43 years of operation. Both had obtained licenses to operate up to 60 years but were closed only two and three years into lifetime extension respectively, mainly for economic reasons. In other words, at least a quarter of the reactors connected to the grid in the U.S. never reached their initial design lifetime of 40 years. On the other hand, of the 99 currently operating plants, 44 units have operated for 41 years and more; thus, half of the units with license renewals have already entered the life extension period, and that share is growing rapidly with the mid-2018 average age of the U.S. operational fleet at 38.1 years.

Many other countries have no specific time limits on operating licenses. In France, where the country's first operating Pressurized Water Reactor (PWR) started up in 1977, reactors must undergo in-depth inspection and testing every decade against reinforced safety requirements. The French reactors have operated for 33.4 years on average, and the oldest have completed the process with the French Nuclear Safety Authority (ASN) evaluating each reactor before allowing a unit to operate for more than 30 years. They could then operate until they reach 40 years, which is the limit of their initial design age. However, the assessments are years behind schedule. The French utility Électricité de France (EDF) clearly prioritizes lifetime extension to 50 years over large-scale new-build. In a recent presentation, EDF states: "Continuing operation beyond 40 years means allowing for the progressive increase of renewable energies while guaranteeing a lasting, quasi entirely decarbonized and continuous electricity supply thanks to the flexibility of the nuclear fleet."[27]

EDF's general approach to lifetime extension is currently under review by ASN's Technical Support Organization, the Institute for Radiation Protection and Nuclear Safety (IRSN) and is now scheduled to be examined by its expert committees (Groupes Permanents) in 2018 and 2019. ASN plans to provide its opinion on the general assessment outline by 2020. In addition, lifetime extension beyond 40 years requires site-specific, time-consuming public enquiries.

Current French energy legislation requires planning to limit the nuclear share in power production to 50 percent by 2025. The implementation of this legislation—even if the 2025 target was to be delayed as is currently discussed—in a context of stagnating electricity consumption, would mean the closure of about one third of the French reactor fleet. In other words, many of the lifetime extensions would become obsolete. No point in investing large amounts of money into upgrading if the plant is shut down shortly after. A particularly difficult aspect of the lifetime management in France is that the units licensed to use plutonium-uranium mixed oxide fuel (MOX) are also amongst the oldest reactors. The criteria for selection of reactors to be closed remain under discussion.

In assessing the likelihood of reactors being able to operate for 50 or 60 years, it is useful to compare the age distribution of reactors that are currently operating with those that have already shut down (see Figure 11 and Figure 12). As of mid-2018, 77 of the world's reactors have operated for 41 years and more, and a total of 81 that have already passed their 40-year lifetime are considered in lifetime extension.[28] As the age pyramid illustrates, that number could rapidly increase over the next few years. A total of 254 units have already reached or exceeded age 31.

The age structure of the 173 units already shut down completes the picture. In total, 60 of these units operated for 31 years and more, and of those, 20 reactors operated for 41 years and more (see Figure 12). Many units of the first-generation designs only operated for a few years. Considering that the average age of the 173 units that have already shut down is about 25 years, plans to extend the operational lifetime of large numbers of units to 40 years and far beyond seemed rather optimistic. However, the operating time prior to shutdown has clearly increased continuously. But while the *average* annual age at shutdown got close to 40 years, it only passed that age in two years so far: in 2014, when the only such unit shut

27 EDF, "Le parc nucléaire en exploitation en France : Exploitation, maintenance et Grand Carénage", 11 January 2018.

28 WNISR considers the age starting with grid connection, and while figures used to be rounded by half-years, as of WNISR2016 they are rounded by the tenth of the year.

down that year (Vermont Yankee in the U.S.) after 42 years of operation; and in 2016, with two reactors shutting down at age 43 (Fort Calhoun, U.S.) and 45 (Novovoronezh, Russia) respectively.

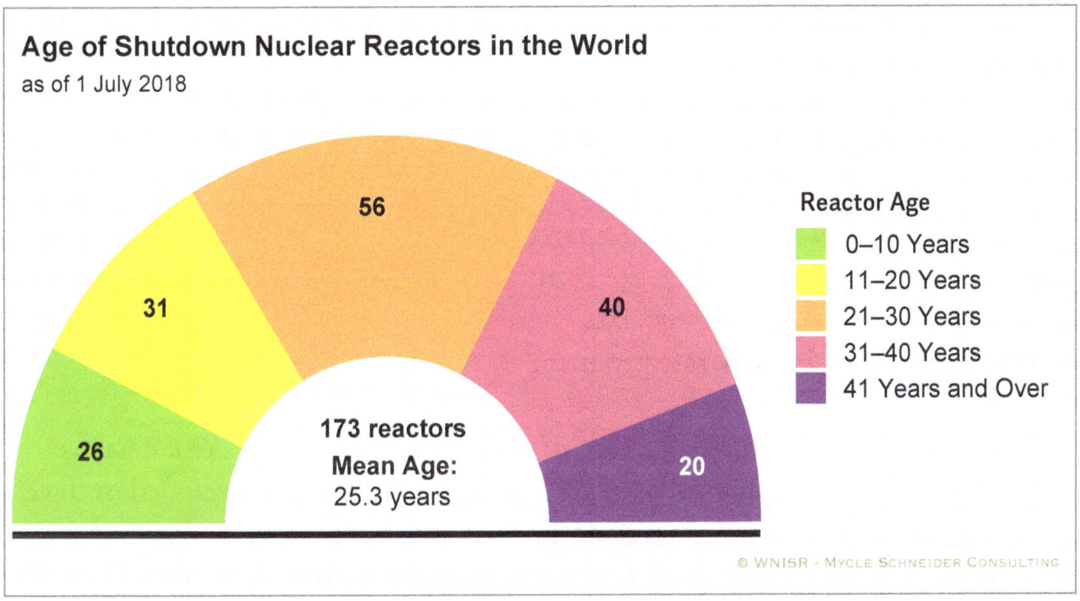

Fig. 12 Age Distribution of Shut Down Nuclear Power Reactors
Sources: WNISR, with IAEA-PRIS, 2018, p. 43

As a result of the Fukushima nuclear disaster, more pressing questions have been raised about the wisdom of operating older reactors. The Fukushima Daiichi units (1 to 4) were connected to the grid between 1971 and 1974. The license for unit 1 had been extended for another 10 years in February 2011, one month before the catastrophe began. Four days after the initial events in Japan began, the German government ordered the shutdown of seven reactors that had started up before 1981 (plus one commissioned in 1983). The sole, decisive selection criterion was operational age. Other countries did not adopt the same approach, but it is clear that the 3/11 events had an impact on previously assumed extended lifetimes in other countries as well, including in Belgium, Switzerland, and Taiwan. And more recently, in the first half of 2017, South Korea's incoming President Moon shut down the country's oldest reactor (Kori-1), explicitly at the age of forty, ruling out lifetime extensions in the future. Sweden also closed its oldest unit, Oskarshamn-1 at age 46. And Germany shut down its then oldest reactor, Gundremmingen-B, at year-end 2017, at age 33. The average age of these three units remains with 39.9 years again just below 40.

8 Potential newcomer countries

The International Atomic Energy Agency (IAEA) assumes that to meet their prediction of more than doubling of current capacity in the higher nuclear scenario, considerable new construction will occur in existing nuclear countries, such as China, South Korea and India, but also envisages significant capacity build-up in newcomer countries.

The industry's lobby organisation WNA suggests that there are 20 countries in which nuclear power is being planned for the first time, with an additional 20, where the nuclear option is under consideration. Notwithstanding any reality check of these assumptions, this is small compared to renewable energy, as at end of 2015, targets had been established in 173 countries at the national or state/provincial level.[29] The WNA further categorizes those countries in which nuclear power is being planned into five separate groups[30]:

- Power reactors under construction: United Arab Emirates (UAE), Belarus.
- Contracts signed, legal and regulatory infrastructure well-developed or developing: Lithuania, Turkey, Bangladesh, Vietnam (but deferred).
- Committed plans, legal and regulatory infrastructure developing: Jordan, Poland, Egypt.
- Well-developed plans but commitment pending: Thailand, Indonesia, Kazakhstan, Saudi Arabia, Chile; or commitment stalled: Italy.
- Developing plans: Israel, Nigeria, Kenya, Laos, Malaysia, Morocco, Algeria.

The following section will look at the countries, in which WNA considers nuclear power plants are at least 'committed plans'.

Under construction

Bangladesh

On 30 November 2017, Bangladesh officially began construction of the first unit of the Rooppur nuclear plant.[31] Unit 1 is now scheduled to begin operation in 2023

---

29 REN21, "Renewables 2016—Global Status Report", Renewable Energy Policy Network for the 21st Century, 2016.

30 WNA, "Emerging Nuclear Energy Countries", January 2018, see http://www.world-nuclear.org/information-library/country-profiles/others/emerging-nuclear-energy-countries.aspx.

31 WNISR, "Construction Start at First Nuclear Power Plant in Bangladesh", 5 April 2018, see https://www.worldnuclearreport.org/Construction-Start-at-First-Nuclear-Power-

followed by unit 2 in 2024.[32] The idea of building nuclear reactors at Rooppur goes back to even before Bangladesh became an independent country, to a 1963 plan by the Pakistan Atomic Energy Commission to build one reactor in West Pakistan and one in East Pakistan.[33]

The current reactor deal dates back to November 2011 when the Bangladeshi Government announced that it was prepared to sign a deal with the Russian Government for two 1000 MW units—the first of which was to start up between 2017 and 2018—at a total cost of US$1.5-2 billion.[34] Since then, although negotiations have reportedly been ongoing, the startup date has been continually postponed and the expected construction cost has risen sharply.

By 2015, the Bangladeshi Finance Minister was quoted as saying the project was then expected to cost US$12.65 billion.[35] A December 2015 agreement was said to be signed between the Bangladesh Atomic Energy Commission and Rosatom for 2.4 GW of capacity, with work then expected to begin in 2016 and operation to start in 2022 and 2023.[36] According to the deal, Russia would provide 90 percent of the funds on credit at an interest rate of Libor plus 1.75 percent. In late May 2016, negotiations were concluded over the US$12.65 billion project, with Russia making available US$11.385 billion.[37] In April 2017, Tass, the Russian news agency, reported that permission to start construction had been granted and that work would commence in the second half of 2017.[38]

Plant-in-Bangladesh.html, accessed 29 June 2018.

32 *NEI*, "Work begins on foundation for unit 1 of Bangladesh NPP", see http://www.neimagazine.com/news/newswork-begins-on-foundation-for-unit-1-of-bangladesh-npp-6107152/, accessed 22 April 2018.

33 IAEA, "Bangladesh", Country Nuclear Power Profiles, 2012, see https://www-pub.iaea.org/mtcd/publications/pdf/cnpp2012_cd/countryprofiles/Bangladesh/Figures/Bangladesh%20CNPP.pdf, accessed 8 May 2018.

34 Srinivas Laxman, "Bangladesh & Russia Sign N-Plant Deal For Two Reactors At Rooppur, Asian Scientist", *Asian Scientist*, 4 November 2011, see https://www.asianscientist.com/2011/11/topnews/rooppur-nuclear-power-project-bangladesh-russia-sign-nuclear-agreement-2011/, accessed 22 April 2018.

35 *WNN*, "Bangladesh, Russia ink $12.65 billion Rooppur plant deal", 29 December 2015, see http://www.world-nuclear-news.org/NN-Bangladesh-Russia-ink-12.65-billion-Rooppur-plant-deal-29121501.html, accessed 22 April 2018.

36 *WNN*, "Bangladesh, Russia Ink $12.65 Billion Rooppur Plant Deal" (ref. 4).

37 *NEI*, "Russia initials credit agreement with Bangladesh for Rooppur NPP", 30 May 2016, see http://www.neimagazine.com/news/newsrussia-initials-credit-agreement-with-bangladesh-for-rooppur-npp-4907672/, accessed 22 April 2018.

38 TASS, "Rosatom plans to launch construction of Ruppur power plant in Bangladesh", 19 April 2017, see http://tass.com/economy/942156.

The project's economics have been widely questioned. Earlier in 2017, a retired nuclear engineer who had been involved in advising the Bangladesh Atomic Energy Commission (BAEC), argued in one of the leading English-language newspapers in Bangladesh that the country was "paying a heavy price" for BAEC not having "undertaken a large-scale programme of recruitment, and training of engineers"; he also charged that Bangladesh was buying reactors at the "unreasonable and unacceptable" price of US$5,500/kW because its "negotiators didn't have the expertise to properly scrutinise the quoted price".[39]

Belarus

Construction started in November 2013 at Belarus's first nuclear reactor at the Ostrovets power plant, also called Belarusian-1. Construction of a second 1200 MWe AES-2006 reactor started in June 2014. In November 2011, the Russian and Belarusian governments agreed that Russia would lend up to US$10 billion for 25 years to finance 90 percent of the contract between Atomstroyexport and the Belarus Directorate for Nuclear Power Plant Construction. In July 2012, the contract was signed for the construction of the two reactors for an estimated cost of US$10 billion.[40] The project assumes liability for the supply of all fuel and repatriation of spent fuel for the life of the plant. The fuel is to be reprocessed in Russia and the separated wastes returned to Belarus. When construction started, it was stated that the reactors will be completed until 2018 and 2020 respectively.[41] In August 2016, the reactor pressure vessel of unit one slipped and fell two meters before hitting the ground, during installation. This lead to an eight-month delay, while it was replaced.[42] In March 2018, the head of the reactor division at the power plant said that it was then expected that electricity would be supplied to the grid in the 4th Quarter of 2019 with the second one online in July 2020.[43]

---

39 Abdul Martin, "The economics of the Rooppur Nuclear Power Plant", *The Daily Star*, 2 March 2017, see https://www.thedailystar.net/op-ed/economics/the-economics-the-rooppur-nuclear-power-plant-1369345, accessed 22 April 2018.

40 NIW, "Belarus, Aided by Russia and Broke, Europe's Last Dictatorship Proceeds With NPP", 28 September 2012.

41 WNN, "Ostrovets plant meets construction safety rules", 7 November 2014, see http://www.world-nuclear-news.org/NN-Ostrovets-plant-meets-construction-safety-rules-07111401.html, accessed 21 April 2018.

42 NIW, "Briefs-Belarus", 7 April 2017.

43 Belarus News, "Belarusian nuclear power plant to give electricity to national power grid in Q4 2019", *Belarus News*, 28 March 2018, see http://eng.belta.by/economics/view/belarusian-nuclear-power-plant-to-give-electricity-to-national-power-grid-in-q1-2019-110418-2018/.

The official cost of the project has increased by 26 percent, to 56 billion Russian roubles, in 2001 prices (US$$_{2001}$1.8 billion).[44] However, the falling exchange rate of the rouble against the dollar significantly affects the dollar price of the project.

The project is the focus of international opposition and criticism, with formal complaints from the Lithuanian government, that has published a list of fundamental problems of the project.[45] The Belarussian government, in order to allay European concerns about Ostrovets submitted the project to a post-Fukushima nuclear stress test and it produced in 2017 a national report, which is currently submitted to peer-review by a commission from the European nuclear regulators group ENSREG and the European Commission. In July 2018, the European Commission announced that the report had been presented to the Belarussian authorities and the executive summary was made public, which concludes that "although the report is overall positive, it includes important recommendations that necessitate an appropriate follow up".[46] The next step is these recommendations need to be incorporated into the next draft of the National Action Plan.[47]

Belarus has historically been an importer of electricity from Russia and Ukraine. But in May 2018, Vice-Premier Vladimir Semashko stated: "In 2018 we stopped electric energy import, because we had upgraded our own power grid. We are self-reliant and can provide ourselves with our own electric energy."[48] In fact, Semashko claims that in the first four months of 2018, Belarus exported 0.4 TWh. The startup of the Ostrovets nuclear plant would significantly increase excess capacity. Lithuania has said it will not accept any electricity from Belarus and is trying to get is neighbours to follow the ban. Currently this has not been successful,

---

44 Charter 97, "Astravets NPP Becomes 12 Billion More Expensive In One Day", see https://charter97.org/en/news/2016/12/30/236059/, accessed 21 April 2018.

45 Bryan Bradley, "Lithuania Urges Belarus to Halt Nuclear Project on Safety Issues", *Bloomberg*, 20 August 2013, see https://www.bloomberg.com/news/articles/2013-08-20/lithuania-urges-belarus-to-halt-nuclear-project-on-safety-issues, accessed 22 April 2018.

46 ENSREG, "Belarus Stress Tests Peer Review – Executive summary", European Nuclear Safety Regulators Group, July 2018, see http://www.ensreg.eu/sites/default/files/attachments/hlg_p2018-36_156_belarus_stress_test_prt_report_-_executive_summary_0.pdf accessed 5 July 2018.

47 European Commission, "Comprehensive risk and safety assessments of the Belarus nuclear power plant completed" (Press Release), 3 July 2018, see http://europa.eu/rapid/press-release_IP-18-4347_en.htm, accessed 4 July 2018.

48 Belarus News, "Belarus ramps up electricity export in 2018", 14 May 2018, see http://eng.belta.by/economics/view/belarus-ramps-up-electricity-export-in-2018-111638-2018/, accessed 3 July 2018.

although there has been an agreement to introducing an electricity import tax.[49] Russia is currently upgrading its grid connection between the Leningradskaya and Smolensk nuclear power stations, thus potentially also enabling a better connection of Ostrovets to the West-Russian electricity grid, circumventing the Baltic States. Vice-Premier Semashko is confident: “Our energy is cheaper, and it will be on demand on this market.”[50]

Turkey

In Turkey, three separate projects are being developed with three different reactor designs and three different sets of financial sources. Despite this, in early 2018, construction formally began on the first of these projects.

Some four decades after the first ideas came up for a nuclear power plant at Akkuyu, in the province of Mersin on Turkey’s Mediterranean coast, construction started in April 2018, a day before President Putin of Russia visited Turkey for the official launch of the project.[51] The power plant is to be implemented by Rosatom of Russia under a Build-Own-Operate- (BOO) model. In February 2018, only two months prior to the official construction start, Rosatom’s Turkish partners pulled out. The consortium of private companies Cengiz Holding, Kolin Insaat Turizm Sanayi ve Ticaret et Kalyon Insaat Sanayi ve Ticaret, which was to hold 49 percent of the shares, quit the project because they expected too little benefits from the project.[52]

A company, JSC Akkuyu Nuclear has been established to ensure construction of the project and has been designated as the Strategic Investor. According to the establishing agreement, at least 51 percent of shares in the finished project should belong to Russian companies and up to 49 percent of shares can be available for sale to outside investors. Negotiations with potential Turkish investors continue

---

49 Reuters, “Baltics to cooperate on Belarus nuclear power tax”, 14 December 2017, see https://www.reuters.com/article/baltics-energy/baltics-to-cooperate-on-belarus-nuclear-power-tax-idUSL8N1OC3QD, accessed 22 April 2018.

50 Belarus News, “Belarus ramps up electricity export in 2018”, op.cit.

51 Tuvan Gumrukeu and Orhan Coskun, “Turkey grants Rosatom construction license for first unit of Akkuyu...”, *Reuters*, 2 April 2018, see https://www.reuters.com/article/us-turkey-russia-nuclearpower/turkey-grants-rosatom-construction-license-for-first-unit-of-akkuyu-nuclear-plant-idUSKCN1H91OY, accessed 22 May 2018.

52 AFP, “Un consortium turc se retire du projet de la centrale nucléaire d’Akkuyu”, see https://www.romandie.com/news/887776.rom, accessed 4 July 2018.

after the three prospective partners withdrew.[53] However, Rosatom has stated that it would be able to complete the project even if it is unable to attract local investors.[54]

An agreement was signed in May 2010 for four VVER1200 reactors (Generation III+), with construction originally expected to start in 2015. At the heart of the project is a 15-year Power Purchase Agreement (PPA), which includes 70 percent of the electricity produced from units 1 and 2 and 30 percent of units 3 and 4. Therefore 50 percent of the total power from the station is to be sold at a guaranteed price for the first 15 years, with the rest to be sold on the market.

After a fresh series of delays, on 3 March 2017, Akkuyu JSC applied for a construction license.[55] Rosatom stated: "According to the Intergovernmental Agreement, the commissioning of the first power unit must take place no later than 7 years after the issuance of all permits for construction by the Republic of Turkey."[56]

In July 2017 the European Parliament adopted a resolution which called on the Turkish Government to halt the plans for the construction of the Akkuyu project due to its location in a region prone to severe earthquakes and called on "the Turkish Government to involve, or at least consult, the governments of its neighbouring countries, such as Greece and Cyprus."[57]

In April 2018, a construction license was awarded, and the first concrete was poured, with first electricity expected to be in 2023 (the 100th anniversary of the founding of the modern state of Turkey), with all four units to be operational by 2025.[58] (See "Contract Signed" section hereunder for information on further projects in Turkey.)

---

53 Rosatom, "JSC Akkuyu Nuclear designated strategic investor in Turkey", 2 April 2018, see http://www.rosatom.ru/en/press-centre/news/jsc-akkuyu-nuclear-designated-strategic-investor-in-turkey/, accessed 20 April 2018.

54 Reuters, "Russia is able to complete Akkuyu nuclear power plant construction: Russian minister", 6 April 2018, see http://www.hurriyetdailynews.com/russia-is-able-to-complete-akkuyu-nuclear-power-plant-construction-russian-minister-129886, accessed 6 April 2018.

55 WNN, "Akkuyu project receives production licence"," see http://www.world-nuclear-news.org/NN-Akkuyu-project-receives-production-licence-16061701.html, accessed 22 April 2018.

56 Coskun, "Turkey's First Nuclear Plant Facing Further Delays – Sources" (ref. 32).

57 European Parliament, "P8_TA(2017)0306, 2016 Report on Turkey European Parliament resolution of 6 July 2017 on the 2016 Commission Report on Turkey (2016/2308(INI))", 6 July 2017, see http://www.europarl.europa.eu/sides/getDoc.do?pubRef=-//EP//NONSGML+TA+P8-TA-2017-0306+0+DOC+PDF+V0//EN, accessed 22 April 2018.

58 NEI, "Construction of Turkey's Akkuyu NPP begins", 4 April 2018, see http://www.neimagazine.com/news/newsconstruction-of-turkeys-akkuyu-npp-begins-6102914/, accessed 22 April 2018.

United Arab Emirates

In the United Arab Emirates (UAE), construction is ongoing at the Barakah nuclear project, 300 km west of Abu Dhabi, where there are four reactors under construction. At the time of the contract signing in December 2009 with Korean Electric Power Corp., the Emirates Nuclear Energy Corp (ENEC), said that "the contract for the construction, commissioning and fuel loads for four units equalled approximately US$20 billion, with a high percentage of the contract being offered under a fixed-price arrangement".[59]

The total cost of the project is at least €24.4 billion (US$28.2 billion). The financing for this was US$16.2 billion Abu Dhabi's Department of Finance, equity financing US$4.7 billion, US$2.5 billion through a loan from the Export-Import Bank of Korea, with loan agreements from the National Bank of Abu Dhabi, First Gulf Bank, HSBC and Standards Charter making up the remainder.[60] In October 2016, Korea Electric Power Corporation (KEPCO) took an 18 percent equity stake in Nawah Energy Company that owns the four reactors, with ENEC, holding the remaining 82 percent.[61]

In July 2010, a site-preparation license and a limited construction license were granted for four reactors at Barakah, 53 kilometres from Ruwais[62] A tentative schedule published in late December 2010, and not publicly altered since, suggested that Barakah-1 would start commercial operation in May 2017 with unit 2 operating from 2018, unit 3 in 2019, and unit 4 in 2020. Construction of Barakah-1 officially started on 19 July 2012, of Barakah-2 on 28 May 2013, on Barakah-3 on 24 September 2014 and unit 4 on 30 July 2015.[63] As late as October 2016, Korean press was reporting unit 1 to be still scheduled for completion by May 2017.[64] In May 2017,

---

59 ENEC, "UAE Selects Korea Electric Power Corp, as Prime Team as Prime Contractor for Peaceful Nuclear Power", Emirates Nuclear Energy Corporation, 27 December 2009, see https://www.enec.gov.ae/news/uae-selects-korea-electric-power-corp-as-prime-team-as-prime-contractor-fo/, accessed 22 April 2018.

60 NIW, "Kepco takes 18% of Barakah", 21 October 2016.

61 NEI, "Kepco and Enec set up joint venture for Barakah NPP", 25 October 2016, see http://www.neimagazine.com/news/newskepco-and-enec-set-up-joint-venture-for-barakah-npp-5647366/, accessed 22 April 2018.

62 Arabian Business, "ENEC Welcomes Regulator's License Approval", 11 July 2010, see http://www.arabianbusiness.com/enec-welcomes-regulator-s-licence-approvals-306150.html, accessed 22 April 2018.

63 Ibid.

64 Lee Hyo-sik, "KEPCO to operate UAE nuclear plant for 60 years", *The Korean Times*, 20 October 2016, see http://www.koreatimes.co.kr/www/news/biz/2016/10/123_216466.html, accessed 22 April 2018.

ENEC announced it had "completed initial construction activities for Unit 1" and the "handover of all systems for commissioning"; the plant as a whole would be 81 percent complete, with Barakah-1 at 95 percent finished. At the same time, ENEC stated: "The timeline includes an extension for the start-up of nuclear operations for Unit 1, from 2017 to 2018, to ensure sufficient time for international assessments and adherence to nuclear industry safety standards, as well as a reinforcement of operational proficiency for plant personnel."[65] In March 2018, the extent of the delay was confirmed with *Nawah* reporting that the startup of Unit 1 would only be in 2019.[66] But only a few months later, in July 2018, a new delay was announced, so that startup would be in late 2019 or early 2020[67], so that commercial operation would not be undertaken until 2020, three years behind schedule.

The UAE released a long-term energy plan in February 2017, which proposes that by 2050 renewable energy will provide 44 percent of the country's electricity, with natural gas 38 percent, "clean fossil fuels" 12 percent and nuclear six percent.[68] The nuclear share is in line with expected output from the Barakah nuclear power plant, so it seems that no further nuclear power plants are envisaged at this point. In September 2017, Government officials confirmed that there were no plans to build a second plant.[69]

---

65 ENEC, "ENEC Announces Completion of Initial Construction Work for Unit 1 of Barakah Nuclear Energy Plant & Progress Update Towards Safety-led Operations", Emirates Nuclear Energy Corporation, 5 May 2017, see https://www.enec.gov.ae/enec-announces-completion-of-initial-construction-work-barakah-unit-1-progress-update/, accessed 22 April 2018.

66 Nawah, "Next phase of preparations for Barakah Unit 1 Nuclear Operations starts" (Press Release), 28 May 2018, see http://www.nawah.ae/en/news/NextphaseofpreparationsforBarakah.html, accessed 29 May 2018.

67 Arabian Business, "UAE further delays launch of first nuclear reactor", 4 July 2018, see https://www.arabianbusiness.com/energy/400041-uae-further-delays-launch-of-first-nuclear-reactor, accessed 8 July 2018.

68 LeAnne Graves, "UAE Energy Plan aims to cut CO2 emissions 70% by 2050", *The National*, 10 January 2017, see https://www.thenational.ae/uae/uae-energy-plan-aims-to-cut-co2-emissions-70-by-2050-1.51582, accessed 22 April 2018.

69 Amena Bahr, "UAE Abu Dhabi Unlikely to Build a Second Nuclear plant", *Nuclear Intelligence Weekly*, 29 September 2017.

Contracts signed

Egypt

In Egypt, the government's Nuclear Power Plants Authority was established in the mid-1970s, and plans were developed for 10 reactors by the end of the century. Little development occurred for several decades. Then, in February 2015, Russia's Rosatom and Egypt's Nuclear Power Plant Authority eventually did sign an agreement that was expected to lead to the construction and financing of two reactors and possibly two additional ones. In November 2015, an intergovernmental agreement was signed for the construction of four VVER-1200 reactors at Dabaa, 130 km northwest of Cairo. In May 2016, it was announced that Egypt concluded a US$25 billion loan with Russia for nuclear construction.[70] According to the Egyptian official journal, the loan is to cover 85 percent of the project cost, with the total investment thus estimated at around US$29.4 billion.

In December 2017, Rosatom Director General Alexey Likhachov and Mohamed Shaker, Egypt's Energy Minister signed a notice to proceed with construction as well as an agreement that "spans the power plant's entire life cycle, i.e. 70 to 80 years".[71] The total cost of the project was now reported to be US$60 billion, of which US$30 billion for the reactor construction. Three other deals were signed to cover the supply of nuclear fuel for 60 years, operation and maintenance for the first 10 years of operation and operating and training of personnel.[72] Russia would supply a loan of US$25 billion, at three percent interest for 85 percent of the construction cost. The Egyptian government agreed to pay back over 22 years starting in 2029. The next two and half years will focus on site preparation and licensing. With construction expected to take five years, the completion of the project is now expected in 2026/27.[73]

---

70 Asma Alsharif, "Russia to lend Egypt $25 billion to build nuclear power plant", Reuters, 1 May 2016, see http://www.reuters.com/article/us-egypt-russia-nuclear-idUSKCN0YA1G5, accessed 23 May 2016.

71 WNN, "'Notice to proceed' contracts signed for El Dabaa", 11 December 2017, see http://www.world-nuclear-news.org/NN-Notice-to-proceed-contracts-signed-for-El-Dabaa-1112178.html, accessed 24 April 2018.

72 Phil Chaffee, "Rosatom Locks in $30 Billion Nuclear Deal in Egypt", *NIW*, 15 December 2017.

73 Dan Yurman, "Egypt's $60 Billion Bet on Nuclear Energy", The Energy Collective, 23 April 2018, see https://theenergycollective.com/dan-yurman/2431718/egypts-60-billion-bet-on-nuclear-energy, accessed 24 April 2018.

Turkey

Besides Akkuyu (see above), Turkey has two other nuclear projects under development – Sinop and İğneada.

Sinop is on Turkey's northern coast and is planned to host a 4.4 GW power plant of four units of the ATMEA reactor-design. If completed this would be the first reactors of this design, jointly developed by Japanese Mitsubishi and French AREVA.[74] In April 2015, Turkish President Erdogan approved parliament's ratification of the intergovernmental agreement with Japan.[75]

The estimated cost of the project was initially US$22 billion and involved a consortium of Mitsubishi, AREVA NP (now known again as Framatome), GDF-Suez (now known as Engie), and Itochu, who between them would own 51 percent of the project, with the remaining 49 percent owned by Turkish companies including the State-owned electricity generating company EÜAS.[76]

The division between the international partners remains in fact undecided. The ongoing financial problems of new-old Framatome after the absorption by EDF are affecting its ability to invest in the project, as does the review by Engie of its involvement in nuclear projects across its portfolio. Furthermore, concerns remain about site suitability given its seismic conditions.[77]

In March 2018, reports from Japan suggest that the expected cost of the project has doubled and is now expected to be US$37.5 billion and that it would be difficult to see completion by 2023.[78] Then in April 2018, press reports from Japan suggested that Itochu would no longer be willing to participate due to the exploding cost estimates, which have risen to more than JPY5,000 billion (US$46.2 billion) from JPY2,000 billion (US$19 billion) in 2013.[79]

---

74 WNN, "Turkish utility eyes large stake in Sinop project", 12 May 2015, see http://www.world-nuclear-news.org/C-Turkish-utility-eyes-large-stake-in-Sinop-project-12051501.html, accessed 22 April 2018.

75 WNN, "Ground broken for Turkey's first nuclear power plant", 15 April 2015, see http://www.world-nuclear-news.org/NN-Ground-broken-for-Turkeys-first-nuclear-power-plant-1541501.html, accessed 22 April 2018.

76 WNN, "Turkish Utility Eyes Large Stake in Sinop Project", op.cit.

77 NIW, "Akkuyu's Prospects Pull Past Sinop", 22 July 2016.

78 Yasuaki Oshika, "Japan's nuclear export to Turkey in doubt as costs estimate doubles", *The Ssahi Shimbun*, 15 March 2018, see http://www.asahi.com/ajw/articles/AJ201803150046.html, accessed 22 April 2018.

79 NEI, "Japan's Itochu pulls out of Turkish nuclear project – Nuclear Engineering International", see http://www.neimagazine.com/news/newsjapans-itochu-pulls-out-of-turkish-nuclear-project-6133206, accessed 15 May 2018.

In October 2015, the Turkish government suggested it was aiming to build a third nuclear power plant, at the İğneada site. The most likely constructors would be Westinghouse and the Chinese State Nuclear Power Technology Corporation (SNPTC). Chinese companies are said to be "aggressively" pursuing the contract, reportedly worth US$22-25 billion.[80] In September 2016, China and Turkey signed a nuclear co-operation agreement, a similar mechanism used to develop the other nuclear projects in the country.[81] However, the financial collapse of Westinghouse, makes their current involvement in the project unlikely.

Vietnam

A decision by the Prime Minster of Vietnam of July 2011 stated that by 2020 the first nuclear power plant will be in operation, with a further 7 GW of capacity to be in operation by 2025 and total of 10.7 GW in operation by 2030. In October 2010, Vietnam had signed an intergovernmental agreement with Russia's Atomstroyexport to build the Ninh Thuan-1 nuclear power plant, using 1200 MW VVER reactors. Construction was slated to begin in 2014. However, numerous delays have occurred and the national electricity development plan, approved by the government in March 2016, envisioned the "first nuclear power plant put into operation in 2028".[82] Vietnam's nuclear power ambitions were severely curtailed in November 2016, when 92 percent of the members of the National Assembly approved a government motion to cancel the proposed nuclear projects with both Russia and Japan, due to slowing electricity demand increases, concerns of safety and rising construction costs.[83]

80 NEI, "Turkey finalizes site for third NPP", 18 March 2016, see http://www.neimagazine.com/news/newsturkey-finalizes-site-for-third-npp-4843161/, accessed 26 April 2017.

81 Herguner Ozeke, "Turkey Looks to China for Third Nuclear Power Plant", Lexology, 3 January 2018, see https://www.lexology.com/library/detail.aspx?g=d0b6672d-14e1-43d0-a5b3-c9750552f521, accessed 22 April 2018.

82 VietNamNet, "Vietnam needs US$148 billion to develop national electricity until 2030", 20 March 2016, see http://english.vietnamnet.vn/fms/society/152739/vietnam-needs-us-148-billion-to-develop-national-electricity-until-2030.html, accessed 22 April 2018.

83 NIW, "Briefs – Vietnam", 28 November 2016.

"Committed plans"

Lithuania

Lithuania had two large RBMK (Chernobyl-type) reactors at Ignalina, which were shut down in 2004 and 2009, a requirement for joining the European Union. Since then there have been ongoing attempts to build a replacement, either unilaterally or with neighbouring countries. However, in October 2012, a consultative national referendum on the future of nuclear power was held and 63 percent voted against new nuclear construction, with sufficient turnout to validate the result.[84] Prior to his appointment as Prime Minister, Algirdas Butkevicius stated that legislation prohibiting the project would be submitted once the new parliament convenes and that "the people expressed their wish in the referendum, and I will follow the people's will".[85] In early 2016, the Energy Minister of Lithuania, Rokas Masiulis, said that the project had been shelved indefinitely, due to unfavorable market conditions.[86]

Jordan

Influential policy makers in Jordan have long desired the acquisition of a nuclear power plant. In 2007, the government established the Jordan Atomic Energy Commission (JAEC) and the Jordan Nuclear Regulatory Commission. JAEC started conducting a feasibility study on nuclear power, including a comparative cost/benefit analysis.[87] And in September 2014, JAEC and Rosatom signed a two-year development framework for a project, which was estimated to cost under US$10 billion and generate electricity costing US$0.10/kWh.[88] After lengthy unfruitful negotiations, in May 2018, an unnamed government official revealed to *The Jordan Times* that the plan to build two 1000 MW "is now over", and that "Jordan is now focusing on small modular reactors".[89] This was confirmed the following month by the Jordan

---

84 Christian Lowe, "Lithuanians send nuclear plant back to drawing board", *Reuters*, 15 October 2012, see http://www.reuters.com/article/us-lithuania-nuclear-idUSBRE89E-0BW20121015, accessed 22 April 2018.

85 NIW, "Lithuania—Prospective PM Wants to Scrape Visaginas", 9 November 2012.

86 The Baltic Course, "Masiulis: Visaginas NPP project has been shelved for now", 20 January 2016, see http://www.baltic-course.com/eng/energy/?doc=115564, accessed 22 April 2018.

87 Mark Hibbs, "Jordan reactor siting study to be done in 2009, JAEC says", *Nucleonics Week*, 27 September 2007.

88 NIW, "Briefs – Jordan", 18 April 2014.

89 Mohammad Ghazal, "Jordan to replace planned nuclear plant with smaller, cheaper facility", *Jordan Times*, 26 May 2018, see http://www.jordantimes.com/news/local/jordan-replace-planned-nuclear-plant-smaller-cheaper-facility, accessed 1 June 2018.

Atomic Energy Commission. The development suggests not only that Jordan was unable to secure financing for the two 1000 MW proposal, but also that Russia was unable to provide low-interest financing.

In the last couple of years, JAEC has signed a series of MOUs and agreements on SMRs.[90] The most recent announcement to come from JAEC Chairman Toukan, in April 2018, is that the organization is in "serious and advanced" talks with China National Nuclear Corporation (CNNC) to build a 220 MW High Temperature Gas-Cooled Reactor (HTR) in Jordan.[91] Although SMRs could be cheaper in terms of total cost, they are expected to be more expensive on a per-MW basis and would pose a range of problems such as siting.[92]

Poland

Poland planned the development of a series of nuclear power stations in the 1980s and started construction of two VVER1000/320 reactors in Żarnowiec on the Baltic coast, but both construction and further plans were halted following the Chernobyl accident. However, on 28 January 2014, the Polish Government adopted a document with the title "Polish Nuclear Power Programme" outlining the framework of the strategy. The plan includes proposals to build 6 GW of nuclear power capacity with the first reactor starting up by 2024.[93]

In January 2013, the Polish utility PGE (Polska Grupa Energetyczna) had selected WorleyParsons to conduct a five-year, US$81.5 million study, on the siting and development of a nuclear power plant with a capacity of up to 3 GW.[94] At that time, the project was estimated at US$13–19 billion, site selection was to have been

---

90 WNN, "Jordan to consider deployment of X-energy SMR", 29 November 2017, see http://www.world-nuclear-news.org/NN-Jordan-to-consider-deployment-of-X-energy-SMR-2911175.html, accessed 24 April 2018; "Russian Sweep in the Middle East", *NIW*, 15 December 2017.

91 Mohammad Ghazal, "Jordan, China in 'serious talks' to build gas-cooled $1b reactor", *Jordan Times*, 28 April 2018, see http://ftp.jordantimes.com/news/local/jordan-china-serious-talks%E2%80%99-build-gas-cooled-1b-reactor, accessed 8 May 2018.

92 M.V. Ramana and Ali Ahmad, "Wishful Thinking and Real Problems: Small Modular Reactors, Planning Constraints, and Nuclear Power in Jordan", *Energy Policy*, 26 March 2016.

93 Lukasz Kuzniarski, "Polish Nuclear Power Programme", Ministry of Economy, 17 March 2014, see https://www.iaea.org/NuclearPower/Downloadable/Meetings/2014/2014-03-17-03-21-WS-INIG/DAY2/COUNTRY/L_Kuzniarski_POLAND_IAEA_workshop_Seoul_2014.pdf, accessed 24 April 2018.

94 NIW, "Briefs-Poland", 8 February 2013.

completed by 2016, and construction was to begin in 2019.[95] However, financing remained a key barrier. And in December 2017, the rating agency Fitch, warned that "if the utilities decide to get involved in building the nuclear power plant and put it on their balance sheets then certainly we will have a close look as this may be negative for the ratings." This is because Polish utilities are already "substantially leveraged" and the massive cost of nuclear investment would be problematic. Furthermore, the agency suggested that offshore wind, with falling technology costs would be more economic.[96]

In late 2017, the Energy Minister, Krzysztof Tchorzewski, said that he would like to see Poland build three nuclear reactors, at five-yearly intervals, the first to operate in 2029, with each unit costing US$7 billion.[97] The Government, in January 2018, announced that it would decide during the year, if it did proceed with nuclear power, with a decision "definitely, in the first half".[98] That did not happen.

Conclusion on potential newcomer countries

The history of potential nuclear newcomer countries is a history of delays, cost estimate increases and abandoned projects before they even get started on the ground. While construction is under way in Bangladesh, Belarus, Turkey and the UAE, projects have been suspended or cancelled in most of the other candidate countries. Two countries, Egypt and Saudi Arabia, both in the Middle East, appear to have made some progress in the deployment of nuclear power but the next few years will determine whether this will result in the actual commencement of construction.

---

95 Economist, "Polish Energy, Going nuclear", 31 January 2014, see http://www.economist.com/blogs/easternapproaches/2014/01/polish-energy, accessed 24 April 2018.

96 Reuters, "Funding nuclear project could hit Polish utilities' ratings: Fitch", 8 December 2017, see https://www.reuters.com/article/us-poland-nuclear/funding-nuclear-project-could-hit-polish-utilities-ratings-fitch-idUSKBN1E21YM accessed 30 June 2018.

97 Reuters, "Poland may have first nuclear power plant by 2029", 6 September 2017, see https://www.reuters.com/article/poland-nuclear/poland-may-have-first-nuclear-power-plant-by-2029-idUSL8N1LN222, accessed 24 April 2018.

98 Reuters, "Poland to decide later this year on building nuclear plant | Reuters", 29 January 2018, see https://www.reuters.com/article/us-poland-nuclear/poland-to-decide-later-this-year-on-building-nuclear-plant-idUSKBN1FI1Q8, accessed 30 June 2018.

Tab. 3 Summary of Potential Nuclear Newcomer Countries

	Site	Proposed Vendor	Initial Startup Date	Proposed Construction Start	
			Under Construction		
Bangladesh	Rooppur	Rosatom	Nov 2017	April 2018	2023
Belarus	Ostrovets	Rosatom	2016/18		2019 (Q4)/2020 (Q3)
Turkey	Akkuyu	Rosatom	2015	2018	2023
UAE	Barakah	KEPCO	2017/18/19/20		2019/2020
			Contract Signed or Advanced Development		
Lithuania	Visegrade	Hitachi	2020	Suspended	-
Turkey	Sinop	Mitsubishi/ Areva		?	-
	Ingeada	SNPTC/ Westing-house		2019	-
Vietnam	Ninh Thuan	Rosatom	2020	Suspended	-
			Committed Plans		
Egypt		Rosatom	2019	2018	2026/2027
Jordan		Rosatom		2019	2024
Poland				?	2029
			Well Developed Plans		
Chile			2024	Suspended	-
Indonesia		Rosatom		Indefinitely Postponed	-
Kazakhstan		Rosatom or Wes-tinghouse		?	-
Saudi Arabia			2020	?	2027
Thailand			2020-8	?	-

Sources: Various, compiled by WNISR, 2018

9 General conclusions

The global nuclear industry is struggling with a combination of factors that severely impact its competitiveness. The continuous ageing of the reactor fleet induces longer maintenance outages and costlier upgrades. In many wholesale markets, the price-level is lower than the operating and maintenance costs of amortized reactors. The incumbent nuclear utilities are facing ferocious competition from new players that enter the market following sector liberalization measures. For example, the largest nuclear operator in the world, the French EDF, is losing 100,000 clients *per month*. In the U.S., many uneconomic reactors are only surviving on the grid with massive direct subsidies on state level. While nuclear generating costs are increasing, costs of competing technologies, in particular solar, wind and natural gas, have been falling dramatically over the past decade. It has turned out impossible to build a new nuclear power plant under market economy conditions and massive government support is indispensable. But even then, the nuclear industry is suffering from excessively long lead times compared to its competitors. So the renewal rate of the fleet is below the minimum necessary for survival. Nuclear power is turning into an endangered species that, in addition, is increasingly threatened by an invasive species, cheap and abundant renewables.

Economics

The Collision of Atomic and Flow Renewable Power in Decarbonization of Electricity Supply

Aviel Verbruggen, and Yuliya Yurchenko[1]

Abstract

Transitions to sustainable, renewable energy supply are the major components of serious climate policy framed by the aims and constraints of sustainable development. The Paris Agreement does not provide the strategy, actions, instruments, or means to boost the transition processes in global North and South. The world's rich countries and people continue to exert rights to pollute the atmosphere with greenhouse gases. A spearhead climate policy can trigger fast elimination of energy-related carbon dioxide emissions, with full de-carbonization of the electricity supply as priority. Atomic power and flow renewable power (wind, solar, running water) are simply juxtaposed as the two major low-carbon supply options. In reality they are mutually exclusive in fully decarbonized power generation systems. They are hard to match technically while their major mutual impact is that they undermine the economic case for each other.

1 Aviel Verbruggen, University of Antwerp, Belgium, aviel.verbruggen@uantwerpen.be; Yuliya Yurchenko, University of Greenwich, United Kingdom, y.yurchenko@greenwich.ac.uk

R. Haas et al. (Eds.), *The Technological and Economic Future of Nuclear Power*, Energiepolitik und Klimaschutz. Energy Policy and Climate Protection, https://doi.org/10.1007/978-3-658-25987-7_4

1 Introduction

COP21 was widely celebrated as a global achievement in tackling climate change certainly is a milestone in climate talks but does not bear enough thrust to spearhead enough action of enough urgency. It carries many endemic problems of previous international climate change agreements i.e., ambiguity of wording, lack of binding obligations, room for interpretation, etc. Too many concessions were made to turn it into text that all parties were ready to sign (Verbruggen, 2015). It came at a sacrifice of drafting the foundations of urgently needed frameworks for governing the global climate commons. The problem of differentiated responsibility for the current state and contamination of those very commons is not addressed in a meaningful way. By putting a price on GHG emissions, environmental destruction is traded, normalized and not addressed; inviting those who can afford i.e., rich countries whose industrialization destroyed the planet in the first place, to continue polluting.

A lot of energy intensive and contaminating production of industrialized countries has been allowed to move to newly industrialized and industrializing ones who are not economically strong to control and mitigate polluting or make polluters pay – something that too will not be solved without coordinated global action and a binding framework. Fast elimination of energy-related CO_2 emission is needed. Two main low-carbon options are seen as a solution here: renewable power (solar, water, wind) and atomic. In this chapter we start from a brief comment on the COP21 Paris Agreement (section 2). In section 3 we focus on the question of rights to emit greenhouse gases and 'the polluter pays principle'. Section 4 presents the headlines of spearhead action in climate policy by fast elimination of energy-related CO2 emissions. It is followed by section 5 where we show that there are but few low-carbon energy supply options, with nuclear and flow renewable energy supply as the main contenders for providing electricity. Moreover, low-carbon is only one of energy supply options, which too need to be assessed for their overall sustainability performance as we discuss in section 6. Nuclear power fails on crucial sustainability aspects and collides with the full expansion of flow renewable energy supply (wind, solar, running water) since, as we show in Section 7, the two low-carbon contenders undermine economic profitability of the other. The Conclusion summarizes our main arguments and recommendations.

2 Paris Agreement: winners and losers

On December 12, 2015, French minister L. Fabius forged the general approval of the Paris Agreement after days and nights of tedious negotiations. Evaluations of the agreement range from 'historical success' to 'epic failure'. The 31 pages text of the Paris Decision & Agreement is grey, vague, and silent about how UNFCCC will govern the global climate commons. It holds boundless opportunities for differentiated interpretation. PwC director J. Grant talked of "*constructive ambiguity, or even woolly wording in some areas*", and L. Fabius stated: "*this allows all countries the ability to take the deal home and declare success*". Ambiguity and woolly wording means that every party can read the text as their success and this speaks of a shaky contract. Without mastering COP's 24-year history and its jargon language, the Paris text is difficult to understand. For satisfying all COP delegations, the text is stripped of content, leaving voluntary efforts, voluntary contributions, and voluntary transfers as fillings for patchwork of voluntary projects. It is not clear who undertakes which projects, how and with whom, as the final decisions are under the discretion of "*all Parties and non-Party stakeholders, including civil society, the private sector, financial institutions, cities and other subnational authorities, local communities and indigenous peoples*". Does an agreement, allowing willing people to set up voluntary initiatives, generate sufficient thrust for drastic and urgent change? In our view it does not.

The unanimous adoption of the Paris text prompted praise, high expectations, and certainly relief for the club of veteran COP participants. The process was widely applauded by most media and societal groups: the many people and organizations concerned about derailing climate change, involved scientists, active governments, social organizations, banks, industrial companies, up to corporates with significant activities and assets related to fossil fuels and to atomic power.

Popular enthusiasm over marginal accord in the light of previous COP failures obscures the danger of agreements where interests of participants are too diverging, even opposite, and not reconcilable in the practical realization of the agreement. One tends to forget that the day-to-day economic priorities and limitation as well as ecological plans – or absence of such – in the participant countries will inevitably sour the optimism of the agreement. Industrialized countries' actions and intentions since the COP21 ovation do not appear assuring of their changing course. And these are the parties who should be global leaders of positive change due to their economic capacity and because of their role in destroying the environment in the first place. Instead, Norway, for example, plans exploitation of all its fossil fuel resources. The EU ETS permit prices per ton CO_2 emitted hovered back to the €5 waiting dock, after a ripple beyond €8 during 2015. France's investment in renewable energy fell from €6.2 billion in 2014 to €2.9 billion in 2015. After post-Chernobyl stalling, atomic power was brought back in as a solution to the inability to effectively reduce CO2 emissions (Mez, Schneider, and Thomas 2009) and is still on the table. The French nuclear conglomerates see the COP21 outcome as a window of opportunity for more atomic power projects. France is an important exporter of atomic power production with the state's heavy involvement in atomic producer ownership – AREVA and EDF are some 90% and 85% state owned and are world leaders in atomic exports (World Nuclear Association 2016). It is then not surprising that the French government promotes atomic energy as a "green option" at home and abroad (Ibid.). The option is, however, neither green, nor cheap in the short and long run alike as some £18 billion budgeted – with a very realistic prospect of adding £2.7 billion (EDF 2016) – Hinkley Point C project in the UK confirms (see also Schneider, Froggatt, and Thomas (2011) on high costs of atomic power). What the atomic option may do is pull urgently needed resources from being invested into truly sustainable, green projects.

COP21 did not establish the elementary conditions and instruments for starting an effective UNFCCC governance of the global commons atmosphere and climate (Verbruggen, 2015). Overall, major industrial, financial, and political interests have swindled good-meaning activists, environmental NGOs, and developing countries. Mass outsourcing of production to low-income countries and internationalization of production chains makes it harder to trace which country and whose companies pollute while it is the governments and end consumers who are being made to pay. IPCC, among other, document that 'a growing share of CO_2 emissions from fossil fuel combustion in developing countries is released in the production of goods and services exported, notably from upper-middle-income countries to high-income countries' (IPCC 2014)

The economic and political ideologies and interests that created the energy and climate problems after World War II, continue to occupy the pole positions, now controlling the sort, price, and pace of the low-carbon energy transitions further jeopardizing possibilities for a progressive change.

3 Rights to CO_2 pollution or applying 'the Polluter Pays Principle'

Addressing the annual 50Gt ton GHG emissions must be prioritised because climate change causes or aggravates the other daunting global problems (UNDP 2007). Governments and companies tend to convert the +2°C limit into a spendable carbon emissions budget, considered and handled as '*rights to emit*'. This practice rises the likelihood of transgressing the +2°C limit to near certainty and is geared towards appropriation by present rich countries and by carbon-intensive lifestyles, and is uncritically propagated by mainstream economists and media as a message that reads: '*by mitigating emissions, present generations deliver efforts and make expenses for the benefit of future generations*'. This unwarranted rights position conflicts with a civilized status of environmental policy. Emitting CO_2 in the atmosphere is an activity of dumping without hindsight or '*gaseous littering*'. Industrialized societies acknowledge the 'Polluter Pays Principle' and polluters face two obligations: pollution must immediately stop and polluters must bear responsibility for the mess caused. Due to the atmosphere being the global commons, it is difficult to impose and enforce the actual implementation of the polluter pays principle. The way in which responsibilities are allocated in terms of 'right to pollute' and 'price of pollution' is ridden with problems. It is undisputable that 'both international and national decision making must aim to take account of income and wealth differentials and regional disparities within as well as between nations' (Newell et al 2015: 239). However, there is also the issue of difficulty to trace who pollutes what in internationalized production chains where countries, not companies, are held responsible for pollution that affects their geographic territory or the global atmospheric commons, while the profits from production more often than not escape those countries' controls.

4 Spearheading climate policy by fast elimination of energy-related CO_2-emissions

Since the UN Framework Convention (1992), over the Kyoto Protocol (1997) and the Copenhagen Accord (2009), yearly global GHG emissions continued to grow, as did the annual use of commercial energy (IEA's yearly Outlook). About $4/5^{th}$ of GHG emissions are the result of present energy supply and use practices. In 2015 CO_2 emissions growth stalled due to a global expansion of renewable energy supply. Presumably more than $4/5^{th}$ of the climate policy studies focus on energy-related CO_2 emissions and their mitigation. Climate policy goes beyond the issue of energy (e.g., other GHG than fossil fuel related CO_2, land-use, adaptation) but is also influenced by fossil fuels use (for example methane emissions, changing land-uses affected by low-priced supply of fossil fuels).

Ongoing climate policy is little effective partly because there are many goals on several aspects that are prioritized at the same time. Contrary to the widespread opinion that UNFCCC must mainstream and simultaneously solve multiple major problems of the world[2], rational climate policy should detect spearhead issues functioning as locomotive in accelerating mitigation or adaptation. Strategic advance requires forcing change via a selected issue for breaking the locks on needed technological, industrial and societal transitions. Thorough transformation of energy supply and use is widely recognized as the predominant change to perform (IPCC 2012).

When COP Parties are serious about not crossing the +2°C as a dangerous, they design and agree on *Individual Parties' Emissions Contraction Scenarios* (IPECS). For this, the focus is on Cpp = the average energy-related CO_2 annual emissions per person in a nation. The Cpp indicator is a well-known Sustainable Development Indicator. Cpp is yearly assessed for all UN members and ranges from less than 100 kg in least developed countries to more than 20,000 kg in a few wealthy, oil intensive economies (IEA 2015).

Decomposing Cpp in three, still highly aggregated, factors provides insight and opens the entry to more detailed, hands-on information. The three indicators can be devolved further to reach detailed groups of actors emitting CO_2 in specific conditions, offering neat hands-on policy targets[3]. Respective Cpp calculation is a multiplication of respectively wealth intensity, energy intensity of wealth, and CO_2 intensity of energy use:

Figure 1 presents a stylized view of Cpp 'contraction & convergence' scenarios for a few typical Parties with the agreed upon upper limit of Cpp, which contracts to a low point in 2050, e.g., a maximum of 500 kg Cpp. Every Party's scenario starts at its recently verified Cpp value. Every Party designs its Cpp path, respecting the constraint of staying below the commonly agreed upper limit. The actual Cpp contraction scenarios for sixteen, major CO_2 emitting nations are documented in the *Deep Decarbonization Pathways Project*[4] 2015 report.

---

2 Paris Agreement p.1: *"Parties should, when taking action to address climate change, respect, promote and consider their respective obligations on human rights, the right to health, the rights of indigenous peoples, local communities, migrants, children, persons with disabilities and people in vulnerable situations and the right to development, as well as gender equality, empowerment of women and intergenerational equity."*

3 The decomposition can go on by splitting GDP in its major composing activities, by identifying actors related to the various activities, by specifying the types of energy used, etc. At UN level the higher aggregate suffices and further detailing is the task of the Parties to design the policies for controlling the values of the aggregate indicators. Agnolucci et al. (2009) and Verbruggen (2011) provide examples and suggestions of deeper decompositions.

4 An international consortium of research centers investigates 'deep decarbonization pathways' for a set of countries, together emitting three quarters of the global energy-related CO_2 tonnage (http://deepdecarbonization.org).

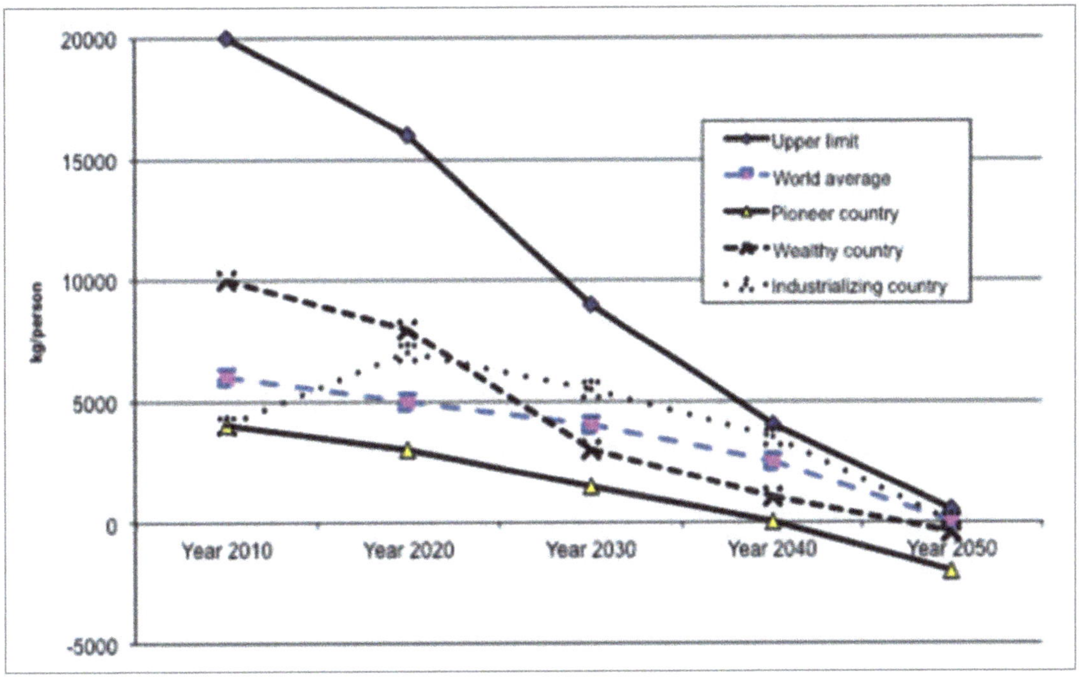

Fig. 1 Individual Parties' Emissions Contraction Scenarios materialize respect for the maximum +2°C average temperature increase; stylized examples of representative cases, selected by the authors.

Decomposing energy-related CO_2 emissions in constituent factors is a widespread practice (IEA 2015). IPCC reports take advantage of this decomposition for explaining the evolution of energy-related CO_2 emissions (e.g. 2014 Assessment report, working group 3, chapter 6). The SE4All initiative of the General Assembly (UN 2011) wants to half the energy intensity (factor 2 of the equation) and double the use of renewable energy (factor 3) in developing countries. Therefore, it is amazing that official COP policy-making neglects the opportunities of decomposition for addressing the 'complex' and 'wicked' policy matters. Also MRV (Monitoring-Reporting-Verification) becomes really practical because numerical indicators are every year available for every factor:

1. The *Budget Reform Index (BRI) for wealth intensity (GDP/person).* The BRI should irrevocably increase year after year. Budget reform is financially promoting sustainable low-carbon activities and charging non-sustainable activities, leading to restructuration of the GDP. The monetary total of the GDP may increase or decrease by the restructuring. The discretionary power of how to practically organize the restructuring remains fully with the Parties. The

BRI only gauges the overall net monetary pressure or thrust of policies for the promotion of sustainable low-carbon technologies and practices.

2. *Energy intensity (energy/GDP)* is a long-time documented indicator (Schipper et al. 1992, 2001; Geller and Attali 2006) and widely used by national and international energy administrations. Energy intensity combines the structure of an economy (how much of which activities take place) with energy efficiency (how much commercial energy is used by one unit of activity). The first factor is affected by budget reform (BRI); the second is mainly technological. Lowering energy intensity is generally high on the list of (proposed) energy and climate policies (IEA, EU, China). More effective progress is part of energy transitions.
3. *Carbon intensity (emitted CO_2 per unit of supplied energy)* is the keystone for controlling CO_2 emissions. Transitions to zero or almost zero carbon emitting energy uses by 2050 is the mission for all nations in the coming decades. Their transitions will be specific, due to differentiated endowment in resources, applied technologies, installed infrastructures, etc. However, all energy transitions are constrained by a small set of energy supply options [Figure 2].

The spearhead approach respects UNFCCC's '*common but differentiated responsibilities and respective capabilities*' in emission reductions. '*Common responsibility*' here is: all countries' Cpp stays below the upper limit scenario. '*Differentiated*' here means: high value Cpp countries must contract first and at a fast rate ('deep cuts'); low value Cpp countries (mostly developing and least developed countries) can grow in Cpp value with the obligation to continue to respect the contracting upper limit values in future years.

5 Few low-carbon energy supply options

For performing activities, the right type and quantity of energy must be supplied at the right place and time. All energy supply are a combination of some energy *source* with particular *technologies* for exploration, generation, conversion, and transmission of energy to the end-users. In sequence of importance, available sources are: renewable flows and stocks in the natural environment, fossil fuel deposits in mines, wells, and shales, and uranium deposits (Figure 2)[5]. The environment supplies for free many energy end-use services with little technology required for extraction or conversion, e.g., phenomena and processes such as

5 The overview does not include the manufacturing of synfuels.

daylight, ambient heat, natural ventilation and drying. Natural processes concentrate diffuse renewable flows (photosynthesis, the water cycle). Over the last decade, the costs of man-made technologies harvesting renewable flows dropped significantly (IPCC 2012). Photovoltaics demonstrate strong performance on cost reduction in comparison with atomic (Haas 2012). Globally speaking, atomic has been showing a slowdown on new installations and decrease of capacity as an industry since 1988 (Schneider, M., Froggatt, A. and S. Thomas 2011). Technological capability announces further cost cuts, for example levelized kWh prices of conventional PV conversion to €ct. 4 to 6 by 2025 and €ct. 2 to 4 by 2050, although dependent on financial and regulatory conditions (Fraunhofer 2015). More innovative technologies can further reduce the costs. Costs distribution however needs to be further examined and reflected in any future policies as EU cross-country research suggests that it is still the households who bear most of the burden 'due to higher costs of direct energy efficiency expenditures in appliances, vehicles and insulation' (Haas et al. 2014).

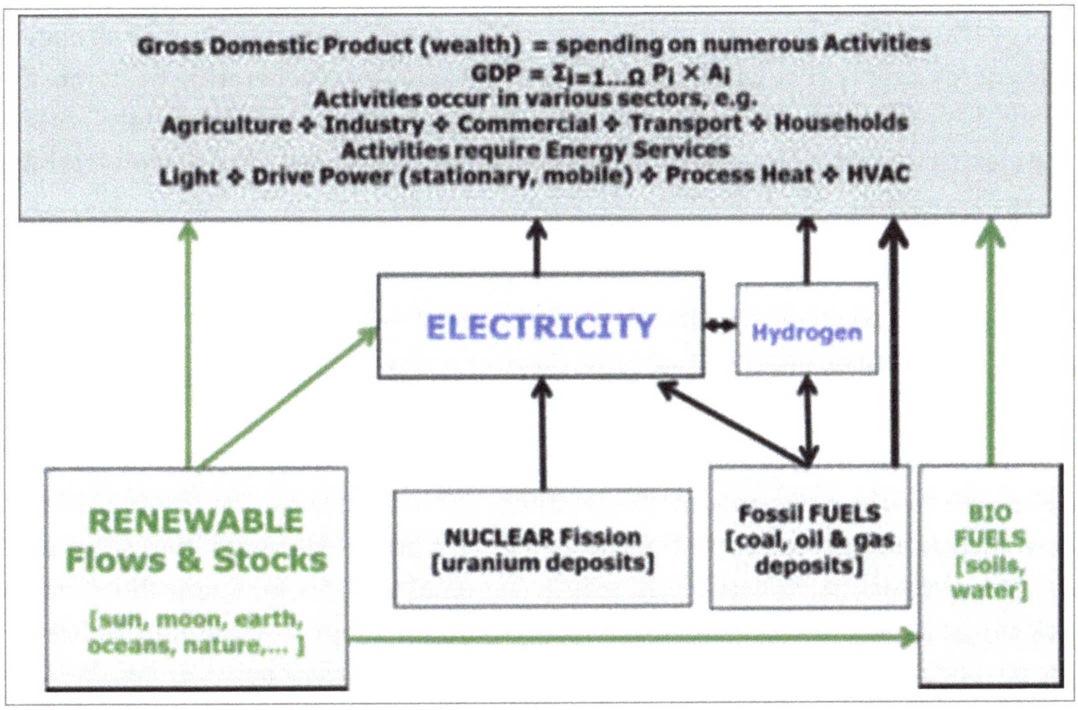

Fig. 2 Overview of energy supply categories, with energy sources in [.] Source: authors' graph.

Nuclear fuel is produced from refined and enriched uranium, dense deposits of which are limited (American Nuclear Society, 2001). Uranium shortage may be overcome by breakthroughs in breeder or fusion technology. Commercial new breeder and fusion plants are not expected before 2050, the year wherein carbon free electricity systems should be operational. Most essential is the high evidence that atomic fission power is not a sustainable power supply option (Verbruggen et al. 2014).

Fossil fuels cover a market share of above 85% of commercially traded energy supply (BP 2015). Their success is due to their versatility, density, (for all scales) divisibility, abundance, storability, and ability to perform on command. However, fossil fuel combustions cause various environmental damages and inevitably fetch CO_2. In a low carbon future their use will be stifled (IEA 2014) but 'carbon lock-in' and related interests are exceptionally strong. A smooth phasing-out of fossil fuels is rather unlikely to happen (Verbruggen and Van de Graaf, 2013). Hydrogen is a carbon free fuel but is not naturally available on earth and difficult to manage safely. New industrial infrastructure may manufacture hydrogen from low carbon electricity but its deployment would be a costly, long-range undertaking. Other non-fossil solutions and options on storage and supply systems integration, some already available, need more investment for the successful and sustainable green energy transition. Electricity plays a central role in that transition as major renewable energy supply (PV, wind, water) and atomic energy is generating heat, mostly medium-pressure steam that is converted to electricity and delivers only power. The electricity sector transformation is the inevitable vanguard of the low-carbon energy transition.

6 Climate change urges sustainable energy transitions

The nuclear renaissance strategy is mostly argued in terms of necessity of atomic power in future low-carbon scenarios. Without public money continuing to subsidize new nuclear power projects, few projects will be started. Also with high subsidies, private investors in industrialised countries remain highly reluctant to invest in risk-prone nuclear projects. In order to obtain public support, the atomic hubris of the past century is shelved by a popular new slogan "*Nuclear power is not the only solution, but there is no solution without*". However, a majority of EU population prefer energy efficiency and renewable power (Eurobarometer 2015). Published statistics prove the fast diffusion of flow renewable power supply by technological innovation resulting in continuously declining generation costs. It has become the survival strategy of the atomic conglomerates to publicly propose co-habitance with renewable power while actually undermining its development because there is no compatibility for lots of renewable flow power and lots of atomic power in the same power system (Verbruggen, 2008). The exclusivity is technically due to physical attributes of electric currents and to the inflexibility of both atomic and flow renewable supply. The options perform differently on sustainable development criteria at the global societal level too.

Notwithstanding many regards towards sustainable development, IAEA skips the true sustainability assessment of atomic power (Verbruggen and Laes, 2015). The UK conservative government substituted the single low-carbon attribute for the spectrum of sustainability criteria covering Planet, People, Prosperity, Politics, and Risks as specific concerns. This reductionist interpretation equals the circumvention and obscuration of sustainable development also observable in the 2014–15 energy and climate policies of the EU (EC 2014a, EC 2015). Sustainability is, however, the primary attribute that future energy supply and use systems should have.

One aspect is the readiness of energy transition pathways for emulation by developing countries that is essential for global CO_2 emissions reductions in the coming decades. Low-carbon energy systems composed of non-sustainable nuclear power and of centralized large-scale, capital-intensive renewable plants are not ready for emulation by the majority of developing countries. For the elimination of energy-related CO_2 emissions, richer countries should develop and deploy renewable energy supply of the kind and size also applicable and affordable by developing countries. Indeed, there has already been progress made on the level of low-carbon energy transition assistance by World Bank and USAid. Power Africa (PA) is an initiative launched by the Obama administration in 2013 that 'works with African governments and private sector partners to remove barriers that impede sustainable energy development in sub-Saharan Africa and unlock the substantial wind, solar, hydropower, natural gas, biomass, and geothermal resources on the continent' (USAID 2013). The program's goal is stipulated as to 'increase electricity access by adding more than 30,000 megawatts of cleaner, more efficient electricity generation capacity and 60 million new home and business connections across sub-Saharan Africa' (USAid 2014). The Beyond the Grid sub-initiative aimed at expansion of rural electrification and providing 'access to small scale and off-grid technology' is also included in PA. One main recent project is the 'Lighting Africa' program – a 'joint initiative of IFC [International Finance Corporation] and the World Bank, [aimed to] help increase access to affordable, clean and safer lighting for more than 30 percent of Nigeria's population who live in rural areas, and have low incomes and no access to grid electricity' (IFC/WB 2015). The program focuses purely on private sector participation in electrification reform and expansion of the sector however the renewably sourced energy focus makes it more hopeful than the Power Africa initiatives.

There are two big problems with the above initiatives. First is the involvement of private sector in delivery – usually associated with less reliability and higher costs (Yurchenko and Thomas 2015). The second is the inclusion of the fossil fuel natural gas as an energy source option, which is not a sustainable option. The two leave us skeptical of the potential final economic costs and effects of the program which seems to mimic typical issues of climate change politics governance discussed by Newell et al (2015) i.e., more of the same conventional approach.

A comprehensive sustainability assessment of nuclear fission power (Verbruggen et al., 2014) reveals that nuclear power fails on most sustainability criteria. The needs of countries with poor grid development and dispersed unsatisfied household electricity demand emphasize the shortfall of nuclear power as sustainability option.

7 Flow renewable power and atomic power supplies are incompatible

There is a growing literature on how integrated power generation systems may embed both flow renewable power (solar, wind, running water) and atomic power. At the outset, this literature adopts the present non-sustainable systems as the default position, with wind and solar power as disturbing newcomers. The incumbent position is: *intermittent and stochastic renewable energy supplies disturb the reliable delivery of power; power on command is the reference.* For a more effective and efficient transition the opposite viewpoint is needed: i.e., the future sustainability goal situation must be treated as a benchmark for assessing present states and required evolutions. Then the overarching guidance in the transition of the electricity sectors must be as follows: *Intermittent and stochastic renewable energy deliver the most sustainable supply and merit priority over the non-sustainable supply; with respect for this sort of lexicographic priority, the supply of reliable power is organized, requiring extended load management capabilities, energy storage facilities, adapted transmission links to convey and match renewable power supply.*

The atomic power and flow renewable supplies are mutually exclusive on five major directions of future power systems. First: atomic power is part and parcel of the expansive "business-as-usual" energy economy since the 1950s. Second, nuclear and renewable power need very different add-ons provided by fossil-fueled or bio-energy power plants, or by dam hydro power; for nuclear the add-on is large and expansive, for renewable power it is distributed, flexible and contracting over time. Third, power grids for spreading bulky nuclear outputs are of another constellation than the interconnection between millions of distributed power sources requires. Fourth, the risks and externalities of atomic power make this technology non-sustainable and therefore without a future. There is no safe or permanent way of disposing of nuclear waste – a problem that requires an international solution (Di Nucci and Losada 2015), effects of potential accidents are insurmountable, climate effects of ore mining are underestimated, emission of radioactive isotopes 'such as tritium or carbon 14 and the radioactive noble gas krypton 85' are not discussed, etc. (Brunnengräber, et al., 2015; Smith, 2006; Mez, 2016); while efficiency/renewable power are still in their infancy particularly in terms of market shares. Fifth, the antagonistic competition for R&D resources and for production capacities and capabilities (e.g., trained experts) will intensify. Nuclear power and renewable power have no common future in safeguarding "Our Common Future" (Verbruggen, 2008).

Fully sustainable renewable energy systems are not just technologically and economically feasible but also the cheapest and only sustainable option for the world's population. Like every successful transition, sustainable energy transitions need profound change in the minds, thinking, beliefs, preferences, etc. to adopt the novel paradigm, perspectives, technologies, and practices. Progressive thinking and actions are unlikely to be delivered by those with vested interests as we mentioned in the case of France, AREVA, and EDF. Although detailed technical analysis of dynamic power systems reveals the incompatibility of flow renewable and atomic

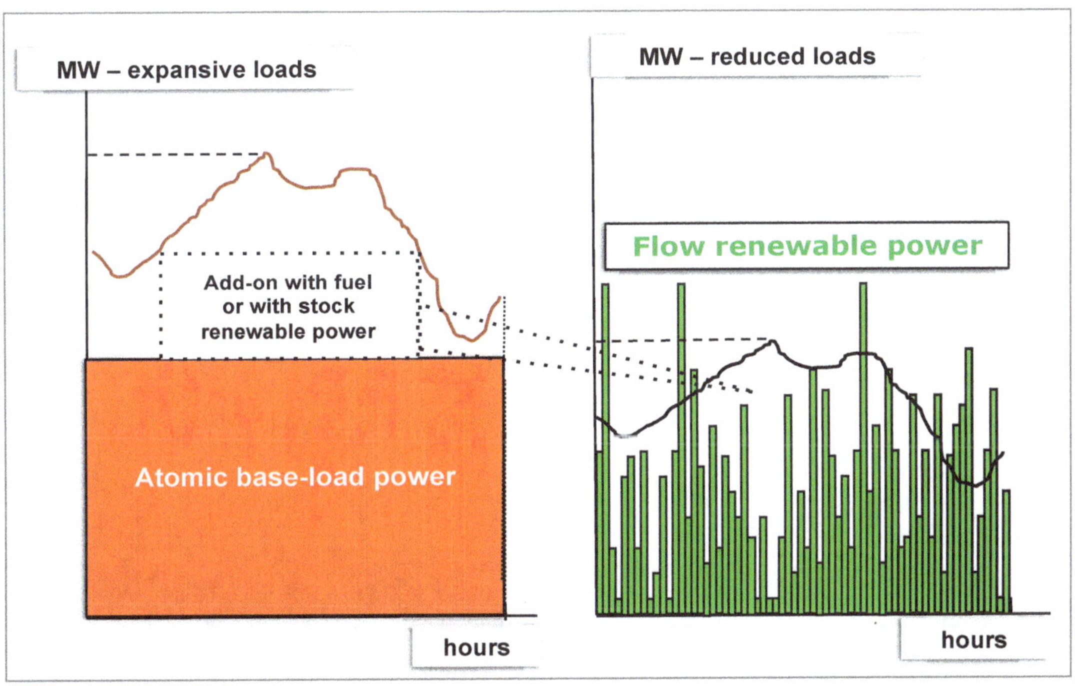

Fig. 3 Atomic versus flow renewable supply as separately serving electric loads
Source: Verbruggen (2008)

power supplies, the prevailing discourse repeats the mantra of a simple juxtaposition of both kinds of supplies (IPCC, 2014; UNFCCC, 2015; Verbruggen and Yurchenko, 2017).

This discourse can be continued when the contenders are embedded in separate power systems with ancillary supplies from fossil fuel based, bio-energy or hydropower dam electricity. However, when 100% carbon-free power in a particular power system is due, flow renewables and atomic power will collide. Both supplies are characterized by inflexibility, although of a different kind and for different reasons. There is a need for add-on current to fill the power loads from the supplied base onwards. Stapling supplies is the principle in merit order loading, not juxtaposing supplies, because electric power is an ephemeral phenomenon switching fifty times per second. Figure 3 shows the juxtaposition of atomic and flow renewable supplies in separately serving electric loads. Figure 4 illustrates that the two contenders will claim the same base-load area when operating in the same system.

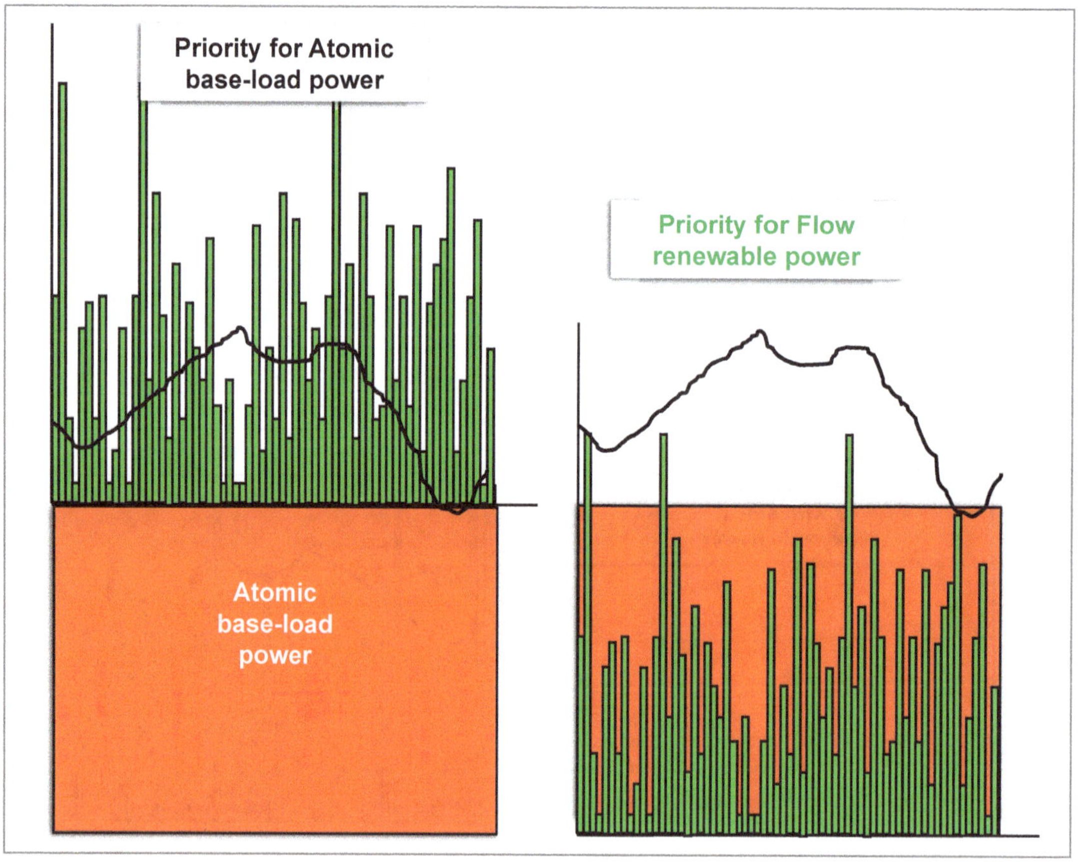

Fig. 4 Atomic and flow renewable supplies mutual impact when operational in the same power system.

Source: Compiled by the authors on the basis of Verbruggen (2008, 2016).

As is shown in Figure 3, atomic power and flow renewable supplies serve separate power loads. They request add-on services from flexible power supply (fossil fuel or bio-energy based power or dam hydropower). Now more flexibility options are added like load management and storage in batteries (IEA, 2014). Supply from other areas in interconnected power systems is considered as a solution, although when the exchange becomes intense and frequent, the power systems become deeper interpenetrated. Figure 4 highlights how atomic and renewable supplies within one electric system ruin the business case for each other as their load factors are eroded when they cannot deliver the base loads. Researchers are looking into possibilities to enhance the load following capability of nuclear power plants, or to cut off sharp peak supply by wind or solar plants, or curtail their outputs for other reasons. Most studies focus on cases of limited annual supply by flow renewables (e.g., 20%, or max.45% of total annual power generated in the system) with a significant share coming from company owned larger scale plants (off-shore wins parks; MW-scale PV fields; concentrated solar power). Our evaluation is that prosumers and cooperatives will become the predominant generators of flow renewable supply. The incompatibility between atomic and flow renewable power is stronger in terms of sustainability, economics, and involved agents than in technical operability.

8 Conclusion

The global climate policy process as deployed by the UNFCCC at the yearly COPs is slow and vague. The participants and most media acclaim COP21, but empirical and theoretical literature on the commons (Hardin, 1968; Ostrom, 1990) and problems of governance (Newell et al., 2015) predicts little positive results (Verbruggen 2015) without a comprehensive reconstruction of the governance regime altogether.

'Energy transition' is a term covering a spectrum of realities, from thorough and sustainable to superficial, deferring and non-sustainable lock-in. One slips in the latter without a clear definition, vision, mission or strategy of sustainable and thorough change. The incurred delays by the slips make the thorough path steeper, and the irreversibility of climate change more probable.

The sustainable renewable energy alternative as such is not costly when fully developed and deployed. Atomic power is and will be more expensive especially in the long run. Evidently, the transition process itself is challenging. Depending on the scores by progressive, viz. reactive strategies, forces, and public support, the transition difficulties and costs will be modest or high. In order to overcome the impasse, urgent transitions bring earlier depreciation of sunk investments. The latter are more significant when incumbent energy companies reacted little or very late to the 1992 Rio summit and ensuing conventions. For example, after 2000, incumbent electricity companies have still built coal power plants in the Netherlands and in Germany (two countries of high exposure in energy transition literature and practice). This happened under the cover of the low CO_2 emission permit prices of the failing EU ETS.

An assessment on nineteen sustainability criteria concluded that atomic fission power is not a sustainable option (Verbruggen et al., 2014). A full expansion of flow renewable and atomic power supplies too are mutually undermining in terms of reliability of supply and economic cost alike. A conclusion thus can be made that sustainable transition and sustainable power supply must focus and rely exclusively on renewable energy and renewable flow energy in particular as the main option to tackle climate change. The shift towards that option must happen quickly and will require coordinated efforts of countries globally. For this to be effective and making all parties responsible, a new, more binding, and clearly formulated framework of governance for the global environmental commons will need to replace the Paris Agreement and the business-as-usual functioning of the COPs.

***Acronyms and Glossary*:** COP=Conference of Parties at the UNFCCC (yearly since 1995); Cpp = a Party's average energy-related CO_2 annual emissions per person; EPR= European Pressurized reactor; IAEA=International Atomic Energy Agency; INDC=Intended Nationally Decided Contributions (by Parties); IPCC=Intergovernmental Panel on Climate Change; PV=Photo-Voltaic; RE=Renewable Energy/Electricity; SD=Sustainable Development; UN-FCCC=United Nations Framework Convention on Climate Change (1992).
This text uses mostly 'atomic' rather than 'nuclear' because splitting (or fusion) of atoms generates other atoms, as recognized in the 1950–60s.

References

Agnolucci, P., Ekins, P., Iacopini, G., Anderson, K., Bows, A., Mander, S., Shackley, S., 2009. Different scenarios for achieving radical reduction in carbon emissions: A decomposition analysis. Ecological Economics 68: 1652–1666

Agora Energiewende, 2013. 12 Insights on Germany's Energiewende. http://www.agora-energiewende.de

American Nuclear Society, 2001. Generation IV Roadmap: Fuel Cycle Crosscut Group. Winter Meeting Reno http://gif.inel.gov/roadmap/pdfs/fuel_cycles.pdf

BP, 2015. Statistical Review of World Energy

Brunnengräber, A., Di Nucci, M.R., Isidoro Losada, A.M., Mez, L., Schreurs, M.A., Eds., 2015. *Nuclear Waste Governance: An International Comparison*. Wiesbaden: Springer.

Di Nucci, M. R. and Isidoro Losada, A. M., 2015. 'An Open Door for Spent Fuel and Radioactive Waste Export? The International and EU Framework', in: Brunnengräber, A., Di Nucci, M.R., Isidoro Losada, A.M., Mez, L., Schreurs, M.A., Eds., 2015. *Nuclear Waste Governance: An International Comparison*. Wiesbaden: Springer. Pp.79-97.

DDPP, 2015. Pathways to Deep Decarbonization. Executive Summary 2015 report. Deep Decarbonization Pathways Project. Sustainable Development Solutions Network (SDSN) and the Institute for Sustainable Development and International Relations (IDDRI). www.deepdecarbonization.org

EC, 2014a. A policy framework for climate and energy in the period from 2020 to 2030. European Commission COM (2014) 15 final.

EC, 2014b. Guidelines on State aid for environmental and energy 2014–2020. European Commission. Official Journal of EU 57 2014/C 200/01

EC, 2015. Energy Union Package. A Framework Strategy for a Resilient Energy Union with a Forward-Looking Climate Change Policy. European Commission COM(2015) 80 final.

EDF, 2016. 'Hinkley Point'. Available at: https://www.edfenergy.com/energy/nuclear-new-build-projects/hinkley-point-c/news-views/cost-clarification

Fraunhofer, 2015. ISE. Current and Future Cost of Photovoltaics

Geller, H., Attali, S., 2005. The experience with energy efficiency policies and programmes in IEA countries. Learning from the critics. IEA Information paper, Paris.

Haas, R., 2012. 'On the dynamics of Photovoltaics vs Nuclear power'. 2012 IEEE Third International Conference on Sustainable Energy Technologies (ICSET).

Haas, R., Liebmann, L., Lemprecht, M., Resch,G., Kahles, M. and F. Pause, 2014. 'Phase out of Nuclear Power in Europe – From Vision to Reality'. Global 2000 Report. Available at: https://www.global2000.at/sites/global/files/Nuclear_Phaseout_Study.pdf

Hardin, G., 1968. "The Tragedy of the Commons." *Science* 162, no. 3859: 1243–48.

IEA, 2014. World Energy Outlook. International Energy Agency, Paris

IEA, 2015. CO2 emissions from fuel combustion. IEA Statistics 2015 edition.

International Finance Corporation/World Bank (10 March 2015) 'World Bank Group Launches the Lighting Africa Program for Nigeria'. Available at: http://ifcext.ifc.org/IFCExt/pressroom/IFCPressRoom.nsf/0/D3C00B7394A8D59E85257E0400319C58?OpenDocument

IPCC, 2012. Special Report on Renewable Energy Sources and Climate Change Mitigation. Intergovernmental Panel on Climate Change, Cambridge University Press; www.ipcc.ch

IPCC, 2014. Intergovernmental Panel on Climate Change, Fifth Assessment Report, Working Group III Mitigation of Climate Change. www.ipcc.ch

Mez, L., Schneider, M., and S. Thomas, Eds., 2009. *International Perspectives on Energy Policy and the Role of Nuclear Power.* Brentwood, UK.

Mez, L., 2016. 'Climate protection through nuclear power plants? Hardly.' In *Bulletin of the Atomic Scientists.* Available at: http://thebulletin.org/commentary/climate-protection-through-nuclear-power-plants-hardly9170.

Newell, P, Bulkeley,H., Turner, K., Shaw, Ch., Caney, S., Shove, E. and Pidgeon, N., 2015. 'Governance traps in climate change politics: re-framing the debate in terms of responsibilities and rights'. *WIREs Climate Change*, 6 (6). pp. 535-540.

Ostrom, Elinor, 1990. *Governing the Commons: The Evolution of Institutions for Collective Action.* Cambridge, UK: Cambridge University Press.

Schipper, L., Meyers, S., Howarth, R., Steiner, R., 1992. *Energy Efficiency and Human Activity: Past Trends, Future Prospects.* Cambridge University Press.

Schipper, L., Unander, F., Murtishaw, S., Ting, M., 2001. Indicators of energy use and carbon emissions: understanding the energy-economy link. *Annual Review of Energy and Environment* 26: 49–81

Smith, B., 2006. 'Insurmountable risks: the dangers of using nuclear power to combat global climate change'. A report of the Institute for Energy and Environmental Research. Available at: ieer.org/wp/wp-content/uploads/2006/05/InsurmountableRisks_2006.pdf.

Schneider, M., Froggatt, A. and S. Thomas 2011. '2010–2011 world nuclear industry status report' in *Bulletin of the Atomic Scientists*, July/August 2011 67: 60–77.

Töpfer, K., et al., 2011. Germany's Energy Turnaround: A collective effort for the future. Ethics Commission on a Safe Energy Supply, Berlin.

UK HM Government 2009. The UK Low Carbon Transition Plan.

UN 2011. Promotion of new and renewable sources of energy. Report of the Secretary-General. General Assembly Sixty-sixth session.

UNDP 2007. Human Development Report 2007/2008. Fighting Climate Change: Human solidarity in a divided world. Summary. United Nations Development Programme, 31p. http://hdr.undp.org/reports

USAid 2013. 'Power Africa Roadmap'. Available at: https://www.usaid.gov/powerafrica/roadmap. 2014. 'Power Africa: About us'. Available at: https://www.usaid.gov/powerafrica/aboutus.

Verbruggen, A., 2008. Renewable and nuclear power: A common future? *Energy Policy* 36: 4036–4047

Verbruggen, A., 2009. Beyond Kyoto, plan B: A climate policy master plan based on transparent metrics. *Ecological Economics* 68: 2930–2937

Verbruggen, A., 2011. A Turbo Drive for the Global Reduction of Energy-Related CO2 Emissions. Sustainability 3, 632–648. www.mdpi.com/journal/sustainability

Verbruggen, A., Van de Graaf, T. (2013). Peak oil supply or oil not for sale? Futures 53: 74–85

Verbruggen, A., 2014. 'Could it be that stock-stakeholders rule transition arenas?' P.119-131 In Brunnengräber, A., Di Nucci, M.R., eds. *Im Hürdenlauf zur Energiewende.* Springer VS, Wiesbaden.

Verbruggen, A., Laes, E., 2015. Sustainability assessment of nuclear power: Discourse analysis of IAEA and IPCC frameworks. *Environmental Science & Policy* 51: 170–180

Verbruggen, A., Di Nucci, M.R., Fischedick, M., Haas, R., Hvelplund, F., Lauber, V., Lorenzoni, A., Mez, L., Nilsson, L.J., del Rio Gonzalez, P. Schleich, J., Toke, D. 2015). Europe's electricity regime: restoration or thorough transition. *International Journal of Sustainable Energy Planning and Management* 5: 57–68

Verbruggen, A., 2015. Self-governance in global climate policy: An essay. Essay EM-1. University of Antwerp. DOI: 10.13140/RG.2.1.1512.7128 (ResearchGate)

Verbruggen, A., 2016. Sustainability aspects of transitions to low-carbon electricity supplies. International Atomic Energy Agency, Technical Meeting. Amsterdam, June 21–23.

Verbruggen A., Y. Yurchenko, 2017. "Positioning Nuclear Power in the Low-Carbon Electricity Transition," Sustainability. January 9/1 1-14.

World Nuclear Association, 2016. 'Nuclear Power in France'. Available at: http://www.world-nuclear.org/information-library/country-profiles/countries-a-f/france.aspx

Yurchenko, Y. and S. Thomas, 2015. 'EU Renewable Energy Policy: Successes, Challenges, and market reforms' with Stephen Thomas. PSIRU Report. Greenwich University: London

The Historical Development of the Costs of Nuclear Power

Reinhard Haas, Stephen Thomas, and Amela Ajanovic[1]

Abstract

One of the major historical arguments of the promoters of the use of nuclear power was its low cost compared to other electricity generation technologies. For a long time, it was argued that a strong nuclear power contribution to electricity supplies was the best way to achieve a reliable and affordable electricity supply. However, from the first wave of nuclear reactors deployed, construction costs have been on an escalation course.

The core objective of this paper is to analyze the historical development of the costs – especially the investment costs – of nuclear power plants. With respect to these in recent years in Western countries there is a strong perception: Realized costs has always been higher than forecast costs and construction times promised have almost never been met. Given the reasons identified for these cost increases – and their irreversibility – we conclude that the time of "cheap" electricity from nuclear power is undoubtedly over if it has ever existed and for the next years there are no signs of a reversal of the current upward cost trend.

1 Reinhard Haas, Technische Universität Wien, Austria, haas@eeg.tuwien.ac.at; Stephen Thomas, University of Greenwich, United Kingdom, stephen.thomas@greenwich.ac.uk; Amela Ajanovic, Technische Universität Wien, Austria, ajanovic@eeg.tuwien.ac.at

R. Haas et al. (Eds.), *The Technological and Economic Future of Nuclear Power*, Energiepolitik und Klimaschutz. Energy Policy and Climate Protection, https://doi.org/10.1007/978-3-658-25987-7_5

1 Introduction

One of the major historical arguments of the promoters of the use of nuclear power was its low costs compared to other electricity generation technologies. For a long time, it was argued that it is impossible to retain a secure and affordable and low-cost electricity supply without nuclear power. However, it has to be debated whether that argument was ever valid and from the first wave of nuclear reactors deployed, construction costs have been on an escalation course.

Looking back to the economic promises of the "nuclear dream" of the 1950s and 1960s, these focused on very cheap electricity to be provided by nuclear power plants (NPP), electricity even "too cheap to meter" (Lewis Strauss[2], Cohn, 2007). This idea was based on the rather low investment costs and short construction times (4-6 years) in the early days of the civilian use of nuclear power. Indeed, at this time NPP generated electricity at costs as low as 2–3 cents_{2010}[3] per kWh (Cohn, 2007).[4]

Yet, over time the costs of nuclear, especially the investment costs, have increased continuously. In recent decades the high and still increasing costs of nuclear power have become a key barrier to the construction of new reactors around the world. It is clear that in the long run nuclear power will only succeed if its generation costs are lower than those of competing technologies (MIT 2003). This is especially true as electricity systems become increasingly exposed to competitive markets in many parts of the world.

The core objective of this paper is to analyze the historical development of the costs – especially the construction costs – of nuclear power plants. Specific derived objectives are to analyze (i) why the investment costs have increased, (ii) why the construction times have increased, (iii) why the construction costs as well as the construction times have been underestimated systematically, and (iv) whether the reasons for construction cost and time increases are irreversible.

In this context it is important to note that there is a difference between actual investment costs and so-called overnight costs (ONC). The major difference is that the investment costs include also the costs for interest and represent the whole capital costs while the ONC represent the expenses for the technology and construction work only (incl. labor and material cost). Overnight costs are useful for analytical

2 Reference to the full text of what Lewis Strauss said: https://public-blog.nrc-gateway.gov/2016/06/03/too-cheap-to-meter-a-history-of-the-phrase/

3 This means value in terms of money of 2010

4 One could argue strongly whether there was ever an era of cheap nuclear power. The perception of cheapness was based either on cost forecasts that were not fulfilled or on a perception that costs would come down over time and make nuclear power cheap.

purposes especially for international comparisons because the interest rate is project and country specific but consumers pay the cost including the interest. The European Commission estimated that Hinkley would cost in total £24.5bn when the overnight cost was estimated at £16bn[5].

Of absolutely core interest is why the investment costs increased by such high rates. So far, there is no sound and comprehensive analytical evidence that explains the skyrocketing of the real costs that have occurred since the beginning of nuclear power. An obvious component referred to above but in little clarity is the longer construction times, leading to ever higher interest accrued and to „natural" cost escalation" of labor and equipment The automatic intuitive assumption is that these extra costs arise from the additional safety requirements resulting from accidents at Browns Ferry, Three Miles Island and Chernobyl. There might also be a need for better quality materials, for example Westinghouse steam generators of the 1970s used a material that corroded too quickly. The reason why the vendors used cheaper material was because costs were too high. This proved a false economy. If raw materials like steel and concrete have gone up faster in real terms than inflation, that would also have increased real costs. Another factor is that reactors seem to have become more prone to cost escalation from the pre-construction forecast, again no analysis to back this up. This raises the issue whether construction costs have gone up because real costs have gone up or because things have gone wrong without the intrinsic cost going up, e.g. how far is the higher than estimated cost of OLK-3 due to the forecast being an underestimate and how far because things have gone wrong, including increases in construction duration.

Another issue is initial price dumping by construction companies. The question would be whether the pre-construction costs have become more realistic again appearing to raise the real cost. Certainly the prices quoted in the 1960s were horrible underestimates (e.g. the 12 US turnkey plants). One problem with pre-construction cost estimates is that unless the vendor gives a fixed price contract (turnkey) and no vendor in its right mind would give a genuinely fixed price contract, the vendors know they can't be held to the pre-construction estimate so they have an incentive to underestimate to get the business.

Other possible reasons for the construction cost increases could be:

- increase in interest rates for financing and an increase in construction duration (which influences the interest costs but not the ONC);

5 See the state aid case verdict

- changes in design generation[6], and changes in the engineering design see Grubler (2014), other than extra safety costs covered above.

In addition, in the past it could be suspected that high subsidies such as public subsidies, financial subsidies (low interest rates that did not reflect the economic risk) and government subsidies to industry could have led to much lower costs than have actually been true (Cohn (1997)). Another reason for present day construction cost increases would be that the pre-construction costs have become more realistic thereby appearing to raise the real cost. Certainly, the prices quoted in the 1960s were dramatic underestimates. After the experience of vendors facing heavy losses with the 12 US 'turnkey' projects of the mid-60s, vendors were only willing to sign 'cost-plus' contracts so there were no direct financial consequences to them when costs overran. Utilities, in turn, were generally able to pass on whatever costs were incurred to consumers. So neither the vendor nor the buyer generally had to bear the additional costs, they fell on consumers.

So far, studies conducted on the costs and economics of nuclear have focused mainly on the analysis of single plants and cohorts of NPP. In this work we take the results of other studies, add own analyses, e.g. on OLK3, FLA3 and the UK's Hinkley Point C project (HPC) and derive major findings. To the best of our knowledge so far, no such a systematic analysis of cost developments has yet been conducted.

Regarding the literature on costs and economics the following work is most relevant. The very first studies on cost analyses were already conducted at the end of the 1970s by Tybout (1975), Mooz (1978), Mooz (1979), Mooz (1982) and Komanoff (1981). They already provide very early, sound analyses on the reasons for cost increases and an early outlook on what is looming today, which is that nuclear power will not become a cheap power source at any time.

Cohn (1997) provides a comprehensive corresponding analysis including cost analyses. He explains from a philosophical and economic point-of-view why the nuclear dream has failed to come true. Cohn is the first to describe, why nuclear costs were systematically underestimated, what were the economic problems already in the early years of nuclear and how the word "market" was systematically misused by the major utilities and vendor companies. This work also provides an interesting summary on nuclear spending and costs of NPP in the early days 1955 to 1969 in the US. He showed that these were financed almost completely by the utilities with the incentive to gain know-how. He also documents that even at that time it was generally expected to have higher generating costs than available fossil fuel alternatives and was undertaken as a technology-promoting investment.

6 See Reinberger et al. in this book explaining the change in generations of nuclear

Another major contribution of Cohn's work is that he was the first to provide a critical discussion on the issue of subsidies and cost deferments. He documents in detail for the time-period from 1950 to 1979 direct expenditure (incl. R&D outlays, uranium supply and enrichment subsidies, and regulatory subsidies) and implicit subsidies (e.g. tax exemptions and tax benefits) as well as the cost deferments due to e.g. neglecting nuclear waste disposal charges (see Cohn 1997, p.79 for more details).

MIT (2003) conducted a sensitivity analysis and showed under which conditions NPP could become competitive again.

The issue of pre-announced construction costs and actual ones was already discussed by Koomey/Hultman (2007). They present a reactor-by reactor analysis of historical busbar costs for 99 nuclear reactors in the US and compare those costs with recent (2007) projections for next-generation US reactors. Their analysis suggests that projections of capital costs, construction duration, and total operation and maintenance costs are quite low – far away from the historical medians and that additional scrutiny may be required to justify using such estimates in current policy discussions and planning.

Grubler (2010) provides a seminal contribution and a very comprehensive analysis on the developments in France. Grubler's major point of criticism is that lack of standardization and new engineering approaches have avoided the learning and standardization effect. He points out that it is worth saying that France is widely, but wrongly seen as having a fully standardised programme. Actually its 58 reactors are spread over at least 3 main designs (900MW, 1300MW, 1450MW) and 7 variants. Hence, many of the plants used a new untested design.. The scope for learning was restricted because the new variants were ordered before there was any operating experience with their predecessors. There was no conscious decision by France not to standardize, design changes were required because of experience elsewhere, e.g. the need to learn lessons from the Three Mile Island disaster and the need to improve the economics, e.g. by scaling up.

Harris et al. (2012) provide cost estimates for nuclear power in the UK. Their motivation is to analyze the actual cost developments in Europe and derive major conclusions for the future of investment costs in the UK: The primary finding of this paper is that the capital cost for an NPP may be higher than recent UK government reports have indicated and may therefore require greater levels of financial support than policymakers might have originally envisaged. As Harris et al. (2012) state further, due to the significant uncertainties that surround cost estimates for NPP in general it is very difficult to give a high level of confidence to levelised cost estimates.

Rothwell (2015) discusses the basics of economics of nuclear power. Lovering et al. (2016) present an overview on overnight costs (ONC) of 58% of the nuclear reactors world-wide. Koomey et al. (2017) heavily criticize the work by Lovering et al. (2016) claiming that they cherry pick data and include misleading data on early reactors.

This work is organized as follows. In the next chapter we look at the basic cost structure of NPP. Then we analyze the development of investment costs. A specific focus is dedicated to the development of Technological Learning (TL). We discuss why it took place for different technologies for electricity generation but apparently not for NPP. Finally, we argue why the argument that nuclear electricity is cheap is not valid. We explain and show the wrong predictions regarding investment costs and how construction times look like for some recent projects. A summary of the major reasons for investment costs increases of NPP and conclusions complete this chapter.

2 The cost structure of nuclear power plants

In principle the cost structure of every power plant consists of investment costs, fuel costs and O&M costs. In addition, for nuclear plants, significant costs for decommission and backend activities have to be considered, see Irrek (2018) and Wealer et al. (2018) in this book. The specific cost structure of nuclear plants is shown in Fig. 1. Specific features are:

- A very high share of capital costs
- High turnkey costs, actually the highest among all types of power plants
- An (unknown) share of decommissioning costs

Figure 1 shows that the largest amount of the costs – about 80%– are capital costs resulting from initial investments. In 2004 the IAEA estimated 60% construction costs, today the share is likely to be even higher because constructions have escalated faster than the other elements. Harris (2012) estimate 80% share of capital costs and also according to Rangel et al. (2013) NPP competitiveness depends on its capital costs representing on average 80% of the levelized cost of electricity. However, from the first wave of nuclear reactors construction costs have been on an escalation course and the share of capital costs in total cost increased.

One might ask whether the cost components in Figure 1 are complete and whether all important components are included. Schneider (2006) suggests it is not, stating "The total costs of a nuclear kWh most likely will never be known. Costs for waste management, decommissioning and clean-up are constantly on the rise and are generally expected to be paid by the taxpayer."

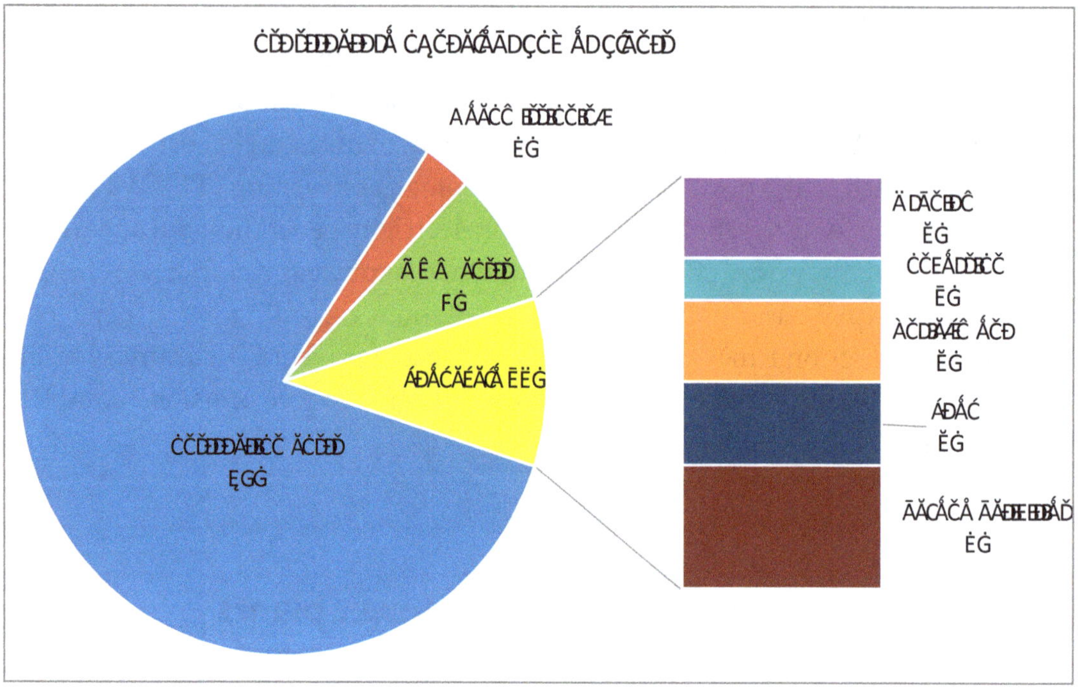

Fig. 1 Typical nuclear electricity generation cost breakdown (based on IAEA (2004) and other sources)

3 The historical development of investment costs

As seen from Fig. 1 investment costs contribute the largest share to the electricity generation costs of NPP. In this chapter we put special focus on the analysis of the historical development of investment costs. We provide a dynamic comparison of the investment costs extracted from different studies, with an emphasis on the explanations for reasons for their increase.

The rising investment cost of building nuclear reactors is a well-established fact. It has been studied in detail for installed capacity in the USA and France. However, sound explanations for these cost increases are difficult to find.

The key literature on investment costs is presented below. The first comprehensive analysis of these was conducted by Koomey (2007). Cohn (1997) describes the development of utilities nuclear investments in the U.S. from the 1950s to the 1970s. He states that utilities nuclear spendings/investments can be divided into four clusters (1) the first investor-owned utilities financed and owned projects (1955-1963); (2) the three rounds of the AEC's Power reactor demonstration program (1955-1963); (3) the turnkey years (12/63-1966); and (4) the bandwagon market (1966-1969). Another important work was done by Harding (2007), who analyzed about 60 plants in the U.S. with respect to their investment costs. His findings show that a rapid growth in investment costs took place already over the period from 1955 to 1995.

For the USA the overnight construction costs (ONC) of the first reactors built in the early 1970s was about $\$1000_{2008}$ per kW. It has increased steadily ever since reaching $\$5000_{2008}$ per kW for the recent reactors built in the early 1990s (Rangel et al. 2013). In other words, a one-to-five ratio in constant USD. The increase in the overall construction costs is even more striking. The average construction duration has increased with time, so interest during construction has increased too. The time taken to build an NPP has risen from between four and six years for the first plants to more than twice as long for the most recent units.

With respect to investment costs an important aspect is the difference between so-called ONC and actual investment costs. It is important to note, that there is a fundamental difference between ONC and the net present value (NPV) of the investment costs, the major difference being that the overnight costs do not include the interest costs of financing depending on the plant construction time. As already said the plant construction time does not in itself increase costs other than the interest costs, although if delays reflect difficulties in construction, these might also increase costs.

Of specific relevance in this context is interest during construction (IDC). During the 80s, there were big battles in the USA between regulators and utilities with utilities trying to get consumers to pay IDC before the plant was on line. Mostly the regulators (rightly) did not give in. US regulation should require utilities to build facilities and only when the facility is complete and the regulator has applied the test of 'used, useful and costs prudently incurred' should the utility be allowed to start to recover its costs from consumers. If a utility fails the test, some or all of the money spent by the utility should not be included in the regulatory asset base and these costs must come out of profits. It was regulators applying this test and threatening to disallow costs that stopped nuclear ordering in its tracks in the USA in 1979 (and led to the cancellation of 100+ orders placed after 1974). So, from an economics point of view this was very important. Without the guarantee of cost pass-through, the banks, credit rating agencies etc made it clear to utilities that building a nuclear plant would be potentially ruinous.

For the financial institutions, delays in construction times and corresponding increases in interest payments play a major role. In addition, almost all modern reactor programs analyzed in detail to date have experienced significantly lengthened construction times particularly in the USA and Europe. Use only of overnight construction costs e.g. by Lovering et al., 2016 means that some of the financial consequences of construction delays is ignored.

ONC has been used in the utility industry for decades (Koomey et al., 2017, EPRI, 1993; Rothwell, 2015), and they attempt to show a cost that is "meant to isolate the cost invariant to construction duration (Koomey et al., 2017) and interest rate, in order to capture the cost intrinsic to the reactor technology", as Lovering et al., 2016 put it. Despite the use of the term "overnight costs" having a long history, there is simply no economic basis for comparing the costs of reactors without including the cost of capital and the construction duration. However, it is not invalid to look at overnight costs. The argument is that adding in interest costs muddies the picture because you don't know if costs have gone up because interest has increased or because the underlying construction cost has gone up.

A key aspect of nuclear reactors that makes them such high-risk investments are that they are large scale, complex, and pre- dominantly site-built. Hence construction takes years (even in the best case) and can extend over a decade or more (Koomey et al., 2017).

Given that financing constitutes a significant part of the electricity generation costs of an NPP, and that the very nature of nuclear power as a large scale, capital-intensive technology makes it particularly sensitive to financial risks, a study that does not take account of interest during construction cannot give a true picture of the costs of nuclear power.

Another major historical analysis on the development of investment costs of NPP has been conducted by Grübler (2010). He investigated all 58 of the French plants in service in 2018 based on an analysis of costs presented in annual reports by the French government. Of specific interest is Civaux in France a N4 type reactor with extremely high costs. However, regarding the four N4 plants (two each at Civaux and Chooz in France) there were clearly design issues that delayed them. They might have been part of the trend but without these plants the trend is still there. There was a trend of cost escalation amongst the other 54 that did not have these design issue problems. For the French case the cost assessment done by Grubler (2010) pointed out that the units completed after 1990 were 3.5 times more costly than the reactors installed in the 1970s. This finding led to the conclusion that cost escalation was inherent to reactors, given that even under the best conditions, as prevailing in France the construction costs have also risen significantly. These favourable factors included more standardization than was achieved elsewhere, predictable series production allowing efficient production line methods to be used to manufacture parts and learning was concentrated in only one reactor vendor and one utility, which also managed the construction process.

Grubler's analysis was a seminal contribution because it led to the publication of the actual costs of the French nuclear power plants by plant. The so-called negative learning found by Grubler for the French case was shocking and led to discussion of this concept. However, the term "negative learning" may lead the reader to think too narrowly. We are skeptical that lost skills account for much of the increased cost and rather think that it is much more about increased scope of the plants, greater complexity which makes the power plants more difficult to build.

Rangel et al. (2013) revisited the French nuclear experience using the actual construction costs of the French nuclear fleet that had been published in a report by *Cour de Comptes* and they found positive learning effects when building the same type of reactors as a result of Grubler's work. With this information they have tried to identify investment costs' main drivers and found some lessons to explain the cost escalation phenomena. Regarding 'same type of reactors' it is important to state that there were 4 different ‚tranches' of 900MW reactors, Programme 1970 (6 reactors), CP900-1 (16), CP900-2 (8), Tranches 900 (4). And there was CP-1300-1 (20) and N4 (4), Thomas (1987).

They stated the importance of recognizing that the centralized nature of the French NPP programme not only allowed a fast deployment of this technology but also shielded its costs from private eyes and public scrutiny. As argued by Rangel et al. (2013) the cost escalation with the Cour de Comptes (CdC) data was less severe. On the basis of the analysis of the Cour de Comptes report there would be reason to believe that the construction cost escalation in France is mainly due to the increase in the labor costs but also due to the scaling-up strategy. No economies-of-scale were observed, rather diseconomies of scale. However, Grubler (2014) argues that the CdC data are heavily biased and omit arbitrily important construction costs components. He compares his estimates with CdC (corrected for omissions) criticizes that Rangel and Leveque (2013) have compared the lowest CdC numbers (73 billion) to the best-guess model estimates 89 billion reported in Grubler (2010), and that Rangel et al. (2013) reached the pre-mature conclusion of a significant overestimation of costs and resulting cost escalation of the Grubler costing model. The low end of the range uses the CdC's original estimate excluding construction engineering and labour costs, and considers the Tricastin 3,4 versus the Chooz 1,2 reactor costs as reported in CdC 2012: 22–23; the high end of the range uses the CdC costs adjusted to include construction engineering and labour, plus the 13 to 23 billion Euro2010 accrued interest during construction

Another specific phenomenon is the issue of economies-of-scale. It has generally been assumed that nuclear power plants would be amenable to scale economies. The bigger the cheaper was a basic approach. However, there has not appeared any empirical evidence to prove this assumption. On the contrary, studies from the 1970s showed no evidence of scale economies. E.g Cantor/Hewlett (1988) calculated that a 1% increase in the size of a reactor resulted in a 0.13% rise in the ONC per kW. Following Leveque (2015), for France the increase in the reactor size was accompanied by greater complexity and lead-times which in turn led to higher investment costs per MW. How far this greater complexity was the result of increased scale and how far it was due to the larger reactors being more recent and therefore requiring additional safety systems, for example to take account of Three Mile Island, is difficult to determine. In addition, a key potential influence is so-called economies of number, ie the more per year you make of a particular item, e.g. a unit of a power plant (same size) the cheaper the production is because the fixed costs of production lines is spread more thinly and more efficient production methods can be used.

In 2016 Lovering et al. conducted an analysis on the ONC of 58% of the nuclear reactors world-wide. In that article the authors purport to show that using this larger dataset yields more representative results than analyses that focus on individual countries explicitly citing Koomey et al. (2007) for the United States and Grübler (2010) for France as examples of country-level treatments. This work was heavily critiziced by Koomey et al. (2017). Koomey et al. argue that construction duration and interest payments are integral parts of the overall construction costs. Another issue with the work by Lovering (2016) raised by Koomey et al. was the reliability of the data they added. They included reactors of several designs, eg heavy water reactors (HWRs) going back a long way to prototype and demo plants and from countries like India, Korea, China where there must be doubts about the reliability of the data.

The big picture with respect to a comparison of major studies on the historical development of investment costs of nuclear power plants is provided in Fig. 2. It provides a descriptive analysis of data of different studies and single plants. As seen over time a considerable uptake took place. An important aspect is that for OLK3 and FLA3 initially much lower costs were expected than reported before e.g. by Grubler (2010) for France in 2000. The latest data suggest that the ONC for Olkiluoto 3 will be about €8bn, the latest estimate for Flamanville is €10.9bn.[7] These are plants where almost everything possible to go wrong has gone wrong yet they are cheaper than Hinkley Point C (HPC) which is only expected to start construction between 2019–21 and whose latest cost estimate is £9.8-10.15bn per reactor or about €12bn. Is that because HPC is really more expensive than FLA3 or OLK3 or because the HPC estimate is so padded to prevent cost escalation falling on the owner?

7 https://uk.reuters.com/article/us-edf-flamanville/edfs-flamanville-reactor-start-again-delayed-to-2020-idUKKBN1KF0VN (Accessed August 22, 2018)

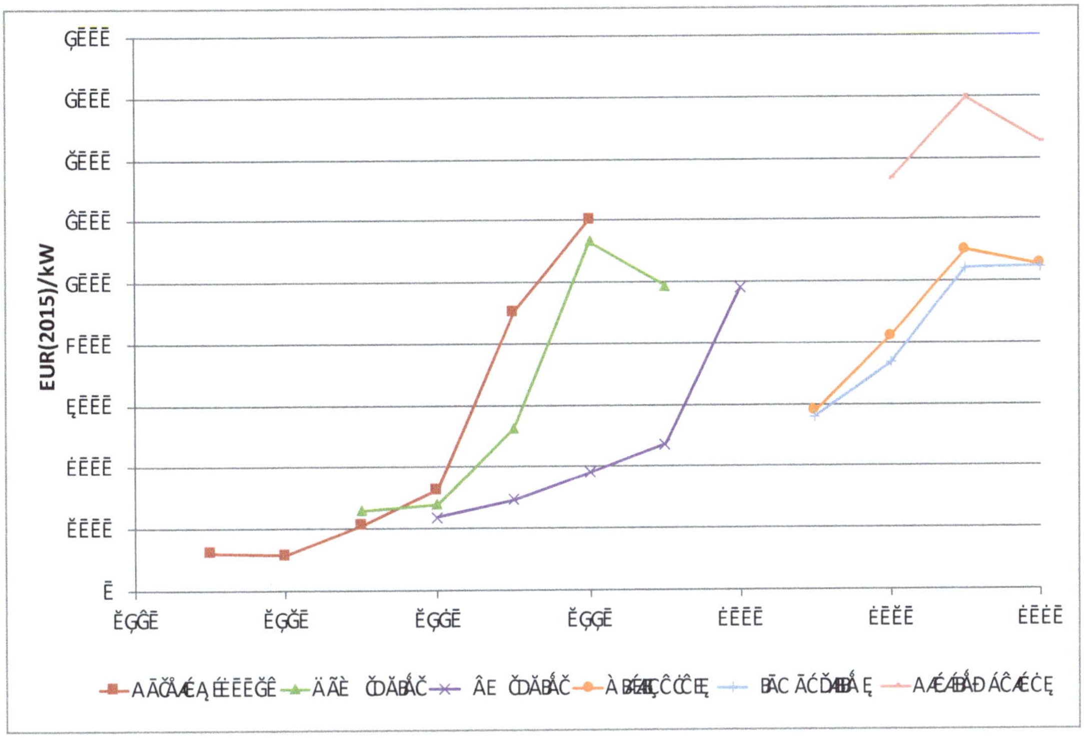

Fig. 2 The big picture: A comparison of major studies on the historical development of investment costs of nuclear power plants

It is important to state that Hinkley Point is unusual amongst nuclear projects because the investment costs and total cost are in advance set relatively high. It is the first time, that the investment costs of a NPP are in advance estimated to be on a higher level than all NPP constructed so far (or under construction).

4 Technological Learning

The next issue we discuss is Technological Learning. It goes in principle along with the dynamic development of investment costs of any technology. The idea is that it is well known that the cost of a technology is expected to drop as it is deployed more widely. That is to say, it is of interest to identify whether with increasing capacities deployed a decrease in investment costs took place. Some major references is in this context are Wene (2000), McDonald/Schrattenholzer (2001), Kobos (2006), Wiesenthal et al. (2010). On learning, one has to be careful. The original Arrow definition (Arrow 1962) was very narrow and encompassed better performance using the same capital stock. In short the workers simply got better at using the equipment. This may be too narrow but it is probably useful to distinguish between design changes that arose because technical change/progress allowed new options to be pursued and design changes that resulted from experience with existing designs.

Nuclear technology displays the opposite trend to reductions in cost. We think that there are four factors that would lead to lower costs for a normal successful technology: economies of scale, economies of number, learning by doing and technical progress. Actually, we are convinced that learning has taken place, but it either hasn't reduced costs or other factors have swamped learning cost reductions. For example, the experience at Three Mile Island was certainly learning but it increased costs. We think it is important to really distinguish and accurately separate these effects which are quite distinct. Of course, this is not easy. A key phrase may be 'a successful technology'. It might be that technologies that do not have scope for these effects fail for that reason. The problem with nuclear is that it was not allowed to fail. In addition, as Leveque states, all other things being equal, the more powerful the reactor, the smaller the number of identical units built.

With respect to Technological Learning for NPP the following is important: Even in the times of booming plant construction in the 1970s and the 1980s nuclear was one of the few exceptions in the sense that additional capacities constructed did not lead to resulting cost reductions. They are mainly that for the early plants no real costs were revealed. Costs were distorted by public subsidies, subsidies from industry (from the constructors of plant to get into the market) and of financing subsides due to very favourable interest rates. Over the course of time these subsidies were gradually removed and costs increased instead of following the classical learning theory. In addition, it is worth mentioning that learning could increase costs, e.g. if a cheap material is not good enough, or existing designs are not safe enough.

5 Historical developments of construction times

One major reason for the increases in nuclear generation costs is the increase of construction times. As an example Grubler (2010) analyzed the historical development of construction times of nuclear power plants in France between 1965 and 2005. His results show that up to 1985 the majority of construction times were between 60 and 84 months. After 1985 the average duration increased in a virtually linear way. In this view for France also the first announcement for the construction time of FLA-3 was included. It is by no means clear what was the intention beyond this cost announcement and the corresponding construction time of five years (see later) because the construction times for the plants built in the years before were already significantly higher, twice as high and more. Obviously, the intention was to convince decision makers that FLA-3 would be economic. The forecast cost (€3.2bn) and construction time (five years) for Flamanville were significantly lower than the most recent experience but more realistic figures would have made the project hard to justify. Fig. 3 shows the increase in estimated construction times for five typical cases world-wide. The graph should be read as follows: On the vertical axis are the construction times indicated in months. The lines show how they increased or remained stable over time. E.g. for OLK-3 in 2004 the estimate for the construction time was 60 months in 2018 it is 200 months (Source: Platts, Power in Europe, various issues).

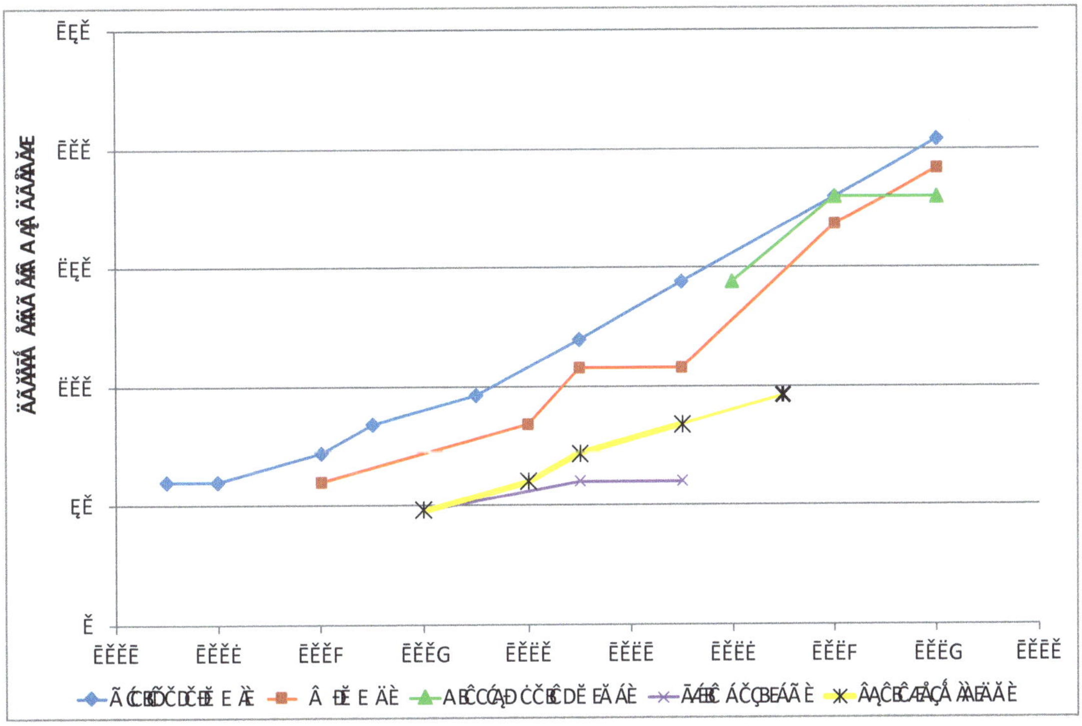

Fig. 3 Increase in the delays of construction times for five typical cases world-wide

6 Lessons learned from the developments of Flamanville and Olkiluoto

Another major question is what caused the increase in construction duration of the NPPs in Finland, France and the US still under the construction. In addition, there was an increase in construction times in China and Russia for China from 2011 onwards and for Russia with the latest design. For instance, the construction of the first European pressurized reactor (EPR) in FLA3 in France revealed that even when this reactor was initially thought as no more costly than its predecessor (the N4) this would not be the case.

At the beginning of 2005 the estimated cost of this project were €3.2 billion. However this figure was revised in 2011, when the state-owned French company Electricite-de–France (EdF) announced that the costs had reached €6 billion. This situation even worsened with the latest press releases stating €8.5 billion in 2012 and 9.5 billion in 2016. The latest estimate is €10.9bn (Platts, 2018).

For the Westinghouse latest design (AP1000) the situation for the two pairs of reactors (Summer and Vogtle) that started construction in the US is very similar. The first cost estimates done both in 2003 were around USD 2400/kW. These costs were later revised stating ONC in USD 2010 of 5100/kW. The pattern is that all three Gen III+ designs that have started construction – EPR, AP1000 and AES2006 – have overruns of time much longer than their predecessors. In Russia and China, contemporary projects using earlier designs were much less delayed. Because Gen III+ has higher design safety, if this leads to greater complexity and makes them more difficult to build, contrary to the claims made for Gen III+, this will increase their costs and increase lead times which, if it reflects construction problems will also increase interest costs.

7 Summarizing the major reasons for investment costs increases of NPP

In the following we summarize the major reasons for investment costs increases of NPP. As stated, so far there is no precise and comprehensive analytical evidence that explains the skyrocketing of the real costs of NPP that have occurred since the beginning of nuclear power use. Our explanations for the major reasons for the cost increases are:

- The intuitive assumption is that these extra costs arise from the additional safety requirements resulting from accidents at Browns Ferry, Three Miles Island and Chernobyl (and of the 9/11 attack). Indeed, as Rangel (2013) and Leveque (2015) have proven for France increases in safety equipment have contributed to about half of the construction cost increases in France between 1970 and 1990;
- There might also be a need for better quality materials, for example Westinghouse steam generators of the 70s used a material that corroded too quickly. If raw materials like steel and concrete have gone up faster in real terms than inflation, that would also have increased real costs (Cohn 1997, Grübler 2010, Thomas (2005)). Increases in labour and material costs is also argued as a major cost driver by Leveque (2015) for France;
- A systematic underestimation of the construction costs as well as the construction duration: One factor is that reactors seem to have become more prone to cost escalation from the pre-construction forecast. Another reason for present day construction costs increases is that the pre-construction costs have become more realistic thereby appearing to raise the real cost. Certainly, the prices quoted in the 1960s were dramatic underestimates.
- Finally, costs could have gone up because things simply have gone wrong without the intrinsic cost going up, e.g. how far is the higher than estimated cost of OLK3 due to the forecast being an underestimate and how far because things have gone wrong, raising costs. In this context also the question is of interest whether Western companies simply do not have the skills anymore to complete huge projects on time. The next obstacle concerns on-site construction and short production runs. Much like other civil engineering projects – bridges, airports e.g in Berlin – NPP are mainly built on-site.
- However, increases in interest rates for financing has not been identified as a driver by any study.

Other possible reasons for the cost increases are:

- the removal of public and industry subsidies;
- increase in construction times: Western companies have lost skills due to the lack of new orders, and are no longer able to construct plants on time, with huge delays leading to construction times two or three times longer than planned;
- changes in generation, e.g from GEN II to GEN III and changes in the engineering design (reduces possible Learning effects);
- The ending of dumping by construction companies;
- The fact that scaling-up appears to have increased, not decreased costs as illustrated in France (Leveque, 2015).

8 Conclusions

Looking back to the years of the nuclear dream in the 1950s and 1960s one of the major historical arguments by the promoters for generating electricity from nuclear power was its low costs compared to other electricity generation technologies. However, from the first wave of nuclear reactors construction costs have been on an escalation course. To date no systematic analysis on the reasons why the costs – especially the investment costs – of NPP have skyrocketed, has been conducted. This work is the first that presents the major reasons for investment costs increases in a systematic structured way.

The major conclusions of this analysis are: There is one core perception regarding the costs of nuclear power plants in recent years: actual costs have always been higher than stated prior to construction and construction times have always exceeded those promised, in most cases considerably. In addition, Western companies have lost skills, and are even less able to construct plants on time, with huge delays leading to construction times two or even three times longer than planned. For nuclear power plants in Western Europe and the U.S. in the last 40 years it can be stated that (i) forecasts of construction times have never been reliable; (ii) forecasts of investment costs have seldom if ever been fulfilled, actual investment costs were always higher than costs announced; (iii) currently there are no signs anywhere of a cost decrease.

What can be stated today is that the economic performance of new nuclear power plants, particularly in the Western countries, has declined substantially compared to their predecessors and to competing power generation technologies. The costs of nuclear power have increased dramatically while on the other hand the costs of wind and photovoltaics – now the major competitors – have fallen the economic performance of nuclear in comparison to these renewable technologies is getting worse.

It will be become much harder for nuclear to recover money in renewable-based electricity markets – much less base load needed – even the pure operation & maintenance costs are difficult to recover in today's electricity markets leading to more and more unfavourable future prospects of nuclear from an economic point-of-view (MIT 2003). In addition, the introduction of competition to electricity markets means the financial risks that were previously borne by the customer must now increasingly be borne by the investors. Because of these risks faced in competitive electricity markets, interest rates have risen and "investors tend to favour less capital intensive and more flexible technologies". Or as an influential interdisciplinary study conducted at the MIT as long ago as 2003 stated "Today, nuclear power is not an economically competitive choice". In addition, as Leveque

(2015) states: "unless nuclear industry moves away from the present model of large, non-modular plants and gigantic construction projects, the investment costs of NPP are likely continue to rise."

Given the identified reasons for the cost increases – and their irreversibility – we state that the time of "cheap" electricity from nuclear power is undoubtedly over – regardless, whether it has ever existed – and for the next years there are no signs of a reversal of current upward going cost trends.

References

Arrow, K., 1962. The Economic Implications of Learning by Doing. Rev. Econ. Stud. 29, 155–173.

Cantor R., Hewlett J., 1988: The economics of nuclear power: Further evidence on learning, economies of scale, and regulatory effects. Resources and Energy, Volume 10, Issue 4, December 1988, Pages 315–335

Cohn, Steven Mark, 1997. Too cheap to meter, State university of New York Press.

EPRI Electric Power Research Institute, 1993. TAG-Technical Assessment Guide:Vol. 1: Electricity Supply-1993. EPRITR-102276-V1R7.

Grubler, Arnulf, 2010. The costs of the French nuclear scale-up:a case of negative learning by doing. Energy Policy. 38, (9), 5174–5188.

Grubler, Arnulf, Wilson, Charles, 2014. Energy Technology Innovation, Learning from Historical Successes and Failures. Cambridge University Press, Cambridge, UK.

Grubler, Arnulf, 2014. The French pressurized water reactor programme. In: Grübler, A., Wilson, C. (Eds.), Energy Technology Innovation, Learning from Historical Successes and Failures. Cambridge University Press, Cambridge,UK, pp. 146–161.

Harris et al (ICEPT), 2012. Cost estimates for nuclear power in the UK. Imperial college center.

Hultman, Nathan E., Koomey, Jonathan G., 2013. Three Mile Island: the driver of US nuclear power's decline? Bull.At.Sci. 69(3), 63–70 http://bos.sagepub.com/ content/69/3/63.abstract.

IAEA, 2004. Nuclear Enery Today. Nuclear Energy Agency.

Komanoff, Charles, 1981. Power plant cost escalation, Van Nostrand Reinhold, New York

Koomey, Jonathan, Hultman, E. Nathan, Grubler Arnulf, 2017. A reply to "Historical construction costs of global nuclear power reactors". Energy Policy 102, 640–643.

Koomey, Jonathan G., Hultman, E. Nathan, 2007. A reactor-level analysis of busbar costs for U.S. nuclear plants,1970–2005. Energy Policy 35 vols. (11), 5630–5642. http://dx.doi.org/10.1016/j.enpol.2007.06.005.

Lévêque, Francois, 2015. The Economics and Uncertainties of Nuclear Power, Cambridge University Press.

Lovering, Jessica R., Yip, Arthur, Nordhaus, Ted, 2016. Historical construction costs of global nuclear power reactors. Energy Policy 91(4), 371–382 http://www. sciencedirect. com/science/article/pii/S0301421516300106.

Kobos P. H., Jon D. Erickson, Thomas E. Drennen, 2006. Technological learning and renewable energy costs: implications for US renewable energy policy, Energy Policy 34, 1645–1658.

McDonald A., L. Schrattenholzer, 2001. Learning Rates for Energy Technologies, Energy Policy, Vol. 29, No. 4, 255–261.

MIT, 2003. The future of nuclear power, An international MIT study.

Mooz William E., 1978 Cost analysis of light water reactor power plants, Rand Corporation, Santa Monica, CA (1978) R-2304-DOE.

Mooz William E., 1979. A second cost analysis of light water reactor power plants, Rand Corporation, Santa Monica, CA, R-2594-RC.

Mooz William E., 1982. An updated data base for light water reactor power plants, Rand Corporation, Santa Monica, CA, R-1899-RC.

Platts, 2018: Power in Europe, Issue 787 / December 3, 2018.

Rangel L. E., Leveque F, 2013. Revisiting the Nuclear Power Construction costs Escalation Curse, The Energy Forum 3, 14–15, 2013.

Rothwell, Geoffrey, 2015. Economics of Nuclear Power. Routledge, NewYork,NY http://amzn.to/1RyHfLx.

Schneider, Mycle, 2006. The nuclear endgame.

Thomas, Steve, 2005. The economics of nuclear power, Heinrich Böll Stiftung, Nuclear Issues Paper No. 5.

Thomas, Steve, 1987. The realities of nuclear power‘ (CUP), Cambridge Energy Studies.

Tybout, Richard, "The Economics of nuclear power", Am Econ Rev, 47, 2, 359–360.

Wene, Claes-Otto, 2000. Experience Curves for Energy Technology Policy, International Energy Agency, OECD/IEA.

Wiesenthal et al., 2012. Technology Learning Curves for Energy Policy Support, EC Joint research centre.

Renewable Energies versus Nuclear Power

Comparison of Financial Support Exemplified at the Case of Hinkley Point C

Gustav Resch, and Demet Suna[1]

Abstract

The energy policy debate in Europe has set (industrial) competitiveness high on the agenda. Support for renewable energies (RE) was in debate and partly suspended. The recent discussion on supporting nuclear power in the UK has, however, demonstrated that renewables are not the only low-carbon option that requires financial incentives under the current framework conditions. The aim of this paper is to compare the costs of state aids necessary for constructing new nuclear power plants for the example of the planned plant at Hinkley Point C in the UK with support incentives for RE.

For doing so, a static and a dynamic approach are followed: The static approach compares today's support incentives for renewable energy with the state aid for Hinkley Point C, whereas for the dynamic approach a model-based assessment of future RE deployment up to 2050 in the EU is undertaken. This is done by use of the Green-X-model (www.green-x.at) and incorporates the impact of technological learning (future cost reductions) as well as aspects of market integration of variable renewables like solar and wind.

1 Gustav Resch, Technische Universität Wien, Austria, resch@eeg.tuwien.ac.at; Demet Suna, AIT Austrian Institute of Technology, Vienna, Austria, demet.suna@ait.ac.at

R. Haas et al. (Eds.), *The Technological and Economic Future of Nuclear Power*, Energiepolitik und Klimaschutz. Energy Policy and Climate Protection, https://doi.org/10.1007/978-3-658-25987-7_6

The assessment is conducted at EU level and for selected EU countries where nuclear power plays a role at present or is considered as a viable future option. From an analytical point of view we undertake an evaluation of effectiveness (i.e. amount of electricity generation stipulated) and economic efficiency of RE and nuclear power support for today and for the future.

1 Introduction

The European Union is divided on the issue of electricity production. While there is consensus that generation technologies need to be low on greenhouse gas- emissions, the question of whether to use renewables or nuclear to meet this power demand is highly controversial. Both options still require financial support and this is not going to change in the near future. This raises the question of where our money should be invested in order to achieve greater economic efficiency: into support for renewable energies (RE) or support for nuclear power plants?

This paper sets out to answer this question. The recent state aid case for the construction of the nuclear power plant Hinkley Point in United Kingdom serves as the model for the nuclear option. After discussing the costs for the nuclear model, we undertake an overview on existing support schemes for renewables in the European Union. Next to that, we conduct the prospective comparative assessment. Here a detailed model-based scenario assessment serves as basis for estimating future cost developments concerning renewable energies. This is then again contrasted with the nuclear model derived from the Hinkley Point case. Finally, conclusions end up this paper.

2 Background – existing and planned support for nuclear power and renewable energies

This section is dedicated to shed light on support schemes for low-carbon energy technologies, specifically nuclear power and renewable energies in the electricity sector. Here the planned support scheme for the new nuclear power plant at Hinkley Point in United Kingdom serves as the model for the nuclear option. After a brief recap on the planned support scheme, classified as state aid, we lay down the resulting costs for the nuclear model case. Next to that we take a closer look at renewable energies, undertaking an overview on existing support schemes for renewables in the European Union.

2.1 New milestone in nuclear state aid: Hinkley Point

The launch of a state aid scheme for a new nuclear power plant at Hinkley Point C in the United Kingdom has been heavily debated across Europe since it has represented a change in paradigm concerning nuclear power. While in the early years of the nuclear built-up there have been arguments for very cheap nuclear electricity in future, new cost figures and specifically the requested support for Hinkley Point C have set an end to that myth. Below we recap some key figures and facts that could be extracted from official documents and public statements on that subject.

The NNB Generation Company Limited (NNBG), part of EDF Energy, plans to construct and operate a new NPP, consisting of two units with an electrical cumulative capacity of 3,260 MW, and an estimated electricity production of 26 TWh per year at the Hinkley Point NPP site (Hinkley Point C 1&2). If constructed, Hinkley Point C would be the UK's first new reactor since 1988.

The construction costs of Hinkley Point C were first estimated to be ca. € 19 billion (EDF, 2013), but were corrected by the EC to € 31.2 billion, and overall capital costs are assumed to be € 43 billion (EC, 2014a). To cover such enormous investments, EDF has undergone lengthy negotiations with the UK government. The start of the operation is supposed to be in 2023[2] with expected operational lifetime of 60 years. The key terms of the final agreement between EDF and the UK government contain the following provisions:

Financial support based on "contract for difference" model

The agreement took the form of a so-called "Contract for Difference" (CfD): if the wholesale prices for electricity fall below an agreed strike price, then the Secretary of State will pay the difference between the strike price and the wholesale price, ensuring that NNBG will ultimately receive a fixed level of revenues. When the wholesale price is higher than the strike price, NNBG will be obliged to pay the difference to the Secretary of State. The duration of the contract is 35 years for each of the two reactors.

---

2 Our analyses are based on the initial operation start time of 2023, and any update on this issue could not be taken into consideration. Nevertheless, the readers should keep in mind that as of September 2013, a delay of start date is expected (Gosden 2015).

The strike price is set at € 108 per MWh (expressed in real terms, as of 2012). If EDF constructs a second nuclear power plant at another site (i.e. Sizewell C) using the same design, the strike price would become €104 per MWh. The strike price will be fully indexed to the Consumer Price Index, meaning that based on current assumptions concerning inflation, this would translate into a nominal strike price of € 329 per MWh in 2058 (as the last year in which the CfD scheme applies).

After the modifications urged by the European Commission, a gain-share-mechanism for the overall profits will be in place for the entire project's lifetime, namely 60 years. If the construction costs are lower than expected, these gains will also be shared (EC, 2014a).

Credit guarantee

The NNBG will also benefit from a credit guarantee issued by the UK Treasury. This guarantee would significantly reduce EDF's risk exposure and therefore the cost of capital. After the modification in 2013, the guaranteed fee that the operator must pay the UK Treasury was significantly raised, resulting in an effective reduction of the subsidy by more than € 1.3 billion. (EC, 2014a). Table 1 summarises the main characteristics of the planned NPP at Hinkley Point C.

Tab. 1 Main characteristics of Hinkley Point C

Capacity per unit	MW_e	1,630
Number of units		2
Total capacity (two units)	MW_e	3,260
Electricity generation	TWh/a	26
Estimated start of operation	Year	2023
Financial support (Contract for Difference / Feed-in Tariff)	GBP_{2012}/MWh	92.5 (89.5)
Duration of support	Years	35

European regulations allow Member States to determine their energy mix within their national competence. However, when public money is spent to support companies, the European Commission must verify that this is done in accordance EU rules on state aid. Therefore the UK's support scheme was investigated in 2013. During this investigation, the UK was required to modify the terms of the project financing. In October 2014, the European Commission concluded that "the modified UK measures for Hinkley Point nuclear power plant are compatible with EU rules" (EC, 2014a).

The October 2014 decision of the European Commission has led to massive protests. The protesters include the Republic of Austria. Based on a legal study, Austria regards subsidies for nuclear power reactors as inacceptable according to EU legislation (BMWFW, 2014).

2.2 EU support for renewable energies

As outlined in detail in the RE-Shaping study (see Ragwitz et al., 2012), the first decade of the new millennium was characterized by the successful deployment of RE across EU Member States – total RE deployment increased by more than 40%. More precisely:

- Electricity generation from RE grew by approximately 40%, RE heating and cooling supply by 30% and biofuels in transport by a factor of 27 during the period 2001 to 2010,
- New renewables in the electricity sector (all technologies except hydropower) increased fivefold during the same period,
- Total investments in RE technologies increased to about € 40 billion annually in 2009 and more than 80% of all RES investments in 2009 were in wind and PV.
- With respect to PV an ongoing trend of achieving impressive cost reductions from year to year has started in the final period close to 2010. .

These impressive structural changes in Europe's energy supply are the result of a combination of strong national policies and the general focus on RES created by the EU Renewable Energy Directives in the electricity and transport sectors towards 2010 (2001/77/EC and 2003/30/EC).

Despite the challenges posed by the financial and economic crisis, RE investments were generally less affected than other energy technologies and partly increased even further over the last couple of years. The European Energy and Climate Package is one of the key factors that contributed to this development. The EU ETS (Emissions Trading System) Directive has introduced full auctioning post 2012, thus exposing fossil power generation to the full cost of carbon allowances, at least in theory. In practice, an oversupply of allowances has however led to a deterioration of prices on the carbon market.

The pathway for renewables towards 2020 was set and accepted by the European Council, the European Commission and the European Parliament in April 2009. The related policy package, in particular the EU Directive on the support of energy from renewable sources (2009/28/EC), subsequently named RE Directive, comprises the establishment of binding 2020 RE targets for each Member State – in line with overall EU target of increasing the RE share to 20% by 2030.

Later on, the EU Energy Roadmap 2050 gave first signals of renewable energy development pathways beyond the year 2020 and identified renewables as a "no-regrets" option. In a next step, Europe's way forward towards 2030 has been discussed intensively. Thus, at the Council meeting of this October (2014) the next step was taken: A binding EU-wide RES target of achieving at least 27% as RES share in gross final energy demand was adopted. This has to be seen as an important first step in defining the framework for RES post 2020. Other steps, like a clear concept for, and an agreement on the effort sharing across Member States have to follow.

Concerning financial support for RE, various policy instruments have been implemented across EU Member States to promote the use of RE (cf. Box 1). Although there are already substantial experiences with the use of support schemes, the dynamic framework conditions have led to a continuous need for reforming the applied policies. Also policy priorities have changed in most Member States. Whilst the policy effectiveness or the ability of support instruments to trigger new investments was a main policy target while the RE-share was still negligible, economic efficiency has become increasingly important in the light of higher shares of RE, rising support costs and the financial crisis. In particular the strong growth of photovoltaics in some Member States has enhanced this change of policy priorities. The stronger focus on cost control mechanisms has led to a revival of tender or auction mechanisms to control the additional RE-capacity eligible for support and to determine support levels in a competitive bidding procedure. Another highly relevant issue regarding renewables support is related to the increasing share of intermittent RE leading to evolving requirements for effective electricity market design. While initially fair remuneration of RE power in the market should be a priority for market design, a more systemic focus on system flexibility should be adopted with a rising share of RE. This is reflected in several market design parameters, e.g. how the system matches temporal profiles of different generation and load types and how it accommodates the spatial profile of intermittent RE generation.

Box 1 Support schemes for electricity from renewable sources

Globally as well as within the European Union, a feed-in tariff (FIT) system is the most common policy instrument for promoting electricity generation from renewable energy sources (RES-E).A quota obligation with tradable green certificates (TGCs) is another widely implemented support scheme. These main instruments for RES-E are often accompanied by complementary instruments like grants offering investment support, fiscal incentives (e.g. tax reductions) or (cheap) loans. The two main support instruments can be characterised as follows:

1. **Feed-in tariffs** offer financial support per kWh generated, paid in the form of guaranteed (premium) prices and combined with a purchase obligation by the utilities. The most relevant distinction is between fixed FIT and feed-in premium systems. The former provides total payments per kWh of electricity of renewable origin while the latter provides a payment per kWh on top of the electricity wholesale-market price (Sijm 2002). In recent years, feed-in tariff systems are also combined with auctions for price determination as well as for having a cost/quantity control on the market. Note that the planned CfD scheme in the UK falls also under the category of a FIT scheme.
2. In a **quota obligation with Tradable Green Certificates** the government defines targets for RES-E deployment and obliges a particular party of the electricity supply-chain (e. g. generator, wholesaler or consumer) with their fulfilment. Once defined, a parallel market for renewable energy certificates is established and their price is set following demand and supply conditions (forced by the obligation). Hence, for RES-E producers, financial support may arise from selling certificates in addition to the revenues from selling electricity on the power market.

3 Method of approach

Renewable energies were compared with the nuclear option by looking at the quantities of power they can both generate and the level of financial support this requires. This mirrors the extra costs which must be borne by the end consumer or society. Five different renewable technologies were analysed: biomass, onshore and offshore wind, small-scale hydropower plants and photovoltaics.

In brief, the static approach compares the current (as of 2013) level of incentives for renewables with the state support mechanism for Hinkley Point. The dynamic approach, in contrast, also considers additional factors including future cost reductions achieved through increasing technological experience and aspects of market integration of variable renewables like solar and wind power. The dynamic approach has been calculated up to 2050; the nuclear option is added from 2023 onwards (planned start-up for Hinkley Point C). The dynamic calculation applies a detailed model-based analysis using the Green-X-model (www.green-x.at). This model takes into account a multitude of factors including costs, potentials, regulatory frameworks, diffusion constraints like non-cost barriers, electricity prices and energy demand, all of which have a strong impact on the economics of power generation. Below we provide for the interested reader further insights on both approaches taken.

3.1 Static approach: comparison of planned support for nuclear with existing RE support

The level of financial support paid to the supplier of nuclear as well as of electricity from the renewable energy sources (RES-E) is a core characteristic of a support policy. Actual support levels are, however, often not directly comparable, and details of the support policy applied, including main instrument like Feed-in-Tarifs (FIT) or quotas as well as complementary incentives, need to be taken into account.

For a comparative assessment of support incentives, the available remuneration level during the whole lifetime of a (RE) power plant has to be taken into account. This is also stated in a detailed assessment report of the performance of RE support policies in EU Member States (Steinhilber et al. 2011). To make the remuneration levels comparable, following the methodology applied in (Steinhilber et al. 2011), time series of the expected support payments per unit of electricity generated are created for each of the assessed options (i.e. biomass, small hydro, photovoltaic (PV) and wind (on- and offshore) as well as nuclear power by country) and the net present value (NPV), representing the current value of overall support payments, is calculated. After that the annualised remuneration level is calculated from the NPV using a discount rate of 6.5% and following under each type of instrument a normalisation to a common duration of 20 years. Below Formula (1) and (2) show further details on the underlying calculation approach.

$$NPV = \sum_{n}^{N} \frac{SL_t}{(1+z)^n} \quad (1)$$

$$A = \frac{z}{(1-(1+z)^{-N})} * NPV \quad (2)$$

where:

NPV: Net present value;
SLt: Support level available in year t;
A: Annualised remuneration level;
Z: Interest rate;
n: Reference year;
N: Payback time

In addition, expected future wholesale electricity prices are normalised over the same time period. In the case of a quota scheme with tradable green certificates (TGCs), it is assumed that the total remuneration level is composed of the conventional electricity price (wholesale electricity prices) and the average value of TGCs. The results on remuneration levels, wholesale electricity prices or net support expenditures are expressed subsequently in real terms, using $€_{2013}$.

3.2 Dynamic approach: a prospective model-based assessment of planned support for nuclear with expected future RE support

The dynamic assessment follows the principles sketched above, assessing effectiveness and economic efficiency (i.e. cost effectiveness) of RE and nuclear power support from a future perspective. The approach taken builds on a model-based assessment of future RE deployment in the European Union and at country level for the UK up to 2050.

A scenario of dedicated RE support is assessed that follows the policy decisions taken, i.e. the binding 2020 RE target (of reaching a share of 20% RE in gross final energy demand), and that reflects the European policy agenda for tomorrow where mitigation of climate change and the built-up of a sustainable energy system are expected to remain as top priorities in the period post 2020. The scenario proclaims the prolongation of establishing enhancing framework conditions at EU level while national (or in future European) RE support instruments aim for setting the corresponding incentives to assure the achievement of European RE targets by 2030 and beyond. Complementary to fine-tuned financial incentives for RE this requires enabling framework conditions and a mitigation of currently prevailing non-economic barriers (i.e. administrative barriers and grid constraints that hinder the upscaling of RE deployment across Europe at present).

To derive the scenario, the Green-X model is used. Green-X is a dynamic simulation model for assessing the impact of energy policy instruments on future RE deployment and related costs, expenditures and benefits at technology-, sector- and country-level, that has been widely used in various studies at a national and European level, e.g. for the European Commission to assess the feasibility and impacts of "20% RE by 2020", and to explore policy options post 2020 – for a detailed description of this model we refer to (Green-X, 2015).

Future requirements concerning support schemes for RE

Generally, the need to incentivise the deployment decreases for RE technologies thanks to technological learning. Technological progress and related cost reductions go hand in hand with the ongoing market deployment of a certain technology. This has been impressively demonstrated for example by the uptake of PV in Germany and other countries and the achieved significant decline of capital cost. But what has been observed for PV is by far not an exceptional case, it is rather an affirmation of a general empirical observation – i.e. the technological learning theory.

On the contrary, with ongoing market deployment of variable renewables like solar and wind we see however also an opposing tendency that ultimately may cause an increase in the need for financial support. This concerns the market value of the produced electricity that is fed into the grid. For these technologies it is becoming apparent that in future years (with ongoing deployment) a unit of electricity produced is less valuable than of a dispatchable RE technology like biomass where the plant may interrupt operation during periods of oversupply (because of massive wind and solar power inflow) and correspondingly may low the wholesale power prices. Accordingly this may increase the required net support, determined by difference between total remuneration and market value.

Whether the cost decrease due to technological learning or the increase in support requirements due to a decreasing market value will be of dominance depends on the country- and technology-specific circumstances. This will be analysed in further detail for all assessed energy technologies for the UK and EU 28 within the dynamic assessment.

Overview on key parameters

In order to ensure maximum consistency with existing EU scenarios and projections the key input parameters of the scenarios presented in this work are derived from PRIMES modelling (EC, 2013) and from the Green-X database with respect to the potentials and cost of RE technologies. Table 2 shows which parameters are based on PRIMES, on the Green-X database and which have been defined for this assessment.

More precisely, the PRIMES scenario used is the reference scenario as of 2013 (EC, 2013). However for this assessment, demand projections have been contrasted with recent statistics (from Eurostat) and corrected where adequate (in order to assure an appropriate incorporation of impacts related to the recent financial and economic crisis). Moreover, mid- to long-term trends have been further modified to reflect an adequate representation of energy efficiency, assuming a proactive implementation of energy efficiency measures in order to reduce overall demand growth. Here we base our demand trends on a detailed study led by Fraunhofer ISI, done on behalf of the European Commission (Braungardt et al. 2014).

Tab. 2 Main input sources for scenario parameters

Based on PRIMES	Based on Green-X database	Defined for this assessment
• Primary energy prices • Conventional supply portfolio and • conversion efficiencies • CO2 intensity of sectors	• RE cost (investment, fuel, O&M) • RE potential • Biomass trade specification • Technology diffusion / Non-economic barriers • Learning rates • Market values for variable RES-E	• RE policy framework • Reference electricity prices • Energy demand by sector*

4 Results

The **static approach** undertaken at country level provides a comparison of planned support for nuclear power at Hinkley Point C in the UK with existing RE support, that is, as implemented in 2013. Key outcomes of that are summarised in Fig. 1, indicating by RE technology the possible annual electricity generation that could be supported with currently implemented RE policies in analysed countries. For doing so, average remuneration and net support levels are taken as given. Note that generally a range of feasible generation volumes is depicted for the assessed RE technologies by country:

- The lower boundary of possible volumes answers the question how much renewable electricity (from different technologies) could be supported in the assessed country, if annual net support expenditures as expected for Hinkley Point C under UK circumstances are taken as given.
- If a new nuclear power plant like the one planned for Hinkley Point C is constructed in another country under similar support conditions as planned for the UK (i.e. same FIT level as set in the UK), the net support level would differ because of different electricity wholesale prices. Thus, the upper range in Fig. 1 is consequently taking into account this difference, using country-specific wholesale prices and corresponding annual net support expenditures, and showing how much electricity generation could be achieved with that for the assessed RE technologies.

In accordance with Fig. 1, key results of the static assessment can be summarised as follows:

Under similar budgetary constraints, a higher amount of electricity generation appears feasible with wind onshore and small-scale hydropower plants compared to nuclear in all analysed

- countries (with the exception of hydro in the Czech Republic). This means, in turn, that small hydro and wind onshore represent "least cost" options from today's perspective across all assessed countries. In those countries where support is offered to that option, i.e. the UK and Poland, co-firing of biomass in fossil-fuel based power plants represents another cost-effective generation option.
- A cross-country comparison indicates a comparatively small benefit for wind onshore in the UK. While in all other countries remuneration levels and net support are significantly lower and, in turn, feasible generation volumes are higher for wind onshore compared to nuclear power. This is the result of an unequal risk perception of two distinct policy instruments that come into play for the UK: Today's support for wind onshore in the UK via a certificate trading regime can be classified as significantly more risky than safe revenues stemming from a "Contract for Difference" scheme as planned for Hinkley Point C.
- Both PV and wind offshore represent the most costly options from today's perspective in the majority of countries.

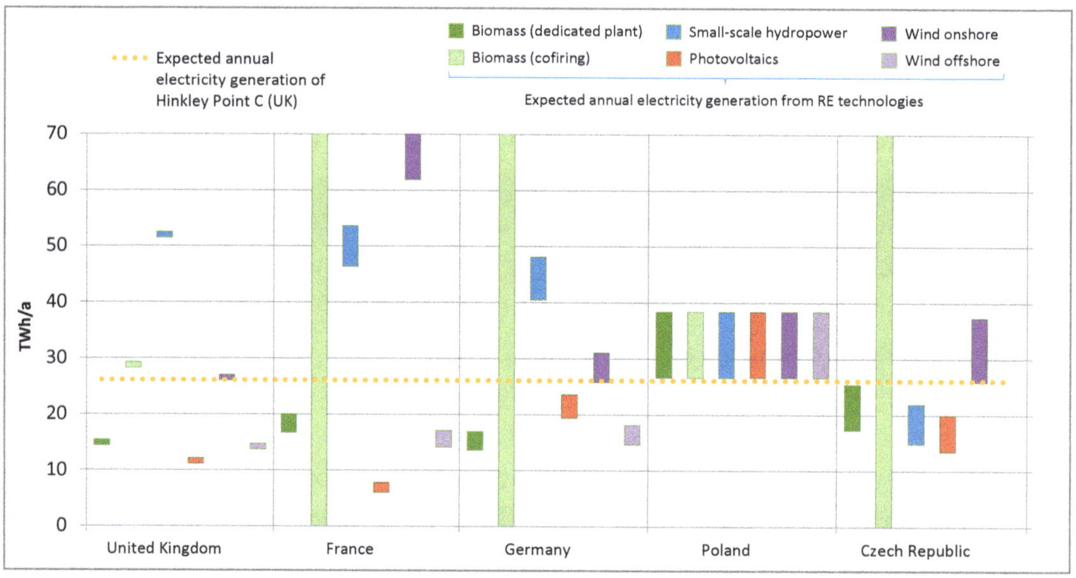

Fig. 1 Comparison of expected annual electricity generation of Hinkley Point C with feasible volumes from assessed RE technologies by assessed country (Source: Own calculations)

The static assessment as discussed above compares today's incentives for RE with a planned aid scheme for nuclear power that may become effective ten years ahead. Since partly significant cost reductions have been achieved throughout the last decade for several RE technologies it can be expected that ongoing technological learning will trigger additional cost decreases and, consequently, reduce the need for RES-E support in forthcoming years. Thus, complementary to the above, a **dynamic approach** is followed within this study: Building on the Green-X scenario of dedicated RE support and the therein sketched deployment of renewables in the EU28, a comparative assessment of future RE support with the planned subsidy for Hinkley Point C is undertaken for all assessed countries. More precisely, the years from 2023 to 2050 form the assessment period whereby 2023 is chosen since this is the year when Hinkley Point C is expected to start full operation. Within that assessment support expenditures for RES-E and nuclear power are contrasted and, finally, the cost-effectiveness of the two distinct pathways is derived.

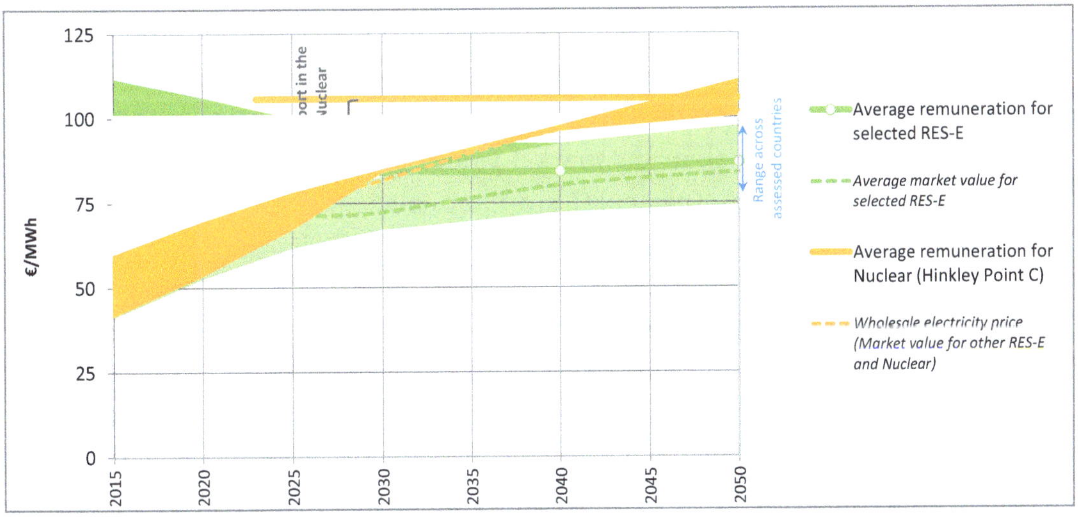

Fig. 2 Future development of remuneration levels and corresponding market values of the assessed RE technologies (as aggregate) and of nuclear power across assessed countries and at EU28 level according to the Green-X scenario of dedicated RE support (Source: Own assessment (Green-X))

Fig. 2 shows the development over time of remuneration levels and the corresponding reference price for the assessed technology options, using weighted average figures to determine market value and the remuneration level for the aggregated RE technology cluster that comprises the basket of assessed individual RE technologies. This graph shows these developments at EU 28 level (i.e. via dotted or solid lines) while shaded areas indicate the ranges of expressed items occurring across assessed countries. Generally, the need for net support for a new installation in a given year can then be derived by subtracting the market value from overall remuneration. Thus, this allows for a first interpretation of cost efficiency:

- For nuclear power it can be observed that during early years of operation a significant gap between remuneration and market value, in this case determined by the yearly average wholesale electricity price, occurs. This is however getting smaller in later years thanks to the expected increase in wholesale electricity prices (that goes hand in hand with an increase of fossil fuel and carbon prices over time).
- For renewables an interpretation appears more difficult since outcomes reflect the over shading impacts of a basket of technologies that come into play: In early years a strong decline of remuneration levels is apparent, reflecting expected technological progress across all considered RE technologies but, thanks to their dominance driven by cost trends for on- and offshore wind as well as photovoltaics. In later years, with increasing deployment the merit-order-effect and the related decrease in market values of variable renewables is applicable. Offshore wind is then mainly responsible for the small gap remaining, where average RE remuneration is higher than the market value at EU28 level as well as in some of the assessed countries. In general, similar to nuclear the need for net support shows a decreasing tendency in the final years up to 2050.

Comparing cumulative electricity generation and corresponding support expenditures that would arise throughout the assessment period (2023 to 2050) an overall conclusion related to the cost effectiveness of the two distinct pathways (i.e. nuclear versus RE) can be drawn next. Results on specific net support as derived by dividing cumulative support expenditures by cumulative electricity generation are shown in Fig. 3. Complementary to that, resulting cost savings at country as well as at EU28 level that would arise if the preferred option is followed are shown in Fig. 4.

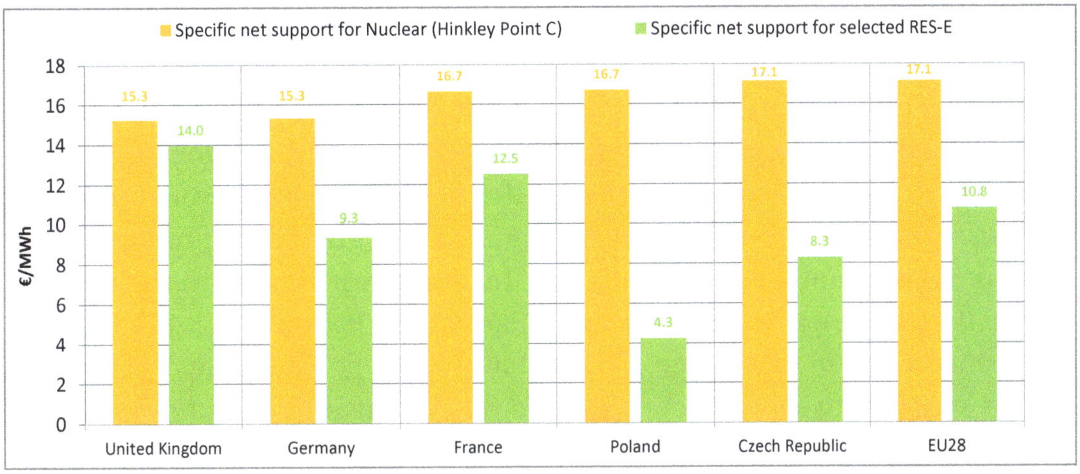

Fig. 3 Comparison of overall cost-effectiveness: Specific net support for assessed RE technologies and nuclear power by assessed countries and at EU28 level according to the Green-X scenario of dedicated RE support (Source: Own assessment (Green-X))

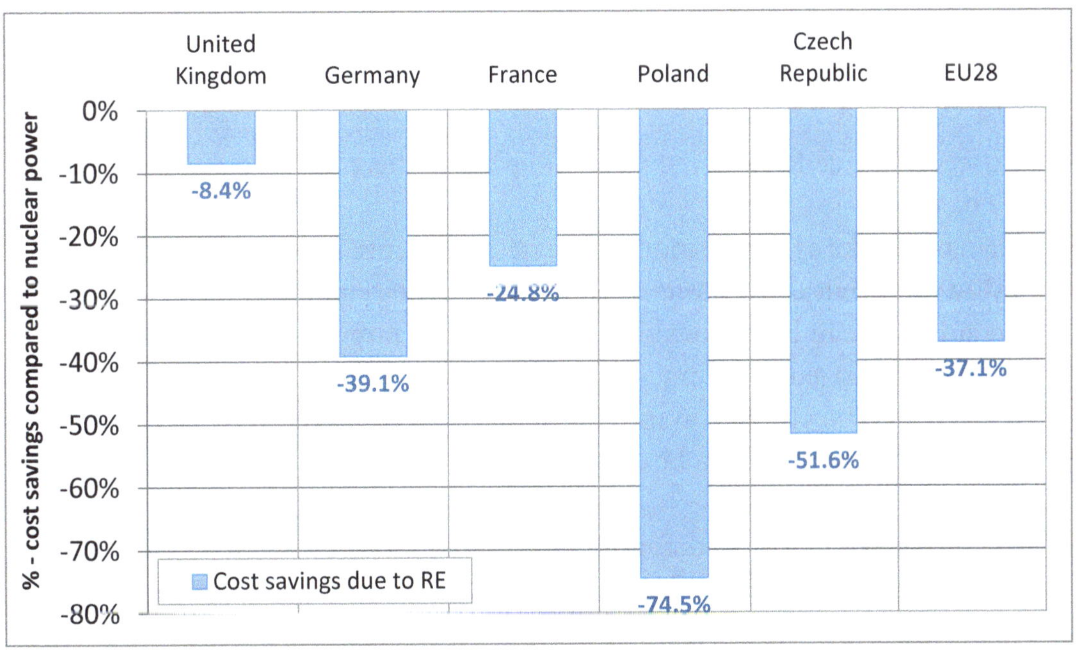

Fig. 4 Comparison of overall cost-effectiveness: Cost savings due to RE compared to nuclear power by assessed country and at EU28 level according to the Green-X scenario of dedicated RE support (Source: Own assessment (Green-X))

We would like to highlight the following As discussed above, net support is generally defined as the difference between total remuneration and the market value of the fed in electricity. If a new nuclear power plant like the one planned for Hinkley Point C is built in another country under similar support conditions as planned for the UK (i.e. same FIT level as set in the UK), the net support level would differ because of different electricity wholesale prices that in the case of nuclear power serve as determinant for its market value. In future years lower electricity prices than in the UK are expected for countries like France, Poland, the Czech Republic and the whole EU28 on average. Thus, a new nuclear power would consequently require significantly higher net support in these countries than in the UK. This would strongly increase the burden for consumer and/or the society, respectively.

Results on specific net support as shown in Fig. 4 point out that supporting a basket of RE technologies as analyzed in this assessment leads to a higher cost-effectiveness than the planned support for the nuclear power plant at Hinkley Point C that served as nuclear comparator throughout this exercise. This statement is valid for all assessed countries as well as for the EU28:

- Highest cost savings due to RE can be observed for Poland where following a RE pathway instead of nuclear would lead to savings in support expenditures of 74.5%.
- On second place follows the Czech Republic where savings due to RE are in size of 51.6%.
- Germany ranks as third among the assessed countries with respect to feasible cost savings that come along with following the renewable pathway. Support expenditures can be reduced by 39.1% through targeting support to RE technologies compared to the nuclear alternative.
- At EU28 level on average savings in support expenditures are in range of 37.1%.
- A slightly lower figure can be observed for France where savings are in magnitude of about 25%.
- Last on the list of assessed countries is the UK. However also in that country following a RE pathway instead of a nuclear appears beneficial – i.e. cost savings of 8.4% are identified for the UK.

5 Conclusions

The level of financial support paid to a nuclear or a RE power plant is a core characteristic of the related policy intervention. Support instruments need to be *effective* in order to increase the penetration of energy sources (in this case RE and/or nuclear) and *efficient* with respect to minimising the resulting public cost, i.e. the transfer cost for consumers (society) over time.

This study assesses the effectiveness and efficiency of support schemes in selected European countries for nuclear and specific renewables (wind, hydro, PV and biomass) using two distinct approaches; a static and a dynamic comparative assessment.

The **static comparative assessment** of the envisaged state aid scheme for the UK's planned nuclear power plant at Hinkley Point C contrasted with today's support incentives for renewables leads to the following conclusions:

Onshore wind and small hydropower plants (with the exception of the Czech Republic) represent the "least cost" option from today's perspective in all the countries analyzed. Consequently, if the planned annual support expenditures for Hinkley Point C were channeled into these RE options, then more carbon-free electricity could be generated. In contrast to above, PV and offshore wind can be classified as the most costly options from today's perspective (with the exception of Poland).

If Hinkley Point C were to be built in the assessed countries and under similar support conditions as those planned for the UK (i.e. same feed-in tariff level), then the net level of support would differ from country to country because of varying electricity wholesale prices. Wholesale electricity prices in the UK are currently among the highest in Europe. Prices in the Czech Republic and Poland are lower. Consequently, under the same feed-in tariff level as set in the UK, a new nuclear power plant would require significantly more net support (i.e. defined as the difference between remuneration and wholesale electricity prices for nuclear) in Poland or the Czech Republic than in the UK. In turn, this would strongly increase the burden for consumer and/or society.

The static assessment, as discussed above, compares today's incentives for RE with a planned aid scheme for nuclear power that may become effective in ten years. As some significant cost reductions in RE technologies have been achieved over the past decade, we can expect that growing technological experience in this field will trigger more cost reductions and, consequently, will reduce the need for RES-E support in coming years. Therefore this study also takes a **dynamic approach**: building on the Green-X scenario of dedicated RE support and its outline for the deployment of renewables in the EU28, future RE support has been compared with the planned subsidy for Hinkley Point C for all the assessed countries.

This analysis leads us to the following conclusions:

- *A constant level of remuneration, as guaranteed for nuclear power at Hinkley Point C in the UK, may lead to a high consumer burden in the early years, but thanks to expected increases in fossil fuel and carbon prices, net support will decrease over time.*
  During the early years of operation at Hinkley Point there will be a significant gap between remuneration level and market value, in this case determined by the yearly average wholesale electricity price. However, this gap will reduce with time thanks to the expected increase in wholesale electricity prices (which goes hand in hand with an increase in fossil fuel and carbon prices over time).
- *Two opposing trends determine the need to support renewables: cost reductions resulting from technological progress lead to decreasing remuneration, whilst increasing deployment of variable RE technologies cause reductions in their market value. The need for net support depends on the country and technology-specific circumstances.*
- *Generally, the need to incentivise deployment of renewables falls thanks to technological learning. Technological progress and related cost reductions go hand-in-hand with ongoing market deployment of a technology. This has been impressively demonstrated, for example, by the uptake of PV in Germany and other countries, and the corresponding, significant decline in capital costs. But the massive cost decline for PV is certainly not exceptional; it affirms a general empirical observation, i.e. technological learning theory.*
  In contrast, the ongoing market deployment of various renewables including solar and wind demonstrates an opposing tendency that may ultimately cause an increase in the need for financial support: the market value of the generated electricity that is fed into the grid. For these technologies it is becoming apparent that in future years (with ongoing deployment) a unit of electricity will be less valuable than that produced by a dispatchable renewable energy technology such as biomass where the plant may interrupt operation during periods of oversupply and wholesale power prices are correspondingly low.
  Thus the net level of required support is determined by the difference between remuneration and market value. Whether the cost decreases resulting from technological learning outweigh the need for increased support as a result of the decreasing market value, or vice versa, depends on the country and technology-specific circumstances.
- *The assessment at country and at EU levels confirms that remuneration for renewables is expected to decline over time. This decrease is strong in the early years, followed by a slowdown and stagnation in later years. Contrarily, market*

values for variable renewables are expected to more strongly decouple from average wholesale electricity prices.

The analysis, which considers selected EU Member States as well as the EU28 as a whole, indicates a strong decline in remuneration levels for renewables in the early years as a result of expected technological progress across all the RE technologies considered. Thanks to their dominance, this positive trend is driven by cost trends for onshore and offshore wind and photovoltaics. With increasing deployment in later years, the merit order effect and the related decrease in market value of variable renewables applies. Offshore wind is then mainly responsible for the small remaining gap, where average RE remuneration is higher than the market value, both at EU-28 level as well as in some of the assessed countries.

- *If we compare cumulative electricity generation and corresponding support expenditures we can draw an overall conclusion regarding the cost effectiveness of the two distinct pathways (i.e. nuclear vs. RE). Results for specific net support clearly indicate that supporting a basket of RE technologies is more cost-effective than the planned support for the nuclear power plant at Hinkley Point C that has served as the nuclear comparator throughout this exercise. This statement is valid for all the assessed countries as well as for the EU28.*

  The highest cost savings achieved through RE can be observed in Poland where following a RE pathway instead of nuclear would lead to savings in support expenditures of 74.5% whereas average savings in support expenditures for the EU28 as a whole are in the range of 37.1%. Finally, the UK comes last in the potential savings ranking, yet even in the UK it is economically beneficial to follow a RE pathway rather than the nuclear option, with cost savings of 8.4%.

Acknowledgment. This paper builds on an analysis conducted by e-think in cooperation with Austrian Institute of Ecology within the project "Renewable Energies versus Nuclear Power", cf. (Mraz et al. 2014). We gratefully acknowledge financial support from Vienna Ombuds Office for Environmental Protection (Wiener Umweltanwaltschaft).

References

Braungardt, S.; Eichhammer, W.; Elsland, R.; Fleiter, T.; Klobasa, M.; Krail, M.; Pfluger, B.; Reuter, M.; Schlomann, B.; Sensfuss, F.; Tariq, S.; Kranzl, L.; Dovidio, S.; Gentili, P.; 2014. Study evaluating the current energy efficiency policy framework in the EU and providing orientation on policy options for realising the cost-effective energy-efficiency/saving potential until 2020 and beyond. Report for the European Commission, Directorate-General for Energy. Available at: https://ec.europa.eu/energy/sites/ener/files/documents/2014_report_2020-2030_eu_policy_framework.pdf.

Černoch, F. & Zapletalová, V., 2015. Hinkley point C: A new chance for nuclear power plant construction in central Europe? Energy Policy, 83, pp.165–168.

EC, 2014a. 2030 framework for climate and energy policies. European Commission Climate Action. Available at: http://ec.europa.eu/clima/policies/2030/index_en.htm [Accessed June 24, 2015].

EC, 2013. EU Energy, Transport and GHG Emissions Trends to 2050-Reference Scenario 2013, European Commission, Directorate-General for Energy, Directorate-General for Climate Action and Directorate-General for Mobility and Transport. Available at: http://ec.europa.eu/transport/media/publications/doc/trends-to-2050-update-2013.pdf.

EC, 2014b. State aid: Commission concludes modified UK measures for Hinkley Point nuclear power plant are compatible with EU rules. European Commission-Press Release Database. Available at: http://europa.eu/rapid/press-release_IP-14-1093_en.htm [Accessed June 22, 2015].

EDF, 2013. Agreement reached on commercial terms for the planned Hinkley Point C nuclear power station- Article from 21 October,2013. EDF Energy. Available at: http://edfenergy.presscentre.com/News-Releases/Agreement-reached-on-commercial-terms-for-the-planned-Hinkley-Point-C-nuclear-power-station-82.aspx [Accessed June 22, 2015].

Gosden, E., 2015. Nuclear delay: EDF admits Hinkley Point won't be ready by 2023 – Telegraph. The Telegraph. Available at: http://www.telegraph.co.uk/finance/newsbysector/energy/11841733/Nuclear-delay-EDF-admits-Hinkley-Point-wont-be-ready-by-2023.html [Accessed July 9, 2016].

Green-X, 2015. Deriving optimal promotion strategies for increasing the share of RES-E in a dynamic European electricity market. Available at: http://www.green-x.at/ [Accessed June 15, 2015].

Held, A.; Ragwitz, M.; Boie, I.; Wigand, F.; Janeiro, L.; Klessmann, C.; Nabe, C.; Hussy, C.; Neuhoff, K.; Grau, T.; Schwenen, S.; 2014. Indicators on RES support in Europe. A report compiled within the Intelligent Energy Europe Project DIACORE (Policy Dialogue on the assessment and convergence of RES policy in EU Member States), Fraunhofer ISI, Karlsruhe, Germany.

Mraz, G.; Wallner, A.; Resch, G.; Suna, D.; 2014. Renewable Energies vs Nuclear Power – Comparing Financial support. A study conducted by Austrian Institute of Ecology and e-think, with support from Vienna Ombuds Office for Environmental Protection. Available at www.e-think.ac.at.

Ragwitz, M., Steinhilber, S., Breitschopf, B., Resch, G., Panzer, C., Ortner, A., Busch, S., Rathmann, M., Klessmann, C., Nabe, C., De Lovinfosse, I., Neuhoff, K., Boyd, R., Junginger, M., Hoefnagels, R., Cusumano, N., Lorenzoni, A., Burgers, J., Boots, M., Konstantinaviciute, I. and Weöres, B. (2012), RE-Shaping: Shaping an effective and efficient European renewable energy market. Report compiled within the European project RE-Shaping, supported by Intelligent Energy – Europe, ALTENER, Grant Agreement no. EIE/08/517/SI2.529243. Fraunhofer ISI, Karlsruhe, Germany.

Resch, G.; Ortner, A.; Panzer, C.; 2014. 2030 RES targets for Europe – a brief pre-assessment of feasibility and impacts. A report compiled within the Intelligent Energy Europe project Keep-on-Track!, coordinated by Eufores and Eclareon. TU Vienna, Energy Economics Group, Vienna, Austria, 2014. Available at: www.keepontrack.eu.

Steinhilber, S.; Ragwitz, M.; Rathmann, M.; Klessmann, C.; Noothout, P.; 2011. Indicators assessing the performance of renewable energy support policies in 27 Member States. A report compiled with the Intelligent Energy Europe Project RE-Shaping (Shaping an effective and efficient European renewable energy market), Fraunhofer ISI, Karlsruhe, Germany.

Financing Nuclear Decommissioning

Wolfgang Irrek[1]

Abstract

While more and more nuclear installations facing the end of their lifetime, decommissioning financing issues gain importance in political discussions. The financing needs are huge along the Uranium value chain. Following the polluter pays principle the operator of a nuclear installation is expected to accumulate all the necessary decommissioning funds during the operating life of its facility. However, since decommissioning experience is still limited, since the decommissioning process can take several decades and since the time period between the shutdown of a nuclear installation and the final disposal of radioactive waste can be very long, there are substantial risks that costs will be underestimated and that the liable party and the funds accumulated might not be available anymore when decommissioning activities have to be paid. Nevertheless, these financing risks can be reduced by the implementation of transparent, restricted, well-governed decommissioning financing schemes, with a system of checks and balances that aims at avoiding negative effects stemming from conflicts of interests.

1 Wolfgang Irrek, Hochschule Ruhr West University of Applied Science, Bottrop, Germany, Wolfgang.Irrek@hs-ruhrwest.de

R. Haas et al. (Eds.), *The Technological and Economic Future of Nuclear Power*, Energiepolitik und Klimaschutz. Energy Policy and Climate Protection, https://doi.org/10.1007/978-3-658-25987-7_7

1 Introduction

By the end of 2017, 166 nuclear power reactors had been permanently shut down, of which 144 were in the process of dismantling or had already been fully demolished (IAEA 2018). Moreover, 64% of the operational nuclear power reactors in the world at this time were 30 years old or older, and are thus candidates for being shut down in the near future (cf. Fig. 1).

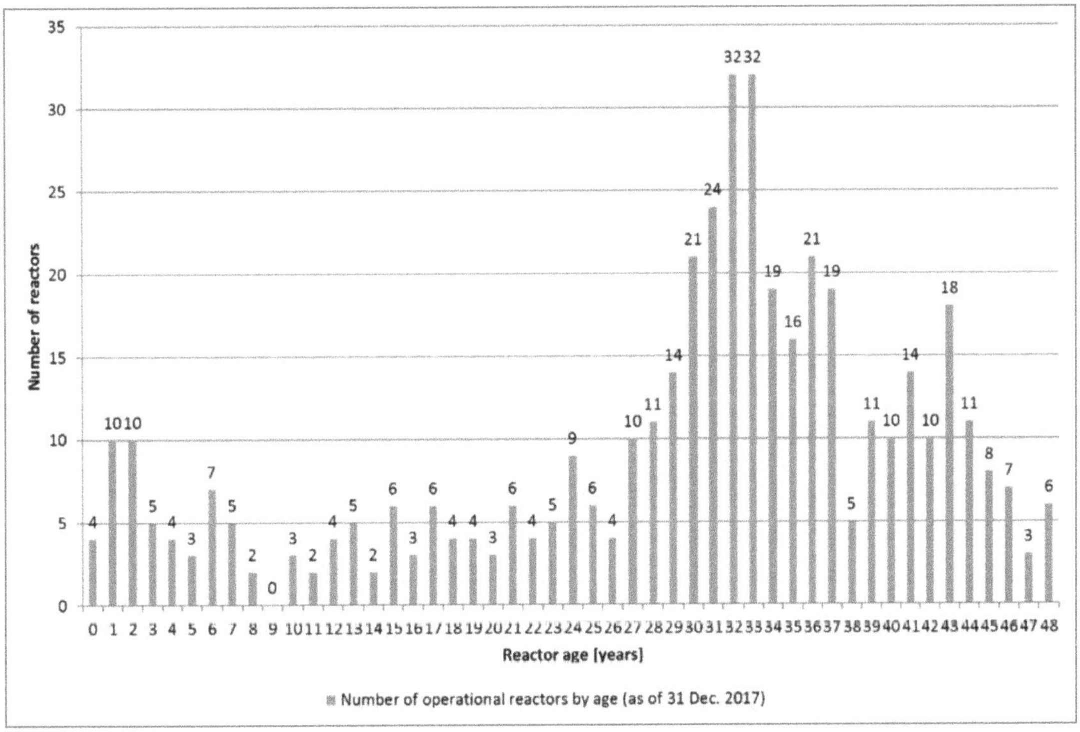

Fig. 1 Number of operational nuclear power reactors by age (as of 31 Dec. 2017) (Based on IAEA, 2018, 78)

This will result into the need to dismantle, decontaminate and demolish these nuclear facilities as well as to undertake processing, conditioning and disposal of nuclear waste and spent fuel ('decommissioning')(Irrek et al., 2007). It is of paramount importance that the funding of these decommissioning activities will be adequate and available when needed in order to avoid negatively affecting the safety of citizens and natural environment. Although this has been principally known since early days of NPP operation (cf., e. g., Lukes et al., 1978), the political pressure to identify and implement respective solutions has only increased within recent years due to changes in the electricity markets in the course of liberalisation and transformation towards a sustainable energy system and the respective impacts on the economic situation of NPP operators and their mother companies, and due to more and more NPPs facing their end of operation.

In general, decommissioning financing needs are huge along the Uranium value chain. Underground uranium mines and mills, open pits, uranium and thorium processing, conversion, enrichment and fuel fabrication facilities, nuclear reactors, nuclear fuel reprocessing plants, interim storages and further nuclear installations have to be decommissioned in a way that human beings, flora and fauna, air, soil, open and ground water sources will be protected against radiation exposure and radioactive contamination. Decommissioning and rehabilitation of the nuclear sites represents a challenge in ecological and economic terms for the former operators. The amount of radioactive waste from all steps of the Uranium value chain adds to the complexity of task (Hagen et al., 2005).

Operators of nuclear installations are expected to accumulate all the necessary decommissioning funds during the operating life of facilities. However:

- The decommissioning process can take several decades;
- In particular, the time period between the shutdown of a nuclear installation and the final disposal of radioactive waste can be very long;
- Therefore, the liable party and the funds accumulated might not be available anymore when decommissioning activities have to be paid;
- Moreover, expected decommissioning costs are only partly assessable since nuclear decommissioning experience is still limited, and technical concepts for important decommissioning steps, particularly for final disposal of higher activity waste (HAW), often do not yet exist;
- Finally, unexpected radiation and contamination problems might lead to additional financing needs even after decades, hundreds or thousands of years after final disposal of radioactive waste.

Therefore, this chapter will analyse

- to which extent the Polluter Pays Principle can be implemented in order to ensure the complete financing of a safe decommissioning by the operators of nuclear installations;
- the different decommissioning financing steps needed;
- who will benefit from the amounts of finance accumulated;
- and compare selected decommissioning financing concepts realised in different countries.

Based on this analysis, conclusions will be drawn and recommendations given for the design of safe decommissioning financing concepts. The chapter will thereby focus on decommissioning financing of commercial nuclear power plants (NPP). However, the general findings can also be transferred to other private or publicly-owned nuclear installations.

2 Basic liability principles

Nuclear liabilities of a NPP operator include liabilities for the dismantling, decontamination, demolition and site clearance of the nuclear facilities at the end of their lifetime as well as for the storage, processing, conditioning and disposal of nuclear waste and spent fuel. They arise with the start of operation of a nuclear facility, i. e. as the first activation or contamination takes place, and usually increase with operation.

A main principle with regard to the distribution of liabilities is the 'Polluter Pays Principle'. In general, the Polluter Pays Principle is broadly accepted in environmental and economic policy. In theory, in a world of perfect information, perfect competition and full rationality of market actors, allocative efficiency will be gained if all external costs are internalised into the market. In order to maximise the net benefit to the society, in principle, the NPP operators should be fully responsible for covering the full costs of decommissioning a NPP. This requires a clearly defined obligation to plan, implement and finance all decommissioning activities including nuclear waste management and final disposal operations. Since the NPP operator does not earn money with the NPP after its shutdown, all necessary financial means have to be collected during operation of the plant via the electricity prices to cover associated decommissioning costs. If decommissioning is not paid by those who have benefited from its operation inter- and intra-generational justice will not be ensured.

However, in contrast to theory, the Polluter Pays Principle is not fully implemented in every country. In some countries like Finland and Sweden, the 'Polluter Pays Principle' is a legal requirement, and there has been made substantial progress during the last ten years in several European countries with respect to better ensuring its implementation (Irrek et al., 2007; European Commission 2013; European Commission 2013a). But still in several countries, the liability of NPP operators ends with transfer of radioactive waste to a state-governed organisation. According to international law, the state has the responsibility for final disposal of radioactive waste. Therefore, financial liabilities for final disposal (and partly waste management, too) are not always fully with the 'polluters' but in some cases at least partly transferred to the state. For example (Irrek et al., 2007; European Commission 2013; European Commission 2013a):

- In Slovenia, the fees the operator pays for dealing with radioactive waste include the cost of final disposal. They discharge the operator from any waste management and disposal liability. However, the levy is periodically reassessed based on available technical data and other inputs.
- In the case of Bulgaria, Lithuania and Slovak Republic, there are agreements between the state governments, the European Union and some Member States about European contributions to finance decommissioning of the nuclear power plants in the context of the countries' accession to the European Union in order to ease an early shutdown of these NPPs for safety reasons (cf. Schmidt et al. ,2013, and Ustohalova/Schmidt, 2014, for recent discussion on this support and its effectiveness).
- In Germany, in mid 2017, NPP operators transferred all liabilities for interim storage and final disposal of radioactive materials to the state against a lump sum of Euro 24.1 bn, with financing regarding these activities managed by a state-governed fund. Any costs exceeding this amount will not have to be borne by the NPP operators. This severely violates the Polluter Pays Principle. The argument behind this was fed by a substantial decrease in electricity wholesale prices which has caused severe losses of the large energy companies. It was argued that it would be better to secure a lump sum paid by the NPP operators now instead of not knowing if the companies will be able to finance radioactive waste management activities in the future. It should be noted that the cost estimate behind this amount is just a rough one, partly based on an extremely rough estimate of 1997 by a German authority (Bundesamt für Strahlenschutz – BfS) for a final disposal site which is not a technically feasible one (Gorleben). Moreover, it does not take into account various problems with interim storages which have already been foreseen by nuclear experts (cf., e. g., Neumann, 2016). In parallel to the transfer of liabilities and money in 2017, the state took over two central interim storages. From 2019 onwards, the state will also be responsible for the decentral interim storages at the NPP sites, which have been in operation by the NPP operators. From 2020 onwards, the German state will be responsible for all interim storage sites as well as for any final disposal activities.

Even in those countries, in which the Polluter Pays Principle is a legal requirement, a NPP operator will not be made financially liable for

- any radiation exposure and radioactive contamination in the Uranium value chain before the fuel arrives at the NPP. While it theory, for this part of the nuclear fuel chain, the respective owners of the Uranium mill or the Uranium processing, conversion, enrichment and fuel fabrication plants should be made financially liable, ethical considerations could lead to the argument, that the NPP operator has to bear some social responsibility also for these parts of the value chain.
- any problems arising after the final closure of the final disposal facility. The responsibility usually ends as soon as all waste has been finally disposed of and the safety authorities have accepted that the final closure of the final disposal facility has been concluded fulfilling the pertinent safety requirements for final closure.

However, the example of the German Asse II mine shows that unforeseen problems can arise after closure of such a final disposal. From 1967 to 1978, 125,787 drums and waste packages containing low-level and intermediate-level radioactive waste were emplaced in this test repository. The legal basis for this was mining law, not nuclear law, and there was no proof of long-term safety before the waste was disposed. Today, the large total volume of open drifts and chambers and the closeness of the chambers to the adjoining rock cause severe problems in the Asse mine. The salt rock and adjoining rock loosen, and clefts have formed through which groundwater flows into the mine. The severeness of this disposal problem has been increased by the fact that the inventory of radioactive waste in this mine is not exactly known, particularly with regard to the amount of plutonium disposed. It is now planned to retrieve the radioactive waste and to dispose it elsewhere (www.asse.bund.de; status: 08 May 2016). The fees collected for the final disposal of radioactive waste during operation of the mine summed up to about Euro 8.25m, while current estimates for retrieval costs are between Euro 4–6bn (Kirbach 2009; N.N. 2013; www.atommuellreport.de; status: 08 May 2016). There has been some discussion in Germany on who should be made financially liable for these extra costs. In 2010, some German politicians (e.g., Kelber et al., 2010) argued that the introduction of the nuclear fuel tax could be justified, among others, by the fact that the NPP operators have benefited from disposal sites like Asse II and thus should contribute to the additional costs that will occur.

Anyway, if such problems arise decades or hundreds of years after the closure of the final disposal facility, the NPP operators might not exist anymore. This could also happen if problems arise during the final disposal activities decades after the NPP shutdown. Therefore, full implementation of the Polluter Pays Principle cannot be 100% secured in any decommissioning financing scheme. Nevertheless, it should become a legal requirement and should be implemented as far as possible in every country with nuclear facilities in order to increase allocative efficiency. Nevertheless, the implementation of the Polluter Pays Principle will only function well if there is a sufficient amount of transparency to the public in the interest of current and future generations of electricity customers and taxpayers. Therefore a requirement to ensure transparency should be added, which should include transparency of the following steps of decommissioning financing (Irrek et al., 2007):

- Determination of decommissioning strategies and time schedules;
- Identification and estimation of decommissioning costs including cost breakdowns by cost items and details of cost estimation methodology,
- Collection of decommissioning funds;
- Management and investment of funds until the money is used for payment of decommissioning activities;
- Use of funds for the original purpose, i. e. for payment of decommissioning activities.

These steps are analysed in more detail in the following chapters.

3 Technical decommissioning strategies and time schedules

In principle, there are three technical decommissioning (decontamination and dismantling) strategies for NPPs (Irrek et al., 2007; OECD/NEA, 2012; OECD/NEA, 2016):

- *Immediate dismantling:*
  Decontamination and dismantling immediately after operation period. All contaminated material is removed or decontaminated to a level at which no more regulatory control is required. In this case, dismantling starts shortly after the permanent shutdown of the NPP and might take between 15–20 years, with no unforeseen incidents happening. In case of unavailability of routes for the spent nuclear fuel, this is kept in an interim storage on-site, which might be decommissioned decades after the demolition of the NPP has been completed.
- *Deferred dismantling (safe enclosure / safe storage):*
  First, spent fuel is removed from the facility. The plant is then kept intact and in a safe and stable state to enable the radionuclides activity to decay until it reaches levels that reduce difficulties of handling. Decontamination and dismantling

then starts several decades after the permanent shutdown of the NPP. Deferral periods range between 20 and around 100 years, e. g. 25 years in Spain, 60 years in the United States, and a century or more in the United Kingdom.

- *Entombment*
  This option involves encasing radioactive structures, systems and components in a long-lived substance, such as concrete. The encased plant would be appropriately maintained, and surveillance would continue until the radioactivity decays to a level that permits termination of the plant's license and end any regulatory control. Most nuclear plants will have radionuclide concentrations exceeding the limits for unrestricted use even after 100 years. Therefore, special provisions would be needed for the extended monitoring period this option requires. To date, no facility owners have proposed the entombment option for any nuclear power plants undergoing decommissioning. In fact, this is more an emergency option than a strategy option, so far used only in the case of Chernobyl.

The choice of strategy depends on several parameters and framework conditions, the decontamination, dismantling and demolition stage aimed at, and the planning for the future use of the site. Operators of nuclear facilities usually take into account the following criteria when deciding on a dismantling strategy (Irrek et al., 2007; OECD/NEA 2016, 46):

- *Radiation protection*
  There is one major argument for deferred dismantling which is radioactivity decay, as it will ensure the reduction of radiation hazard during dismantling and a reduction of volume of radioactive waste. Moreover, new techniques (e. g., robotics) might be invented that could further reduce radiation harzards. Finally, deferred dismantling might ease disposal routes for radioactive waste if a final disposal site is available by this time. On the other hand, with immediate dismantling, radiological characterisation is much easier and less costly. Moreover, there will be reduced effects of deterioration and ageing like corrosion.
- *Know-how of employees*
  Immediate dismantling ensures the availability of qualified and experienced workforce with internal knowledge on the design and history of the facility from its operation. On the one hand, immediate dismantling might ease a socially acceptable reduction in employed and contracted staff at the site of the nuclear facility after the installation has been shut down (cf. Irrek, 2005). On the other hand, workers might not be motivated to demolish a plant where they had been working. Moreover, deferred decontamination and dismantling might make it easier to outsource dismantling activities at cheap labour costs

because of existing wage differentials between employees in the nuclear sector and employees of contractors.

- *Reuse of site*
  Immediate dismantling allows an earlier reuse of the site, with respective economic effects for the region.
- *Costs*
  A thorough evaluation and comparison of different strategies is needed in order to assess which strategy will be the least-cost while fulfilling all the nuclear safety obligations. For example, for the deferred decontamination and dismantling strategy, it has to be taken into account in how far existing ancillary equipment can be used for decommissioning activities decades after the end of operation as well as the costs for maintenance and surveillance. Moreover, economies of scale could be achieved if several similar plants are dismantled in co-ordinated manner. A general question is how much cheaper is it to outsource decontamination and dismantling activities or to carry them out in-house with existing know-how. Furthermore, total costs also depend on the availability of waste management, storage and disposal options, and on the decision about the use of the site for other purposes after its release from radiological restrictions. However, such cost estimation is not an easy task. Different assumptions on the underlying decontamination and dismantling processes, on the technical feasibility of possible technical solutions and technical developments, on person-years needed, on labour, material/equipment and capital costs, on time horizons, and on developments of the labour market and the general economic environment can lead to different strategic choices.
- *Financial risks*
  Risks and uncertainties of changes in benefits and costs have to be adequately taken into account because of all these possible influences, time horizons of several decades are considered. For example, a long period of deferment not only gives the chance to yield interest over a longer period of time, but includes also a higher risk that the funds will be lost or will significantly lose value.

The preferred decommissioning strategy can differ from case to case, even within the same country. In most cases, economic arguments with respect to expected financial benefits and costs as well as perceived financial risks are the decisive arguments for the operators to choose a specific decommissioning strategy, particularly for privately owned facilities. However, there are also strategic or tactical arguments for particular decommissioning strategies of the NPP operators in the course of political discussions on the distribution of liabilities (cf., e. g., discussion in Germany on the possible final repository site). Nuclear safety authorities are mostly in favour of immediate dismantling, particularly in recent years in the European Union. The reasons given for this include the consideration that the risk of the loss of memory on the conception and operation of a facility will be significant.

4 Identification and estimation of decommissioning costs

Based on the decommissioning strategy and time schedule determined, cost planning starts with an identification and estimation of costs. The direct comparison of decommissioning cost estimates generated for different plants by different cost estimate providers is limited due to different cost structures, different combinations of individual cost items, different methodologies applied and different ways of dealing with uncertainties. This holds true, even if the results are presented in a similar manner, e. g., by using the International Structure for Decommissioning Costing (ISDC). Following the ISDC, costs can be split up for principal activities as follows (OECD/NEA, 2012):

01 – Pre-decommissioning actions.
02 – Facility shutdown activities.
03 – Additional activities for safe enclosure and entombment.
04 – Dismantling activities within the controlled area.
05 – Waste processing, storage and disposal.
06 – Site infrastructure and operation.
07 – Conventional dismantling, demolition and site restoration.
08 – Project management, engineering and support.
09 – Research and development.
10 – Fuel and nuclear material.
11 – Miscellaneous expenditures.

For each of these principal activities, on a second level, several activity groups can be distinguished. For example, according to ISDC (OECD/NEA 2012), pre-decommissioning actions consist of decommissioning planning, facility characterisation, waste management planning and further activity groups, each differentiated on a third level into single activities (e. g., strategic planning, preliminary planning and final decommissioning planning). For each activity, labour costs, investment costs (capital, equipment, material costs), expenses (consumables, taxes, etc.) and contingencies (a specific provision for unforeseeable elements of costs within the defined projects scope) should be identified.

In many cases, cost estimates are bottom-up ones taken into account the specific decommissioning activities required. However, there are also cost estimates by specific analogy to similar past projects, by parametric estimating based on historical databases on similar systems and subsystems, by cost reviews that just look at those cost items to be updated or by rough expert opinion when other techniques or data are not available (OECD/NEA, 2015). As with other complex construction or deconstruction or other engineering projects, cost estimates are based on a number of technical and economic assumptions, and on assumptions influenced by the political-administrative framework conditions. Therefore, there are financial risks to be taken into account in any ex ante-evaluation of decommissioning project costs. In practice, there are different methodologies how to take into account uncertainties and risks, for example (Irrek et al., 2007; Däuper et al., 2014; OECD/NEA, 2016):

- *Risks and uncertainties ignored*
  In Switzerland, in 2014, the Swiss Federal Audit Office claimed that cost estimates were based on an ideal scenario leading to too low contributions to the decommissioning funds and a high degree of risk borne by the Federal Government. However, in the future, following a new ordinance implemented in 2014, a 30% contingency for unexpected costs should be included according to a new ordinance.
- *Conservative estimates*
  In France, national regulation demands dismantling cost estimates to be 'conservative' ones;
- *Cost estimates with flat or specific contingency factors*
  In the Slovak Republic, based on risk analysis and risk assessment, contingency factors between 0.2 and 16.5% are considered;
- *Scenario calculation with sensitivity analysis of major cost drivers*
  In Lithuania, scenario calculations take into account different wage levels;
- *Probabilistic and deterministic cost estimates*
  In Sweden, probabilistic cost estimates performed in Sweden in addition to deterministic ones;
- *Complex modeling*
  Monte Carlo analysis and other quantitative modeling approaches can be used to simulate possible deviations from assumptions taken, if there are many independent variables with significant uncertainties. In the United Kingdom a combination of computational modeling with Monte Carlo simulation and

management judgement based on experience of previous projects leads to contingencies in the range between 1–24%.

In general, optimism bias can cause a NPP operator to believe that the respective NPP is less at risk of experiencing a future cost increase compared to others. Therefore, regulation has to ensure that adequate cost estimation methodologies are applied that properly take into account possible risks of cost increases. Moreover, international organisations and national authorities should ensure that information on costs of past decommissioning activities are widely spread and could be used for calculation of future costs. Information on past decommissioning activities and improved methodologies have led to substantial increases in cost estimates in various countries during the past 15 years (cf. OECD/NEA, 2016).

5 Collection of decommissioning funds

After the costs have been properly estimated, it has to be determined, if, when and how funds should be set aside at the beginning of and/or during plant operation. In general, any financing scheme should ensure and be managed and periodically reviewed in a way that sufficient funds will be collected during the lifetime of a nuclear facility and will be available at the time decommissioning and waste management expenses occur. Basically, the funding schemes can be differentiated into (cf. also Irrek et al., 2007; OECD/NEA, 2016):

- Payment of decommissioning activities from the current budget of public authorities (e. g., for decommissioning of Uranium mines in Germany): Provisions might be collected during lifetime of a plant via a levy or taxes, but very often there is no collection of funds during lifetime of the plant.
- Internal unrestricted fund of a private company (e. g., for dismantling and demolition and waste processing of NPPs in Germany): On the liabilities side of the balance sheet, the liable company discloses the amount of provisions accumulated by the respective year. However, it is not required that any assets are separated and reserved or earmarked for decommissioning purposes. Therefore, if decommissioning activities have to be paid it might happen that there will not be any financial means available. In case of insolvency of the company, the state has to step into the breach.

- Internal restricted fund of a private liable company with public regulation (e. g., for NPPs in France): In contrast to the internal unrestricted fund, there is an enhanced insolvency protection because assets are separated and earmarked for decommissioning purposes and restrictions on investment of funds are imposed.
- External restricted fund (in most of the European countries, e. g., in Switzerland, Finland and Sweden): The funds are managed externally, i. e. not within the liable company, but by a dedicated body that may be a private or state-owned entity, and with respective transparency. This dedicated body has to follow specific restrictions with regard to the investment of financial means in order to enhance insolvency protection. In most cases, although there is an external fund installed the liabilities remain with the NPP operators. Thus the Polluter Pays Principle will be followed. For example, if there is an increase in costs, NPP operators will have to make additional payments to the funds. However, this is not the case with the new external fund for radioactive waste management and final disposal installed in Germany recently, where NPP operators have just paid a lump sum to the fund.
- External unrestricted fund (e. g., the 'Cassia conjugation per ill set tore electric – CCSE' in Italy, which allows surcharges on the electricity price for several purposes, among others, for nuclear decommissioning).

Accruals to an internal fund or contributions to an external fund are usually set up in regular installments or according to the electrical energy produced. For this, costs are usually inflated up to the year they will incur, and then discounted to its current value to determine the size of the accrual. Since discounting rates are usually higher than inflation rates, this leads to the sum of accruals or contributions being lower than the cost estimates. This, in turn, demands to yearly provide funds not only for the regular installment, but also for the difference between the present values of the actual year and the past year. The determination of the inflation and discounting rates is of central importance in any of these funding regimes. Only in few countries, provisions are based on undiscounted costs.

However, there are also funding regimes, where the full amount of costs has to be provided for from start of operation (Irrek et al., 2007; Däuper et al., 2014; OECD/NEA, 2016): For example, in France, since 2006, with a transition period until 2010, provisions for dismantling and decontamination of a NPP have to be fully collected already with start of operation. In Finland, a special requirement exists which, in principle, demands to cover the full nuclear liability already at the start of operation by special financial securities. In Sweden and in the Netherlands, with start of operation, NPP operators have to provide a guarantee for early shutdown.

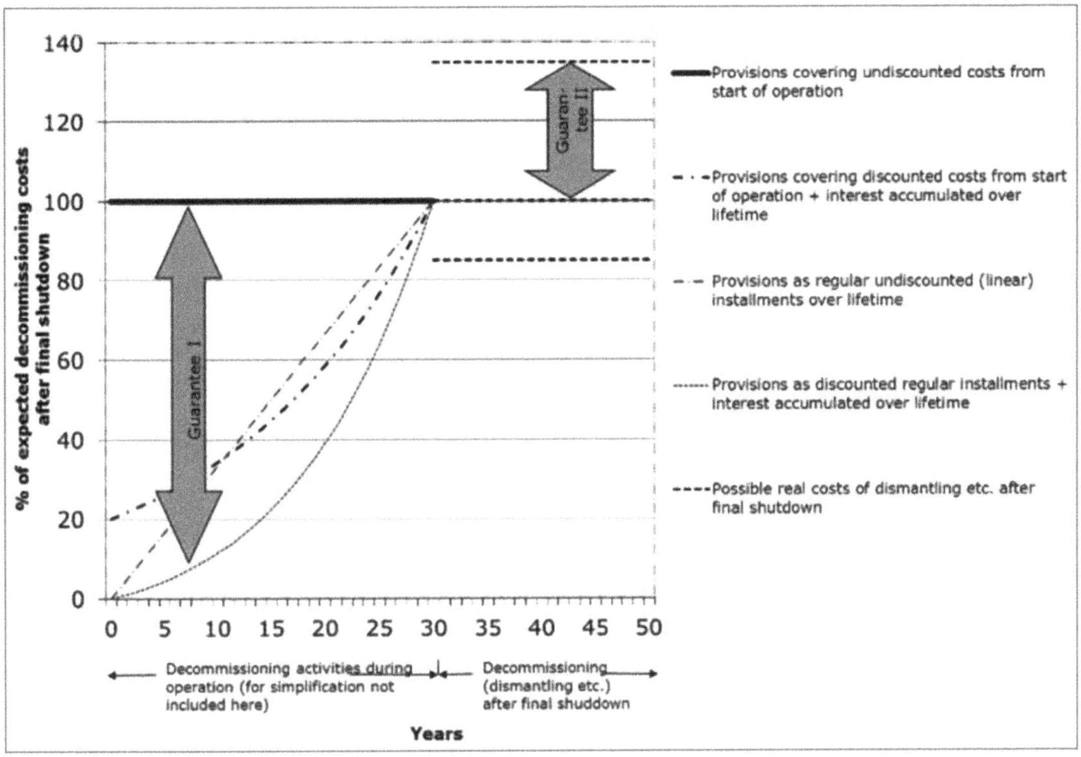

Fig. 2 Guarantees covering financial risks related to decommissioning costs occurring in case of early shutdown (Guarantee I) or after final shutdown of the plant (Guarantee II) (Irrek et al., 2007, 152)

Such schemes substantially reduce the risk that the state will have to pay for early shutdown of plants and contribute to implementing the Polluter Pays Principle. In addition to such guarantees for early shut down (Guarantee I in Fig. 2), in some financing schemes, NPP operators have to provide additional guarantees for additional costs exceeding the cost estimates that might occur after final shutdown (Guarantee II in Fig. 2). For example, in Finland, the state has the possibility to demand such guarantees up to 10% of the sum of liabilities. In Sweden, such guarantees amount to 16–17% of estimated costs. And in Switzerland, there partly is a joint liability of all operators in case one of the NPP operators cannot pay its contributions.

6 Management and investment of funds

A well-designed governance regime of the fund and a high level of quality of fund management is crucial for maintaining or even increasing the value of the funds so that sufficient funds will be available at the time decommissioning activities have to be paid. Funds can be managed by public or private fund managers. Specific restrictions beyond general accounting principles and general accounting law can be imposed on managers of internal or external funds, e. g., with regard to (Irrek et al., 2007):

- the way funds have to be accumulated;
- the investment of the financial means collected until they are used to pay for decommissioning activities;
- the payment for decommissioning costs;
- the regular reporting on funds and fund management;
- the control by the public (e.g., government, parliament, special boards, information rights of the public).

Typical examples of investment restrictions and guarantees required for internal or external funds in practice are (Irrek et al., 2007; Däuper et al., 2014; OECD/NEA, 2016):

- Restrictions regarding the degree of risk allowed to be taken, e. g. by limiting the share of asset classes with higher risks. For example, in France, assets of the internal restricted funds have to present a sufficient degree of security and liquidity. However, since 2010, diversification into real assets has been increased. In Slovenia, it has to be invested not less than 30% in state bonds, not more than 5% in stock, not more than 15% in other bonds, not more than 5% in stocks of one issuer, and approximately 10% in securities on foreign financial markets. In the US, there are just simple guidelines such as using a "prudent investor" standard, with restrictions on self-investing and on foreign investment.
- Restrictions that do not allow investment in companies associated with the legally obliged parties or that have invested the majority of their assets in nuclear facilities.
- In Finland, there is the special situation that the lincensee can borrow back up to 75% of the capital of the fund against securities and at current interest rates.

Actual performance of the funds differs depending on the investment restrictions imposed, i. e. on the degree of risk taken, and on the general economic situation.

Tab. 1 Yield on investment of the Swiss decommissioning and waste management funds (quarterly reports)

		Nuclear Decommissioning Funds		Radioactive Waste Management Funds		
Quarter	Year	Nominal Yield on Investment per Quarter	Real Yield on Investment per Year	Nominal Yield on Investment per Quarter	Real Yield on Investment per Year	Inflation
2	2017*	0,11%/0,17%		0,20%		
1	2017*	2,81%/2,84%		2,82%		
4	2016	1,24%	6,85%	1,25%	6,78%	-0,43%
3	2016	2,38%		2,33%		
2	2016	2,60%		2,59%		
1	2016	0,07%		0,06%		
4	2015	3,28%	0,61%	3,23%	0,66%	-1,14%
3	2015	-2,44%		-2,42%		
2	2015	-2,75%		-2,78%		
1	2015	1,52%		1,62%		
4	2014	1,38%	11,54%	1,47%	11,52%	-0,02%
3	2014	3,57%		3,69%		
2	2014	2,78%		2,58%		
1	2014	3,54%		3,87%		
4	2013	4,87%	7,39%	4,93%	7,58%	-0,20%
3	2013	-1,61%		-1,60%		
2	2013	1,30%		1,33%		
1	2013	2,55%		2,62%		
4	2012	4,38%	10,19%	4,45%	10,33%	-0,70%
3	2012	0,51%		0,52%		
2	2012	3,44%		3,41%		
1	2012	0,85%		0,87%		
4	2011	1,27%	-0,32%	1,31%	-0,34%	0,22%
3	2011	-2,41%		-2,43%		
2	2011	-3,45%		-3,57%		
1	2011	4,69%		4,78%		
4	2010	3,47%	3,29%	3,57%	3,35%	0,69%
3	2010	-3,39%		-3,42%		
2	2010	2,91%		2,86%		
1	2010	1,08%		1,12%		

Investment Strategy 2010-2017**

Category	Strategy	Lower Bound	Upper Bound
Liquidity (Cash)	0%	0%	5%
Bonds (CHF)	25%	15%	35%
Bonds (Other Currencies; hedged)	15%	10%	20%
Shares	40%	30%	50%
Real Estate	10%	7%	13%
Other Investment	10%	7%	13%

* Since 1 April 2017, there is a separated investment strategy for the NPP Mühleberg (KKM) compared to the investment strategy for the NPPs Beznau (KKB), Gösgen (KKG), Leibstadt (KKL) and for the interim storage Würenlingen AG (Zwilag); therefore the first number is relevant for KKM, the second fo KKB, KKG, KKL, Zwilag.

** not including KKM

While the European Commission (2013a) recommends that a secure risk profile should be sought in the investment of the assets, ensuring that a positive return is achieved, a 100% security of a positive return over any given period of time cannot be guaranteed over the many decades of lifetime of such a fund. Moreover, there is a general tradeoff between security and the yield on investment.

The Swiss example in Table 1 shows that an investment strategy with up to 50% of funds invested into shares at the stock market can lead to comparatively high returns in some quarters of a year, but also to a decrease in funds value in others. If the liable company feeds the fund with discounted contributions, the fund will have to yield positive returns to make up for the difference between the cost estimate and the discounted value, or additional contributions by the liable company will be needed. This will be also required with a secure investment strategy, if there are negative real interest rates on safe investment into bonds of solvent states as it could be observed in some states recently.

The internal unrestricted fund differs from the restricted solutions in one important aspect: The provisions accumulated on the liabilities side of the balance sheet do not guarantee that there will be financial means available when decommissioning activities have to be paid. The cash flow from the financial equivalent of the set-aside provisions can be freely used by the companies as a portion of corporate revenue.

In Europe, Germany is the only country where such an internal, completely unrestricted fund still exists for the dismantling and demolition of reactors and for the conditioning of radioactive waste. Here, no information is available on how nuclear power plant operators or their parent groups have invested the unrestricted funds from the nuclear provisions. A direct link cannot be drawn between individual liability items and individual asset items on a group's balance sheet. Just because provisions are set up does not necessarily mean that the funds are being invested to finance dismantling and disposal. Groups can employ any type of financing whatsoever to provide future funding for dismantling and disposal activities (cf., e. g., Perridon et al., 2012, for the general financing options). Using the German groups E.ON and RWE as examples, the financing options can be investigated as follows:

- With regard to financing from current cash flow, one needs to consider that operating margins have fallen sharply in recent years and have been negative at times, while revenue and EBITDA have also fallen year-over-year in most of E.ON's and RWE's business divisions. There is no guarantee that the cash flow generated during the next years will be adequate to finance higher provisions or pay for activities related to nuclear dismantling and the long-term storage of radioactive material.

- The options that E.ON and RWE have for using debt to finance activities related to nuclear dismantling and disposal are becoming more limited in light of their declining credit ratings and relatively high debt-to-equity ratios.
- For financing through asset restructuring, one needs to take a closer look at the groups' assets, especially tangible assets, shareholdings, financial assets and – to the extent that they are not required to cover current liabilities – liquid assets. Assets available in the short-term are not sufficient to cover net nuclear provisions. The sum of the values of E.ON's and RWE's plant and machinery assets and shareholdings declined significantly from 2013 to 2016. There is a risk that this trend will continue.

While observing the availability of the groups' current financial resources to cover their obligations in the nuclear sector, one must also bear in mind that the groups must use their assets and cash flow not only for their nuclear provisions, but also to cover other obligations. Therefore, it is necessary to compare the groups' total financial resources with all of their assumed obligations. For example, for RWE, based on annual reports it can be calculated that the long-term financial resources at RWE's disposal at the end of 2016 were hardly sufficient to cover all of RWE's long-term obligations (own calculation based on annual company reports and Irrek / Vorfeld 2015).

7 Use of nuclear decommissioning funds

In general, decommissioning funds should be used only for the purpose for which they have been established and managed, i. e. to pay for decommissioning and radioactive waste management activities. Therefore, the degree of independence between the operator of a nuclear installation as the liable polluter and contributor to the funds, the company carrying out decommissioning activities and thus using decommissioning funds, the funds management and the position disposing of the power of authorising payments is a key issue in any decommissioning financing system.

In general, market actors in nuclear decommissioning business making use of nuclear decommissioning funds are the following:

- Operators of nuclear facilities, who benefit already during operation as well as after shut-down, depending on the degree they are involved in the decommissioning activities. During dismantling it is important to make use of existing know-how of the personnel of the NPP operator;

- National or international firms specialised in nuclear decommissioning;
- Local firms without any specialisation in nuclear decommissioning, e. g. craftsmen, scaffolders, unqualified staff that can be trained for decontamination activities, etc.

Fig. 3 Past and expected future yearly payment for decommissioning activities in Germany, following groups' balanced sheets and cost estimates by the NPP operators of 31 December 2014 including cost estimates for final disposal of HAW although there is no technical concept or site available yet (balanced sheets of E.on, RWE, EnBW and Vattenfall, 2003–2015; own calculation based on Warth & Klein Grant Thornton, 2015)

Tab. 2 Profits and losses of selected companies active in the German decommissioning market (companies' income statements of 2014, 2015 and 2016 as far as available)

Company	Year	Profit / Loss [1,000 Euro]	Year	Profit / Loss [1,000 Euro]
Nukem Technologies GmbH [Russian Rosatom group]	2014	-26,960	2015	
Siempelkamp Ingenieur und Service GmbH [Siempelkamp Nukleartechnik GmbH]	2014	-2,683	2015	2,176
Celten Service GmbH	2015	105	2016	141
Eckert & Ziegler UmweltdiensteGmbH	2014	-1,367	2015	-4,266
SAT Kerntechnik GmbH	2014	32	2015	
EWN Entsorgungswerk für Nuklearanlagen GmbH [Energiewerke Nord GmbH; German state-owned company]	2015	0 (per definition)	2016	0 (per definition)
Safetec Entsorgungs- und Sicherheitstechnik GmbH	2014	791	2015	793
EnBW Kernkraft GmbH [NPP operator]	2015	0 (per definition)	2016	0 (per definition)
Vattenfall Europe Nuclear Energy GmbH [NPP operator]	2014	-173,100	2015	-459,4000
GNS Gesellschaft für Nuklear-Service mbH	2015	27,400	2016	30,000
AREVA GmbH [incl. losses from NPP construction and modernisation]	2014	-764,164	2015	632,392

In Germany, following the shutdown of NPPs after the Fukushima accident, decommissioning activities have increased during recent years (Fig. 3). It can be expected that the level of expenses will continue in the coming years when further NPPs will be shut down. With regard to the age of NPPs in the world (Fig. 1), a growth in international decommissioning expenses can be expected, too. In 2013, Nukem Technologies GmbH estimated, that the international market volume until 2030 could sum up to more than Euro 250bn. However, in general, as stated in AREVA's financial report of 2014, the international decommissioning market is only slowly developing yet. The available income statements of selected companies active in the German decommissioning market show that several of them are still facing losses (Table 2).

Nevertheless, there are several attempts to reduce costs and thus losses. For example, Nukem Technologies expects that a 15–20% decrease in costs of decontamination and dismantling might be possible (Kutscher 2015). Possible optimisation includes the following:

- Portfolio planning: co-ordination of and synergy effects from parallel activities at different sites including specialisation and centralisation of specific dismantling activities;
- Standardisation and modularisation of decontamination and dismantling activities;
- Mobilisation: Mobile equipment for conditioning of nuclear waste;
- Increase in decontamination and in the share of radioactive waste that does not have to be stored in a final disposal site but can be used with or without any restriction for other purposes if remaining radiation does not exceed specific levels, which can be a problem from the nuclear safety perspective if concerns and new medical knowledge with regard to risks from low-dose level radiation are ignored;
- Lean management and professional logistics, project management, process management, risk management and cost management: reduction in overhead costs by concentration of administrative activities, mergers and acquisitions; optimised planning and control.

It can be expected that international companies specialising in back-end activities like companies of the Russian Rosatom group like Nukem Technologies GmbH will be the first benefiting from such developments. This might lead to market concentration processes, which will be a challenge for governments aiming at getting insight and control of activities with respect to nuclear safety.

8 Comparison of selected decommissioning financing concepts

Existing European and international analyses of decommissioning financing schemes (Irrek et al., 2007; European Commission 2013; European Commission 2013a; OECD/NEA, 2016) show that financing schemes in practice substantially differ with regard to

- Cost estimates and accounting procedures;
- Collection and investment of funds;
- How risks and uncertainties have been addressed;

- Implementation of Polluter Pays Principle;
- Use of funds: Incentives for reducing costs;
- Different degree and ways of public control – differing public information rights.

Table 3 compares the financing schemes in Switzerland, Sweden and Finland from which can be particularly learned for future design of decommissioning financing concepts. In order to implement the polluter pays principle as far as possible and to ensure that governments will be able to control decommissioning activities with regard to nuclear safety, the following central questions have to be addressed when designing the governance scheme for nuclear decommissioning financing (Irrek et al., 2007):

- Who defines or regulates decommissioning (financing)?
  In most cases, this task is assigned to public licensing authorities (government level). A key issue thereby is the independence of the authority, which has to align different objectives from different stakeholders. Employees of the authority should dispose of sufficient personal independence from the operators and, if the operators are public entities, from the government.
- Who is liable or who has to pay the decommissioning activities?
  Due to the polluter pays principle assumption, the operators of nuclear installations should have to carry all decommissioning costs. They should pay through a decommissioning funding system, which urges them to financially contribute to a designated decommissioning fund. Guarantee schemes like the ones in Sweden and Finland aim at ensuring the implementation of the polluter pays principle even in case of an early shutdown of the nuclear installation or in case of increasing decommissioning costs after the end of operation.

Tab. 3 Comparison of decommissioning financing schemes in Switzerland, Sweden and Finland (Däuper at al., 2014; Kaberger/Swahn, 2015; Irrek/Vorfeld, 2015; OECD/NEA, 2016; quarterly reports of Swiss funds)

Country	Switzerland	Finland	Sweden
Legal form	Two separate public law foundations for decommissioning and radioactive waste management.	One single public law foundation.	One single public law foundation.
Nuclear installations included	All 5 NPPs and an interim storage.	All NPPs with their on-site interim storages.	All NPPs.

Country	Switzerland	Finland	Sweden
Obliged party	Operator of nuclear installations remains responsible for decommissioning activities.		
Governing bodies	Operational fund management by an agency at an industrial organization. Board with a cost committee and an investment committee: 4 of 9 members of the board and 4 of 7 (2 of 7) members of the investment (cost) committee are representatives from NPP operators. Revision by certified auditor.	Fund governed by the Ministry for Labour and Economy.	Operational fund management by a state-owned agency. Board of Governors and Administration, with 2 of 7 members being representatives from E.on and Vattenfall.
Cost estimate	Every 5 years.	Every 3 years.	Every 3 years.
Fund allocation	Regular installments over 50 years of operation.	Regular installments over 25 years of operation or depending on the amount of waste produced via a ca. 10% surcharge on the electricity price	Regular installments over 40 years of operation via a surcharge on the electricity price.
Fund investment	Cash, currencies 0–9%, Bonds 38–55%, Shares 30–50%, Real Estate 7–13%, Others 7–13%.	75% of the fund can be borrowed back by the NPP operator who has to invest it in a productive way. 25% state bonds.	Assets with low risks only. In 2013: Covered bonds 62%, Index-based securities 24%, Cash, currencies 14%.
Payment of decommissioning activities	Operators of nuclear installations hand in bills for decommissioning activities. They receive respective payment after bills have been checked by fund management.		
Common advantages	External restricted funds (public law foundation) with a specific degree of public transparency and control in order to ensure a safe decommissioning. Economic risk remains with NPP operator, who has additional payment liabilities in case of short-fall of funds aiming at fully implementing the polluter pays principle.		
Common problems	Cost estimates probably too low and thus the funds. More realistic cost estimates are partly planned but controversially discussed.		

Country	Switzerland	Finland	Sweden
Special disadvantages	Strong influence of NPP operators on funds.	No specific asset constitution strategy determined by legislation.	Weak economic situation of NPP operators might lead to less restrictive governance.
Special advantages	Reform of 2014 foresees a 30% surcharge on estimated costs to cover possible uncertainties. Joint liability of NPP operators in case one of them cannot pay.	100% funding guaranteed from start of operation. NPP operators are allowed to borrow back up to 75% of funds against guarantees. Up to 10% additional guarantee covering cost uncertainties can be required by the state.	Public cost estimates. Prudent investment strategy so far. Guarantees covering short-fall of funds: Guarantee I covering the case of early shut down and plant-specific Guarantee II covering cost uncertainties (16-17% on average).

- Who is the entity, which holds the fund in its general accounts?
  In practice, various solutions are implemented from internal unrestricted to external restricted. However, only a restricted fund has the objective to ensure that assets will be available when needed to pay for decommissioning activities.
- Who outlines the investment policy and the investment guidelines?
  In general, the prudence principle should be followed in order to ensure that sufficient financial means will be available for a safe decommissioning. For this, the independence and competence of all involved stakeholders is important. The example of Switzerland shows that the strong influence of the NPP operators on the fund management can lead to a more risky investment strategy, which, in turn, can lead to substantial returns in some years, but also to a substantial decrease in asset value in others. The incentive to finance part of future decommissioning costs through a high investment performance is evident. A high performance on its part can conflict with the prudence principle. However, the professional application of asset and liability management allows managing a slightly higher risk. Special attention has to be paid to lending practices to related parties like in the example of Finland. In particular, lending can be beneficial for the NPP operator, but should be backed up by respective guarantees. Furthermore, means of finance should be invested in such a way that correlations between the investment and the development of the nuclear industry are avoided. It is recommended to develop guidelines, which describe the required qualifications of investment managers and which give a basic investment policy frame also

defining the acceptable risk levels. A kind of oversight board or decommissioning financing committee could provide such guidance.

- Who manages the fund?
  A sufficient degree of independence between the operators of nuclear installations (as polluters and thus contributors to the funds) and the investment managers is a key issue.
- Who disposes of the power of authorising payments for decommissioning?
  As mentioned already before, the degree of independence between the operator of a nuclear installation as the liable polluter and contributor to the funds, the company carrying out decommissioning activities and thus using decommissioning funds, the funds management and the position disposing of the power of authorising payments is a key issue in any decommissioning financing system. The conflict potential always remains with the entity that has access to the decommissioning funds (power of authority). If the NPP operator solely disposes of the power of authority (e. g., in internal unrestricted funds), he could be tempted to defer payments in periods, where he wishes to use the financial resources for other purposes, or where he has general liquidity problems. If the government solely disposes of the power of authority and the liable NPP operator has to contribute to the fund for any additional costs occurring, the government could be tempted to use the financial resources in an inefficient way or for additional activities not really needed.
- Who monitors or controls decommissioning (financing) and who has the authority for sanctions in the case of non-compliance?
  Like the definition and regulation of decommissioning financing, this task usually is assigned to public licensing authorities on the government level. The independence of the authority from the interests of the various stakeholders is of vital importance here, too.

9 Conclusions and recommendations

In conclusion, with regard to the financial consequences and risks involved in the different nuclear decommissioning financing schemes, existing schemes could be improved by (Irrek et al., 2007; OECD/NEA, 2016):

- Measures that establish a system of checks and balances in the governance chain in order to avoid negative effects stemming from conflicts of interests;

- Measures that increase transparency including regularly reviewed, realistic, well-founded, published cost estimates. Within this context, a much better understanding of the uncertainties affecting decommissioning activities and how to best take them into account in cost estimations should be aimed at;
- Measures that set incentives to cost reduction while at the same time maintaining the level of radiation protection needed;
- A system of rules for regular contributions to the fund and to cover cases of short-falls like the guarantees in the Swedish system in order to ensure the full implementation of the polluter pays principle from the first until the last day of operation of a nuclear installation and beyond;
- Measures that ensure that fund assets will be separated from other assets and liabilities and invested according to the prudence principle so that they are available at the appropriate time and used only for their original purpose.

References

Balance sheets and income statements, annual financial statements and further annual reports, information and documents from websites and further publicly available information from operators of nuclear facilities and decommissioning funding organisations.

Däuper, O.; Fouquet, D.; Irrek, W.; et al., 2014. Finanzielle Vorsorge im Kernenergiebereich – Etwaige Risiken des Status quo und mögliche Reformoptionen, expert opinion of the law firm BBH [Becker Büttner Held] in co-operation with Prof. Irrek prepared on behalf of the German Ministry for Economy and Energy (BMWi), Berlin.

European Commission, 2013. Communication from the Commission to the European Parliament and the Conuncil on the use of financial resources earmarked for the decommissioning of nuclear installations, spent fuel and radioactive waste, COM (2013) 121 final, Brussels.

European Commission, 2013a. Commission Staff Working Document "EU Decommissioning Funding Data" accompanying the document Communication from the Commission to the European Parliament and the Conuncil on the use of financial resources earmarked for the decommissioning of nuclear installations, spent fuel and radioactive waste, SWD (2013) 59 final, Brussels.

Hagen, M.; Kunze, C.; Schmidt, P., 2005. Decommissioning and Rehabilitation of Uranium and Thorium Production Facilities, Kerntechnik (Nuclear Technology) 70, 1–2, pp. 91-99.

IAEA [International Atomic Energy Agency], 2018. Nuclear Power Reactors in the World, Reference Data Series No. 2, Vienna.

Irrek, W., 1996. Volkswirtschaftliche Vorteile und höhere Finanzierungssicherheit durch einen Stillegungs- und Entsorgungsfonds: Eine Untersuchung der Bildung und Verwendung von Rückstellungen für Stillegung, Rückbau und Entsorgung im Kernenergiebereich,

Wuppertal Paper No. 53, Wuppertal: Wuppertal Institute for Climate, Environment and Energy.

Irrek, W., 2005. Development of jobs and supporting measures at nuclear power plant sites in the context of decommissioning, in: Tagungsbericht: Jahrestagung Kerntechnik 2005; 10.-12. Mai 2005, Meistersingerhalle Nürnberg. – Berlin:INFORUM-Verl. u. Verwaltungsges., 594–599.

Irrek, W., 2009. Improving financing schemes for nuclear decommissioning and radioactive waste management in European member states and at EU level, in: Davies, Christophe (ed.): Euradwaste '08: Seventh European Commission Conference on the Management and Disposal of RadioactiveWaste – Community policy and research & training activities, Luxembourg: Publications Office of the European Union, pp. 527–531.

Irrek, W.; et al., 2007. Comparison among different decommissioning funds methodologies for nuclear installations", study by the Wuppertal Institute for Climate, Environment and Energy et al., conducted on behalf of the European Commission, service contract TREN/05/NUCL/S07.55346, Wuppertal.

Irrek, W.; Vorfeld, M., 2015. Liquidity and valuation of assets in unrestricted funds from provisions set up for nuclear decommissioning, dismantling and disposal, Brief study on behalf of the Alliance 90/The Greens parliamentary group in the German Bundestag, Wuppertal.

Kaberger, T.; Swahn, J., 2015: Model or Muddle? Governance and Management of Radioactive Waste in Sweden, in: Brunnengräber, A.; et al.: Nuclear Waste Governance, An International Comparison, Wiesbaden.

Kelber, U.; et al., 2010. Brennelementesteuer – Windfall Profits der Atomwirtschaft abschöpfen. Antrag der Abgeordneten Ulrich Kelber et al., Deutscher Bundestag, Drucksache 17/2400 (06 July 2010), Berlin.

Kirbach, R., 2009. Das Lügengrab, Die Zeit, 38 (10.09.2009) [http://www.zeit.de/2009/38/DOS-Asse/komplettansicht; Status: 08 May 2016]

Kutscher, U., 2015. Challenges, Strategies and Approaches of NUKEM to meet the Changing Boundary Conditions and Needs of Nuclear Decommissioning Tasks, Presentation at the International Conference on Decommissioning of Nuclear Installations, 9–11 November 2015, Moscow.

Lukes, R.; Salje, P.; Feldmann, F.-J., 1978. Finanzielle Vorsorge für die Stillegung und die Beseitigung kerntechnischer Anlagen, Energiewirtschaftliche Tagesfragen, 28, 11, pp. 680–689.

N.N., 2013. Marodes Endlager Asse: Bundestag beschließt Gesetz zu Atommüll-Bergung. Spiegel online, (28.03.2013) [http://www.spiegel.de/politik/deutschland/marodes-endlager-asse-bundestag-beschliesst-gesetz-zu-atommuell-bergung-a-886236.html; Status: 08 May 2016].

Neumann, W., 2016. Mängel beim Sicherheitsstandard von atomaren Zwischenlagern in Bayern, Präsentation beim Atomkongress der bayerischen Landtagsfraktion von Bündnis 90/Die Grünen, 16 April 2016, Augsburg.

OECD [Organisation for Economic Co-operation and Development] / NEA [Nuclear Energy Agency], 2012. International Structure for Decommissioning Costing (ISDC) of Nuclear Installations, Issy-les-Moulineaux.

OECD [Organisation for Economic Co-operation and Development] / NEA [Nuclear Energy Agency], 2015. The Practice of Cost Estimation for Decommissioning of Nuclear Facilities, Issy-les-Moulineaux.

OECD [Organisation for Economic Co-operation and Development] / NEA [Nuclear Energy Agency], 2016. Costs of Decommissioning Nuclear Power Plants, Boulogne-Billancourt.

Perridon, L.; Steiner, M.; Rathgeber, A., 2012. Finanzwirtschaft der Unternehmung, 16., überarbeitete und erweiterte Auflage, München: Vahlen.

Schmidt, G.; Ustohalova, V.; Minhans, A., 2013. Nuclear Decommissioning: Management of Costs and Risks, Study by Öko-Institut Darmstadt on behalf of the European Parliament's Committee on Budgetary Control, IP/D/CONT/IC/2013_054, Brussels.

Swiss Federal Audit Office, 2014. Stillegungs- und Enstorgungsfonds, Prüfung der Governance, EFK-14172, Bern

Ustohalova, V.; Schmidt, G., 2014. Klare Zuständigkeiten und Strukturen für Rückbau von Kernkraftwerken: Herausforderungen des EU-mitfinanzierten Rückbaus von Kernreaktoren in Osteuropa, Präsentation bei der Tagung „Sicherheitsmanagement in der Kerntechnik", 16.-17. Oktober 2014, München.

Warth & Klein Grant Thornton, 2015. Gutachterliche Stellungnahme zur Bewertung der Rückstellungen im Kernenergiebereich im Auftrag des BMWi, Düsseldorf.

Legislation

Nuclear Policy in the EU from a Legal and Institutional Point-of-View

Dörte Fouquet[1]

Abstract

The regulation of nuclear issues dates back as far as the foundation of the European Community. The Treaty establishing the European Atomic Energy Community (EURATOM) was one of the founding treaties of the European Communities next to the Treaty establishing the European Coal and Steel Community (ECSC) and entered into force on January 1, 1958. Since then and unlike the other founding treaties, the EURATOM Treaty has never been significantly amended or reformed. Why?

To be able to answer this question, one must look at the legal conditions and political key messages for the setting up of the EURATOM Community. The latter not only from within Europe but also from abroad. Due to the Members States' original compromise and the limits between what is controlled and regulated under the treaty and what remains in the discretion of the Member States, the EURATOM Treaty is clearly limited in its scope.

The EU has no competences in regulatory fields such as operational safety of nuclear power plants, management and safe disposal of radioactive waste, storage or disposal facilities and decommissioning of installations. All these

1 Dörte Fouquet, BBH, Brussels, Belgium, doerte.fouquet@bbh-online.be

R. Haas et al. (Eds.), *The Technological and Economic Future of Nuclear Power*, Energiepolitik und Klimaschutz. Energy Policy and Climate Protection, https://doi.org/10.1007/978-3-658-25987-7_8

crucial objectives remain the sole responsibility of national authorities and are co-guided by standards adopted at the international level, especially under the framework of the International Atomic Energy Agency (IAEA). The EURATOM Treaty then remains as the only sectoral energy policy impeding the integration of policy towards a democratic energy Union. This is reflected by the European Parliament's role being rather that of an opinion-giving onlooker than a co-decision maker in matters related to nuclear regulation.

1 Introduction: EURATOM – a treaty without harmony

The EURATOM Community and its Treaty establishing the European Atomic Energy Community (EURATOM) are marked by their inherent failure to achieve a joint common nuclear policy and establish a common interest of the European Communities, and later the European Union, in the development of nuclear energy. Instead, the EURATOM treaty has set the European energy landscape on a distortive pathway, which was foreshadowed by its limitations and conflicts back in 1945.

In October 1945, the provisional French government under President Charles de Gaulle issued *ordinance* N° 45–2563 establishing the French "*Commissariat à l'énergie atomique*" in order for France to keep its place in the field of nuclear research.[2] The Commissariat and its Committee were, at that time, directly attached to the Government and under direct tutelage of the French President.[3] De Gaulle established this Committee weeks after the nuclear bombing of Hiroshima and Nagasaki in August 1945. France was the only founding member of EURATOM that, alongside the civil use of nuclear energy, sought hegemony in military nuclear power and competed against the dominant position of the United States of America.

Article 1 of the *ordinance* clarifies : "*Le Commissariat à l'Energie Atomique poursuit les recherches scientifiques et techniques en vue de l'utilisation de l'Energie Atomique dans les divers domaines de la science, de l'industrie et de la défense nationale*".

2 See « Ordonnance N° 45–2563du 18 Octobre 1 945 (J.O. du 31 .10.45) et rectificatif du J.0. du 3.11 .45) modifiée par la loi n° 47–1497 du 13 Août 1947(J.O. du 14.8. 47) et par le décret n° 51 -7. du 3 Janvier 1951 (J.O. du 4. 1 . 51) ; copy online in

3 See exposé des motifs : « ...L'autorité de l'Etat sur la marche du Commissariat est d'ailleurs la contrepartie nécessaire de la liberté, tout à fait exceptionnelle dans notre droit public, qui' lui est donnée dans sa gestion. Pour assurer une consécration indiscutable à cette autorité, il est prévu que le Comité est présidé par le Président du Gouvernement Provisoire.... ».

France's continued research in the military use of nuclear energy resulted in its first bomb tests in 1960, eight years after the United Kingdom's test in 1952. The peaceful use of nuclear energy in France was as important as its wish to remain and grow as an independent nuclear military force in the World. Today, France, together with the United Kingdom, the United States of America, Russia and China, is one of the five nuclear weapon states (NWS) under the Nuclear Non-Proliferation Treaty (NPT).[4], with four undeclared or unrecognized states, with possibly more on their way.

When it came to purchase and choice of nuclear energy technology, none of the other founding states even showed strong interest in purchasing French civil nuclear technology. From the beginning of EURATOM, and even ahead of its ratification, these other founding Member States- Belgium, the Federal Republic of Germany, Italy, Luxemburg and the Netherlands- were more interested in ordering US technology for new nuclear power plants. A specific agreement with the United States – the EURATOM Cooperation Act followed in 1958[5], immediately after the establishment of EURATOM.

Jean Monnet and others in Europe together with eminent non-EU politicians such as US President Eisenhower *"hoped that EURATOM would foster European integration"*.[6] As was outlined by the European Parliament's research unit in 2002: *"Sectoral integration of nuclear energy was chosen by Spaak and Monnet because it was viewed as less ambitious than a general common market or customs union, and therefore less likely to fail. Nuclear energy was an appealing prestige technology, especially after the liberalisation of US nuclear policy. The Euratom proposal was*

---

4 The Principle of the NPT is that other states which were signatories eschewed the nuclear weapons option and in return were promised assistance in civil nuclear power development by the weapons states.

5 EURATOM Cooperation Act of 1958 is a United States statute enabling a cooperative program between the European Atomic Energy Community and the United States. Following the US Atomic Energy Act of 1954, the cooperative program was designed to be an international agreement cleared the way for the United States to build nuclear power plants within the European Atomic Energy Community territory.

6 In fact, Monnet after the first failures for more European institutions tried to push for a EURATOM treaty as integrating force and peace enforcing similar to the ECSC approach. Only, especially countries such as Germany did not want any EURATOM treaty without a Common market: "To Monnet's great surprise the German government was not interested in the new (EURATOM) Community, while Erhard, already credited with the German Wirtschaftswunder, was openly opposed to it. A meeting with German officials established the basic fact of the coming negotiations, namely that Germany would not swallow EURATOM without a general common market..." Werner Kamppeter: Lessons of European Integration, April 2000, http://www.fes.de/analysen/kamppeter1_1.html

expected to appeal to the French, since France had a strong interest in the development of [...] nuclear energy"[7]

On the other hand, the United States was a keen competitor of France in view of the supply of nuclear power plants, having a competitive advantage at the time of the EURATOM negotiations.[8] In addition, relations between France and the US were quite frail after the Suez crisis in 1956.

This made France turn to its European neighbors rather than towards the United States[9] for cooperation in nuclear issues but without wanting to give up dominion on its own nuclear policy.

France wanted to close the widening gap between itself and the United States' prolific construction and sale of nuclear power plants. For France, EURATOM was the vehicle to stop American dominance in the supply of nuclear energy in Europe.

As was distinctly pointed out by Louis Armand, President of the French CEA (Comité de l´équipement industriel au Commissariat à l'énergie atomique (CEA)), during the EURATOM debate in the French National Assembly on 5 July 1956: "*Il est inutile d'attendre pour se demander si telle solution (EURATOM, d.Verf.) serait la meilleure. Je ne sais comment vous faire partager mon angoisse quant à l'urgence. Le temps presse et, si nous voulions faire de l'effet, je vous dirais que, pendant que*

---

7 O' Driscoll, Mervyn, European Parliament, Directorate-General for Research, Working paper, The European Parliament and the Euratom Treaty: past, present and future, 2002,(ENER 114 EN), p.36.

8 Sebastian Wolf; EURATOM, the European Court of Justice, and the Limits of Nuclear Integration in Europe, German Law Journal; 12 (2011), 8. – S. 1637–1657, p. 1653; Dwight D. Eisenhower: " I am especially pleased to approve the EURATOM Cooperation Act of 1958, which enables the United States Government to begin active preparation for the Joint United States-EURATOM Program to develop nuclear power in Europe. EURATOM (The European Atomic Energy Community), which came into being on January 1, 1958, was formed by six of our European friends--Belgium, Germany, France, Luxembourg, Italy and the Netherlands--in order to combine their efforts in developing the peaceful uses of atomic energy. It holds great promise, not only as a means to this end, but also as a means of furthering European unity. Our Joint Program, which is EURATOM's first major program, is designed to achieve the construction in Europe of about six nuclear power reactors with a total installed capacity of about one million kilowatts of electricity and to improve power reactor technology through a research program of great scope. This Joint Program should prove highly beneficial both to Europe and to the United States."-The EURATOM Cooperation Act of 1958 is Public Law 85–846 (72 Stat. 1084). Dwight D. Eisenhower: "Statement by the President upon Signing the EURATOM Cooperation Act." August 29, 1958. Online by Gerhard Peters and John T. Woolley, The American Presidency Project http://www.presidency.ucsb.edu/ws/?pid=11203 .

9 Grégoire Mallard, Crafting the Nuclear Regime Complex (1950-1957): Dynamics of harmonization of opaque treaty rules; EJIL (2014) , Vol. 25 No. 2, 445–472, p. 455.

nous parlons, un ensemble d'alternateurs équivalent à ceux, réunis, de Génissiat, de Donzère, d'Eguzon, de Gennevilliers, soit tout l'ensemble des alternateurs français, travaillent en Amérique à alimenter les usines de séparation isotopiques, donc à augmenter la distance qui nous sépare de l'industrie américaine. Tout cela va tellement vite que, si nous ne nous dépêchons pas, nous ne rattraperons plus. Sans EURATOM, c'est bien simple, tous les pays européens iront s'adresser aux colosses. L'Italie, nous le savons, achète dès maintenant une installation américaine de 75.000 kilowatts, du même type que celle que les Belges ont acquise et qu'on inaugurera l'année prochaine à Bruxelles. Et voilà le commencement de ces accords bilatéraux, le commencement de cet achat de matériel. ...Aussi, ce que je puis vous dire, c'est que le travail des experts à Bruxelles a été un travail fonctionnel. On l'a dit et c'est vrai. Ce n'est pas institutionnellement que le problème a été étudié. Il a été défini en fonction des besoins, et comme M. Perrin l'a dit, en faisant juste le nécessaire, mais en le faisant bien, car on n'a pas besoin de tout associer. Mais ce qu'on associe, il faut l'associer avec des règles telles que l'ensemble travaille mieux que ne le feraient les mêmes éléments, mais séparés. C'est ainsi que nous n'avons pas pensé à la création d'un commissariat européen de l'énergie atomique.... Nous avons respecté tous les programmes nationaux, comme le demandait M. Perrin, et limité l'association à un minimum, mais au minimum vital, c'est le cas de le dire, au minimum nécessaire pour une large irradiation. Les experts de Bruxelles ont envisagé que l'association ne porterait que sur 20 p. 100 des équipements. C'est ainsi que les recherches resteront indépendantes".[10]

And, on the other hand, France insisted during the negotiations for the EURATOM Treaty that "*equality of rights of the Members*", as was claimed by the other founding Member States to EURATOM, was not in any way pursued in order to hinder its own nuclear weapon programme.

During the months before final consent on the EURATOM treaty text was reached, conflict and discussion continued. Disagreement was rife over Article 77 on the balance to be drawn when it came to defining the single jurisdiction that would have control over fissile material on the EURATOM territory.

At the London Conference of nine powers on 28th of September to 3rd of October 1954, German Chancellor Konrad Adenauer declared that his country would not produce nuclear weapons in Germany and thus enabling his country for

10 ARMAND, Louis; PERRIN, Francis. L'EURATOM : Exposés faits à la tribune de l'Assemblée nationale le 5 juillet 1956 par Francis Perrin et Louis Armand. Paris: 1956. 31 p. p. 19-31. http://www.cvce.eu/obj/expose_de_louis_armand_sur_la_politique_nucleaire_de_la_france_et_sur_l_EURATOM_5_juillet_1956-fr-9ee0da8e-928a-4e5e-af43 -a3abfeef683e.html

re-armament.[11] Nonetheless, in January 1957 in Algeria, France and Germany, the latter being a deliberate opponent so that France would not alone have a military use for nuclear fissile material[12], signed a secret military cooperation agreement in a meeting between their defense ministers (Maurice Bourgès-Maunoury and Franz Josef Strauss). This agreement was later extended to Italy in 1957. These secret agreements laid the ground for a biased application of the control articles of EURATOM. In consequence of these agreements, EURATOM never clearly governed material destined for military use. Only in that way did France accept that the EURATOM inspectors would *"control the conformity between the 'real' and 'declared' uses of nuclear fuels (be they civil or military uses)"*[13]

Therefore, the inspectors could only verify that the material in France was for military use but could not prevent this use. This weakened EURATOM's role as a backbone of the international Non Proliferation Treaty (NPT) and its objectives and immediately hamstrung its status as the governing supranational body. Moreover, it created a situation from the beginning where there was a split vision between France and the other EURATOM founding members. This in effect weakened any hope for unity under EURATOM. One can thus conclude that divergence was an integral part of EURATOM from its beginning. Also, to the outside world and especially to the United States, this biased approach prevailed. Monnet briefed his assistant Max Kohnstamm[14] before they went to negotiate details for the EURATOM-US

---

11 For the extent of this policy in order to ensure France's acceptance of Germany having an own army (die Bundeswehr) again, see Soutou Georges-Henri. Les accords de 1957 et 1958 : vers une communauté stratégique nucléaire entre la France, l'Allemagne et l'Italie? In: Matériaux pour l'histoire de notre temps, n°31, 1993. Penser et repenser les défenses (suite). pp. 1-12;doi : 10.3406/mat.1993.404096 http://www.persee.fr/doc/mat_0769-3206_1993_num_31_1_404096

12 See for background of a quite lively s dispute between the ministers of foreign affairs of France and Germany, Grégoire Mallards, a.a.O.

13 Grégoire Mallard, a.a.o.

14 Max Kohnstamm: From 1948 to 1952 Max Kohnstamm was appointed to the Dutch Ministry of Foreign Affairs, as Diplomat under the direction of Minister Hirschfeld, where he was responsible for German affairs and in particular the Ruhr Authority and later under Minister Stikker, he was responsible for European concerns. He helped negotiating for the Schuman Plan where he met Monnet. After this, he participated in the negotiations for EURATOM and in the negotiations for the US-EURATOM Agreement of 1958, where he was Secretary to the three "Wise Men". He later became Vice-President of the Action Committee for the United States of Europe, 1956–1975. He was a close supporter of Monnet in his endeavor for creating an integrated Europe. A fascinating wealth of information are his archives at the European University Institute (http://archives.eui.eu/en/fonds/155415?item=JMDS.A-09.06).

treaties on Nuclear to *"avoid explaining how the system of EURATOM control would work"* and to maintain *"the appearance"* of equivalence between the American and the EURATOM control and thus of strictly non-military civic use of fissionable material belonging to and controlled under EURATOM.[15]

2 EURATOM ltd.

The problematic start thus led to a future where no real Atomic Energy Union was ever established in Europe. Instead, EURATOM has served as a basis for gigantic and continuing benefits for nuclear research over the last decades in comparison to other energy research. Additionally, there never was a unified movement to promote a European policy framework for nuclear.

The EURATOM treaty remains a treaty with limited liability.

There seems to have been a recent move to attach an aspect of exclusive supremacy to EURATOM over the European Union treaties, at least when it applies to the internal energy market and to the observance of procurement and competition rules. Under the current Commission and its predecessor, EURATOM was re-invented as a way of underlining a common European interest to support the creation of new nuclear power plants with public money and guarantees, as will further be shown.

National preferences or antipathies concerning a powerful EURATOM Community prevail until today.

The above, almost fascinating, national obstinacy of Member States to not create a true Atomic Energy Union when agreeing to the EURATOM Treaty in the past is well documented by numerous examples, the first of which being the unwillingness of France in the years after EURATOM entered into force to sign the NPT. The NPT, as joint initiative of the Soviet Union, the United States of America and the United Kingdom has been open for signature since 1968.

The signature of the Community did not lead to an automatic assent of France to the NPT since the EURATOM treaty is only binding internally, with its own safe guard clause under Art. 77, on the regional Member State level.

In 1973, the seven non-nuclear weapon States within EURATOM formally signed, together with the European Community, an agreement with IAEA for the implementation of NPT safeguards.

Concerning non-military nuclear installations, a similar agreement was concluded in 1976 among the IAEA, the Community and the United Kingdom. The

15 Grégoire Mallard, a.a.O.

United Kingdom deliberately had offered to accept the IAEA safeguards for the UK non-military nuclear plants.[16]

France only acceded to the NPT in 1992, in the same year as China did. Before that and after the NPT entered into force, France not being a party to the NPT, signed a similar three party agreement with the EURATOM Community and the IAEA which foresaw the application of IAEA safeguards in a manner similar to the above agreements but, "*limited, however, to those materials which France wants to put under such safeguards. (…)Thus we now find in Western Europe a unique situation in the field of safeguards due to the fact that there exists a regional safeguards authority invested with supranational rights which at the same time not only fulfils, within the framework of the IAEA system, the tasks normally assigned to a national system of accounting and control, but also collaborates with the IAEA in inspections in a way which permits the latter to draw its own independent conclusions.*"[17]

3 Activities and sectors governed by EURATOM

When screening EURATOM following modern, scientific and established rules on circular economy legislation, EURATOM again fails to deliver further underlining its status as a "*failed community*".[18] An example is its limited provisions on planning and permitting procedures which fail to include operation, safe operation, ore supply, dismantling and safe final storage of radioactive waste.

Existing secondary legislation under EURATOM, which covers issues such as waste, were in some cases helped by a progressive interpretation of the European Court rather than by the EURATOM treaty itself with the limited scope of its concerned Chapter (Chapter 3 on Health and safety), as will be reflected on below.

In consequence, EURATOM, following its beginnings and its undercurrent of opposing interests, particularly those of the one Nuclear Weapon State (NWS) (France) and the other EURATOM founding members, emerges as a 'skeleton treaty' unable of consequently regulating the diverse elements of nuclear power including

16 Schleicher, H.W. (1980), (former Director of the EURATOM Safeguards Directorate, Commission of the European Communities). Nuclear Safeguards in the European Community – a Regional Approach, IAEA Bulletin 22 (3/$) p 45.

17 Schleicher, H.W., ibid. page 45.

18 See Wolf, Sebastian, EURATOM, the European Court of Justice, and the limits of Nuclear integration in Europe, German Law Journal 12 (2011), 8, p. 1638 and referring to Weilemann, Peter, die Anfänge der Europäischen Atomgemeinschaft, p. 157.

uranium mining, supply, shipment, use and processing and final dismantling of nuclear installations.

EURATOM's objective is prominently established in the EURATOM treaty:

> "Article 1
> By this Treaty the HIGH CONTRACTING PARTIES establish among themselves a EUROPEAN ATOMIC ENERGY COMMUNITY (EURATOM). It shall be the task of the Community to contribute to the raising of the standard of living in the Member States and to the development of relations with the other countries by creating the conditions necessary for the speedy establishment and growth of nuclear industries."

Art. 2 lists the responsibilities and tasks for the EURATOM Community:

Major objectives are to promote research and ensure the dissemination of technical information and to establish uniform safety standards to protect the health of workers and of the general public and ensure that they are applied; facilitate and ensure investment, particularly by encouraging ventures on the part of undertakings; Chapter 3 EURATOM, entitled, 'Health and Safety', outlines content and limitation of the power of the Community with its Art. 30:

According to EURATOM, basic standards shall be laid down within the Community for the protection of the health of workers and the general public against the dangers arising from ionizing radiations. However, this provision does not give any right to the Community to directly safeguard Nuclear Power stations: "*It essentially provides for the Community to establish a series of dose limits for exposure of human beings to radiation. (This has been done, usually following the line established by the International Commission for Radiological Protection (ICRP)). But it does not provide any competence to EURATOM either with respect to possible damage to the natural environment caused by radiation, and perhaps even more remarkably, it provides no EURATOM Community competence with respect to the safety of nuclear reactors*".[19]

The main European objectives under EURATOM outlined in Art. 2 and detailed for safety and health aspects in Chapter 2, are:

- the establishment of the basic installations necessary for the development of nuclear energy in the Community;
- to ensure that all users in the Community receive a regular and equitable supply of ores and nuclear fuels;
- to make certain, by appropriate supervision, that nuclear materials are not diverted to purposes other than those for which they are intended;

19 O' Driscoll, Mervyn, The European Parliament and the EURATOM Treaty: past, present and future, p. 17.

- to exercise the right of ownership conferred upon it with respect to special fissile materials; details of the property ownership in this sense are further laid down in Chapter VII of the EURATOM treaty. However, while the Community has the property rights, Art. 87 makes it clear that Member States, persons and undertakings shall have the unlimited right of use and consumption of fissile materials in lawful possession.

In reality, these fissile material property rights of the Community have proved of little practical consequence within the European Community, as will again be outlined below.

EURATOM should, under Art. 2, ensure wide commercial outlets and access to the best technical facilities through the creation of a common market in specialised materials and equipment, through the free movement of capital for investment in the field of nuclear energy and through freedom of employment for specialists within the Community.

Another point which was to be a future point of contention as to its scope was the provision under Art. 2 that EURATOM should establish with other countries and international organizations similar relations as it will further progress on the peaceful uses of nuclear energy.

4 The EURATOM Supply Agency

Chapter 6 of Title II EURATOM establishes the EURATOM Supply Agency (ESA) and confers upon it an exclusive right to conclude contracts relating to the supply of ores, source materials and special fissile materials coming from inside or from outside of the Community. Chapter 8 established the property ownership rules where -as laid down in Art. 86 Para 1 EURATOM- the special fissile materials shall be the property of the Community.

The ESA conceived would own and control the supply of all fissile materials in the Community. Art. 86 and 87 EURATOM are perfect examples of a bizarre and redundant legal position of the Community: the provisions under Art. 86 EURATOM declare the Community as being holder of the property right on special fissile materials, and defines a "*right of ownership*" to "*all special fissile materials which are produced or imported by a Member State, a person or an undertaking*" and which are subject to the safeguard clauses under Chapter 7 EURATOM. Despite this principle, the Community has no **real** right of ownership since use is clearly curtailed via the provisions under Art. 87 EURATOM, according to which Member

States, persons or undertakings have unlimited rights over use and consumption of this material. The European Supply agency reflects the ambiguous and incomplete community framework of EURATOM.

It is the EURATOM Community / the Commission that controls the distribution of patent rights and production licenses for a series of reactor designs and fuel cycle technologies to be developed by the Joint Nuclear Research Centre (JNRC). Important conditions are set for research (Art. 7) and international agreements (Art. 101), e.g. on ensuring access to fissile materials.

These tasks and responsibilities by the Community are, to a large extent, kept away from serious control by the European Parliament in the sense of modern legislative power and supervision over the executive or, as was put pointedly: *"Control by democratically elected Parliaments was not exactly a significant feature of the nuclear sector in the 1950s"*.[20] And this democracy-excluding oversight was never reformed by a modernisation of EURATOM as has been the case in other areas via the various reform treaties leading to the Lisbon treaty.

When it comes to uranium supply, the reality of nuclear supply substantially mirrors the existing situation on gas import dependence in Europe. All sourcing is done outside the EU and mostly from former Soviet Union countries, e.g. from Russia. One might question the sustainability of this supply route, a worrying fact also outlined by the European Commission itself.[21] Another point of unease is the fact that Russia is nowadays pushing to supply new nuclear power stations to Europe, such as the one just recently constructed in Hungary named Paks II – a new installation with procurement and state aid authorised by the European Commission.

---

20 O' Driscoll, Mervyn, The European Parliament and the EURATOM Treaty: past, present and future, p. 6.

21 The EU Commission outlined in its Communication COM (2014) 330 final "European Energy Security Strategy" of 28th of May 2014, p. 16: " The worldwide uranium supply market is stable and well diversified but the EU is nonetheless completely dependent on external supplies. There are only a few entities in the world that are able to transform of uranium into fuel for the nuclear reactors, but EU industry has technological leadership on the whole chain, including enrichment and reprocessing. "...However, Russia is a key competitor in nuclear fuel production, and offers integrated packages for investments in the whole nuclear chain. Therefore, particular attention should be paid to investments in new nuclear power plants to be built in the EU using non-EU technology, to ensure that these plants are not dependent only on Russia for the supply of the nuclear fuel: the possibility of fuel supply diversification needs to be a condition for any new investment, to be ensured by the EURATOM Supply Agency. Furthermore, an overall diversified portfolio of fuel supply is needed for all plant operators."

4.1 The reality of supply channels and long term supply contracts

Chapter 6 of Title II EURATOM opens with mention to, and a request for, a common supply policy concerning the sourcing of source and special fissile materials and conveys to the supply agency an exclusive right to conclude contracts relating to the supply of ores, source materials and special fissile materials coming from inside or outside the Community. Strangely enough, Chapter 6 seems to be respected or applied very little and "*many articles have apparently either not been implemented at all, or only partially implemented or applied.*"[22]

The same is valid for the Special Fissile Materials Financial Account as a transaction balancing and auditing tool: Art. 89, para. 4 under Chapter 8 of EURATOM foresees that the Agency could undertake transactions for its own account and shall in this context "be deemed to be an undertaking". This provision for a European nuclear undertaking has never been implemented.

The Supplies Agency exists, but rather worked as an observer than as an active buyer and supplier of fissile material. Until at least 2005 the Agency never itself used its purchase power and there is no publicly documented evidence for direct activities in purchasing, supplying and stockpiling of material.

In 2005, a report on Security of supply, published by an advisory task force to the EURATOM Supply Agency, parted from this past experience, that security of supply concerns in the nuclear fuel cycle result from the fact that primary production of natural uranium covers only some 60% of world demand while the remaining part comes from historical production (inventories and weapons dismantling) and from the re-enrichment of tails of depleted uranium resulting from the enrichment process.[23] This task force of the major Western nuclear operators at that time and the nuclear energy and fuels company, British Nuclear Fuels plc (BNFL), several times outlined the question of stocks but always clearly and self-evidently defined and described the stocks as those kept by the nuclear operators. The report mentioned the possibility that the Agency could act under Art. 72 EURATOM and establish a stockpile, but in no way referred to this as an option near to reality. This is clearly illustrated in the following analysis in the report, underlining that there "*is a decrease of stocks in many countries as part of this reduction of operational costs for utilities.*

---

22 O' Driscoll, Mervyn, The European Parliament and the EURATOM Treaty: past, present and future, p. 13

23 See Analysis of the Nuclear Fuel Availability at EU Level from a Security of Supply Perspective, EURATOM Supply Agency – Advisory Committee Task Force on Security of Supply Final Report of the Task Force, June 2005, page 5.

The appropriate level of stocks and the entire fuel procurement policy depends on the size and electricity generation pattern of each utility".[24] Also, the recommendation in case of shortages does not mention in any way a role for the Agency to execute its right and even objective under EURATOM in stockpiling capability.[25]

The European Court of Auditors has apparently regularly asked what the Supplies Agency could actually be doing. France, over many years, has displayed an attitude of not caring at all about the Agency.[26]

When it comes to national policies and contracts with third countries, an international organisation or a national of a third State (e.g. an energy company or supplier), the line between competencies of the State and of EURATOM remained unclear and was several times subject to European Court decisions. The Member States clearly remain sovereign to bilaterally conclude those international agreements. EURATOM steps in when "*matters within the purview*" of EURATOM are concerned, as laid down under Art. 103 EURATOM.[27] This wording creates ground for uncertainty and thus underlines the reality that there is no unique EURATOM competence, even in matters where the ownership of the fissile material is explicitly given to the EURATOM community under the treaty but where use and

---

24 See, Analysis, ibid. page 5.

25 See, Analysis, ibid, page 16: "Against long term risks (not susceptible to happen in less than a year or two) remedies are more mixed, e.g. exploration and investments in new production facilities, diversification, long term contracting, and partnerships. In any case, an effective monitoring of the supply and demand situation at EU and world levels and its likely evolution would be a very important tool for the Commission's analysis in view of the EURATOM Community's responsibilities in the EU energy security of supply, as well as for the nuclear industry. The ESA which already has a unique insight into the market through its concurrence privilege, could be this monitoring tool, but would benefit from accurate information from all nuclear industries operating in the EU, on their sources and commitments as well as their stocks of nuclear material (including their quantity, form and location)."

26 O' Driscoll, Mervyn, The European Parliament and the EURATOM Treaty: past, present and future, European Parliament, Directorate-General for Research, Working paper, 2002,(ENER 114 EN), page 17: "The Court of Auditors has regularly asked what the Supplies Agency actually does. France appears in the past to have sometimes largely ignored the very existence of the Agency, considering that France is exempt from most of the provisions of Chapter 6 (which it has also challenged the legitimacy of in the European Court of Justice – so far unsuccessfully)."

27 See for example: European Court, Ruling 1/78 of 14. 11. 1978 following an application of the Kingdom of Belgium pursuant to Art. 103 EURATOM, asking as to whether in the absence of the concurrent participation of the Community, Belgium may adhere to the Convention on the Physical Protection of Nuclear Materials, Facilities and Transport under the IAEA regime.

management is kept with the Member States. This limitation is also important in questions around who can sign which kinds of agreements at international level.

In 2016, a recommendation by the Commission was needed on the interpretation of Art. 103 EURATOM. The recommendation tries to establish more clarity around fuel supply contracts but outlines a dilemma: "*In the event that a draft agreement or contract concerned by the present Recommendation includes also provisions on supply matters, its assessment by the Commission pursuant to Article 103 should be without prejudice to the exclusive right of the Agency to conclude supply contracts; co-signature of supply contracts by the Agency on the other hand is without prejudice to the assessment by the Commission of the compatibility of Member States' draft agreements or contracts with the provisions of the Treaty and its secondary legislation.*"[28]

The EURATOM Supply Agency is hesitant, when it comes to agreeing to long-term supply contracts with nuclear fuel.

In a recent decision, the Commission acting as the EURATOM Supply Agency refused to co-sign a Nuclear Fuel Supply (NFS) Contract on nuclear fuel supply and spent fuel storage and treatment for future Units 5 and 6 to be built at the site of the existing Paks nuclear power plant in Hungary. The Hungarian company Paks II and the Russian company Nizhny Novgorod Engineering Company «Atomenergoproekt» had signed the NFS Contract on 9 December 2014. Following its signature, the NFS Contract was submitted to the "Agency" on 23 December 2014 pursuant to Art. 52 of the EURATOM Treaty and screened under the procedure of Art. 103. By letter of 13 January 2015, after extensive discussions with the Paks II consortium and Hungary, the "Agency" notified its refusal to co-sign the NFS Contract, reasoning that the clauses of the contract "*de facto prevent diversification of fuel supply*".[29] Under this contract, Rosatom, the Russian State Atomic Energy Corporation, is mandated to provide the technology and the supply of nuclear fuel (and assemblies) and offered the option of managing the spent nuclear fuel for at least 20 years through its subsidiary "NIAEP". The financing of the project is aimed to be ensured mainly via a loan granted by the Russian Federation to the Hungarian State for which the state gives various guarantees. The state aid part of the deal was recently authorised by the European Commission in view of Art. 2 (c) EURATOM[30]

28 See C(2016) 1168 final COMMISSION RECOMMENDATION of 4.4.2016 on the application of Article 103 of the EURATOM Treaty, para 13.

29 See (declassified part of) COMMISSION DECISION of 2.3.2015 (CÇ2015) 1398 final) relating to a procedure in application of the second paragraph of Art. 53 of the EURATOM Treaty, Para 4.

30 Art. 2 c) EURATOM: (In order to perform its tasks, the Community shall, as provided in this Treaty:..) "(c) facilitate investment and ensure, particularly by encouraging ventures

and its view that there is a common European interest in promoting the creation of new nuclear power stations.[31] On the fuel supply and waste management contract, the Agency was more wary. A problematic reality has appeared: Hungary is to complete a new state aid-supported construction of the Paks II nuclear power plant, with supply and management provided by Rosatom where the Commission itself has acknowledged an almost full import of fissile material from Russia.

Concluding, it is obvious that major executive provisions for the Agency under Chapter 6 and 8 of EURATOM are not mirrored in reality even though they were designed to represent its pillars. They seem to never have really been destined to have a dominant role enshrined to the EURATOM Community and its control agencies: France and Western Germany at the time of the Spaak Committee[32] and the preparation of a Treaty gave a green light for a somewhat inventive definition, concerning "property" and "control" .

To recall: In view of France's insistence as sovereign NWS, the Spaak Committee explained that "*control*" for the Community when it comes to fissile material means a "*control of conformity*". "*The EURATOM Community would be granted "property" of all nuclear fuels used for peaceful ends within the EURATOM territory (therefore leaving the property of fuels used for French military ends to the French government); but this notion of property was defined "as a "property sui generis," an old notion which differed from the common notion in Roman Law, since the property of fuels by the EURATOM Community granted no rights to its owner during normal times*".[33]

And in line with the above task force's report, not only France but none of the nuclear operators in any Member State attributed any real power of property and supreme control to the Community.

This mismatch once again underlines the already drawn conclusion from the late 1990s that "*EURATOM never got off the ground*".[34]

---

on the part of undertakings, the establishment of the basic installations necessary for the development of nuclear energy in the Community; ".

31 SA.38454 – 2015/C (ex 2015/N) Commission decision of 6.3.2017 on state aid which Hungary is planning to implement for supporting the development of two new nuclear reactors at Paks II nuclear power station.

32 The Spaak Committee was an Intergovernmental Committee set up by the Foreign Ministers of the six Member States of the European Coal and Steel Community (ECSC) as a result of the Messina Conference of 1955.

33 Mallard, Grégoire, The European Nuclear Force, An Obscure Object of Desire, Princeton University and Université Paris-Est (LATTS) (2009), page 19.

34 Trachtenberg, Marc, 1999. A Constructed Peace: The Making of the European Settlement. Princeton: Princeton University Press. Page 205.

5 The relationship of the European treaties towards each other

The Merger Treaty, or "Brussels Treaty", which entered into force on 1 July, 1967 created a Single Commission and a Single Council of and for the three European Communities, meaning the executive bodies of the European Coal and Steel Community (ECSC), the European Atomic Energy Community (EURATOM) and the European Economic Community (EEC). The institutions of the EEC would manage all institutional responsibilities under EEC, ECSC and EURATOM. All three were, after the Brussels treaty, denominated as the European Communities but from a legal point of view all three continued to exist separately under their respective treaties. As a consequence, since 1967, the Council and the Commission of the EEC replaced the Commission and Council of EURATOM and the High Authority and Council of the ECSC. Although each Community remained legally independent, they shared common institutions.

In 1993, the Maastricht Treaty created the European Union, which absorbed the three Communities, yet EURATOM and the ECSC kept their legal personality.

The Maastricht treaty was repealed by the Amsterdam Treaty, which was signed in 1997. With the Lisbon Treaty of 2009, the European Union replaced the original European Communities but the structures and legal conditions at the communities' bases as specific treaties remained unchanged. This was due to lack of support in many EU Member States to create a European Constitution as had previously been conceived by the Maastricht treaty. Therefore, the European law is still established by the two international Treaties, minus the ECSC, which had a deadline and expired after 50 years in 2002. The Treaty establishing the European Community was renamed the "Treaty on the Functioning of the EU" (TFEU).

During the finalised 1995 enlargement of the European Union, Austria, Finland and Sweden acceded to the European Union (EU). All these States had already established links to the Union with the European Free Trade Association (EFTA).

6 The role of the European Parliament and of the European Court of Justice

A major headache for the European Parliament and the democratic community in the EU in general is its very limited role in all EURATOM based legal projects since its position is restricted to that of an advisory institution under Art. 31 EURATOM rather than as a legislator. Back in 1988, and two years after the Chernobyl disaster, the European Parliament contested in an annulment procedure before the European Court that Council Regulation (EURATOM) No. 3954/87 of 22 December 1987 laying down maximum permitted levels of radioactive contamination of foodstuffs and of feeding stuffs following a nuclear accident or any other case of a radiological emergency was wrongly based on Art. 31 EURATOM, which provides that the Parliament is to only be consulted, whereas it should have been based on Art. 100a of the EEC Treaty which requires cooperation with the Parliament implementation of the procedure.[35]

The European Court of Justice (ECJ) declared the case as admissible despite an opinion of the Council asserting that the European parliament did not have the legal personality under (former) Art. 173 EEC treaty or Art. 146 EURATOM since it is not included among the institutions which, like the Member States, bring an action for annulment against any measure of another institution before the Court.[36]

The Court accepted the fact that the Parliament is not mentioned in both articles and thus the Court could not include the Parliament among the institutions, which may bring an action per se as institution but "*being required to demonstrate an interest in bringing an action*".[37]

But the Court saw its "*duty to ensure that the provisions of the Treaties concerning the institutional balance are fully applied and to see to it that the Parliament's prerogatives, like those of the other institutions, cannot be breached without it having available a legal remedy, among those laid down in the Treaties, which may be exercised in a certain and effective manner*".[38]

According to the Court, such a procedural gap cannot prevail over the fundamental interest in the maintenance and observance of the institutional balance laid down in the Treaties establishing the European Communities, thus referring to all Treaties. Therefore and insofar as the Parliament disagrees with the Council's choice of legal basis for the contested Council regulation, this led to a breach of Parliament's prerogatives by denying it the possibility of participating in the drafting more actively than it could under the EURATOM consultation procedure.

In consequence, the Court dismissed the Council's objection of inadmissibility and allowed the proceedings to be continued with regard to the substance of the case.

On the substance, though, the Parliament lost the case. The Court did not follow the arguments of the European Parliament that the regulation was also a harmonisation measure within the meaning of Art. 100a of the EEC Treaty. For the Court

35 ECJ Case C-70/88, (interlocutory judgment of 22.05.1990), Para 39.

36 Case C-70/88,(Interlocutory judgment) Para 13.

37 Case C-70/88, (Interlocutory judgment) Para 24.

38 Case C- 70/88, (Interlocutory judgment) Para 25.

the prohibition of marketing provided for in Art. 6(1) of that Regulation was "*only one condition for the effectiveness of the application of the maximum permitted levels. The regulation therefore has only the incidental effect of harmonising the conditions for the free movement of goods within the Community inasmuch as, by means of the adoption of uniform protective measures, it avoids the need for trade in foodstuffs and feedingstuffs which have undergone radioactive contamination to be made the subject of unilateral national measures.*"[39]

Over the ensuing years, the European Parliament has had some success nonetheless in this respect, but it is restricted to very few decisions.[40]

An interesting case in this context may also be the ECJ judgment in C-490/10 of 06.02.2012 concerning an Action for annulment in view of Regulation (EU, EURATOM) No. 617/2010, especially concerning the notification to the Commission of investment projects in energy infrastructure within the European Union and the choice of the legal basis of this Regulation, meaning Article 187 EURATOM versus Article 337 TFEU, in view of Article 194 TFEU.

The case concerned an action introduced in October 2010 by the European Parliament against the Council, where France and the Commission intervened on the side of the Council. The Parliament demanded the ECJ to annul Council Regulation (EU, EURATOM) No 617/2010 of 24 June 2010 concerning the notification to the Commission of investment projects in energy infrastructure within the European Union and repealing Regulation (EC) No 736/96.[41] In 2009, the Commission had

39 CASE C-70/88, JUDGMENT OF 4. 10. 1991 Para 17.

40 Recently, the European Parliament voted a resolution of 13 September 2017 on the draft Commission Implementing Regulation amending Commission Implementing Regulation (EU) 2016/6 as regards feed and food subjected to special conditions governing the import of feed and food originating in or consigned from Japan following the accident at the Fukushima nuclear power station (D051561/01 – 2017/2837(RSP)) urging the Commission not to loosen restrictions on imports. The Parliament argues that it is very difficult to verify whether the measures proposed are sufficient to protect the health of Union citizen, especially in view that the responsible Japanese utility Tokyo Electric Power Company (TEPCO) has officially requested permission from the Japanese Government to dump into the Pacific Ocean almost one million tonnes of highly radioactive water related to the clean-up of the nuclear accident. The Parliament sees the draft Commission implementing regulation inconsistent with Union law in that it is not compatible with the aim and general principles laid down in Regulation (EC) No 178/2002 of providing the basis for ensuring a high level of protection of human life and health, animal health and welfare, the environment and consumer interests; the Parliament calls on the Commission to go back to the drawing board and present together with a new proposal.

41 Council Regulation (EU, EURATOM) No 617/2010 of 24 June 2010 concerning the notification to the Commission of investment projects in energy infrastructure within

submitted a proposal for this Regulation to the Council. The proposal was based on Art. 284 EC and Art. 187 EURATOM.[42] These provisions did not provide for any involvement by the Parliament in the decision–making process, but the Council decided to consult it as it had done when adopting previous Regulation No 736/96. Since the entering into force of the Lisbon treaty in December 2009, the European Parliament asked for a review of the legal basis for the draft Council Regulation so that it would henceforward be based on Art. 194 TFEU[43], instead of Art. 284 EC (which became Art. 337 TFEU) and Art. 187 EURATOM.

The Parliament argued that Art. 337 TFEU and Art. 187 EURATOM were now substituted with Art. 194 (1) and (2) TFEU as legal basis to be observed for this draft Regulation. The Regulation focused on data collection in the energy market and the newly introduced shared responsibility for energy under the Lisbon treaty would make Art. 194 TFEU the relevant bases thus ensuring the Parliament's legislative role as defined in the new co-decision procedure. The Court's judgment not only agreed with the view of the European Parliament, that Art. 187 EURATOM would in this case not take precedence over Art. 194, but that the latter remained the sole basis for the envisaged Regulation, underlining that the contested Regulation con-

---

the European Union and repealing Regulation (EC) No 736/96 (OJ L 180, 15.7.2010, p. 7).

42 Art. 187 EURATOM: "The Commission may, within the limits and under the conditions laid down by the Council in accordance with the provision of this Treaty, collect any information and carry out any checks required for the performance of the tasks entrusted to it."

43 Art. 194 TFEU:
"1. In the context of the establishment and functioning of the internal market and with regard for the need to preserve and improve the environment, Union policy on energy shall aim, in a spirit of solidarity between Member States, to:
(a) ensure the functioning of the energy market;
(b) ensure security of energy supply in the Union;
(c) promote energy efficiency and energy saving and the development of new and renewable forms of energy; and (d) promote the interconnection of energy networks.
2. Without prejudice to the application of other provisions of the Treaties, the European Parliament and the Council, acting in accordance with the ordinary legislative procedure, shall establish the measures necessary to achieve the objectives in paragraph 1. Such measures shall be adopted after consultation of the Economic and Social Committee and the Committee of the Regions.
Such measures shall not affect a Member State's right to determine the conditions for exploiting its energy resources, its choice between different energy sources and the general structure of its energy supply, without prejudice to Art. 192(2)(c).
3. By way of derogation from paragraph 2, the Council, acting in accordance with a special legislative procedure, shall unanimously and after consulting the European Parliament, establish the measures referred to therein when they are primarily of a fiscal nature."

cerns the notification by all Member States of the aggregated data and information relating to all investment projects in energy infrastructure.[44]

The Court came to the conclusion, that it is "*appropriate to find that the contested [R]egulation, in so far as it was based on Art. 187 EURATOM, was adopted on an incorrect legal basis and it should have been based solely on Art. 194(2) TFEU*".[45] The Court annulled the contested Regulation on this ground.

This leads to the conclusion that EURATOM is "lex specialis" only for those objectives which are explicitly and specially regulated by it and only if clear restrictions respectively excluding nuclear matters from general energy policy or other policy fields in the Union with consequence to the energy sector are outlined.

The European Parliament has the right for co-decision in areas that touch upon EURATOM issues when the legal basis is predominately the general TFEU. But the conflict often arises about the legal bases for energy legislation which affects nuclear matters in the energy market. In a recent decision by the ECJ it did not follow a similar annulment request by the European Parliament concerning the legal basis in view of Council Directive 2013/51/EURATOM[46] on the protection of the health of the general public with regard to radioactive substances in water intended for human consumption, underlining EURATOM's specific sphere of application.[47] The choice of legal basis in this case excluded application of the co-decision procedure and thus any substantial involvement of the Parliament.

As has been outlined, the main provisions of the EURATOM Treaty have never been amended since it entered into force on 1 January 1958, which is a sign of a monolithic or static situation, strangely detached from national energy reality and the European objective of a single European Energy market, which was defined in the mid-eighties of last century.

Clear sign of this stagnation is the limited number of judgments of the European Court of Justice on cases with respect to the EURATOM Treaty, most concerning violation of the safeguard provisions under Art. 77. EURATOM. Based on specific research by Wolf, there seem to be no more than 31 cases until 2009 before the European Court of Justice which have secondary legislation based on EURATOM as its object.[48] Most of these cases refer to infringement problems in view of the

---

44 ECJ, C-490/10, European Parliament v. Council, para 84.

45 Ibid, Para 86.

46 O.J. L296/12/2013.

47 Case C 48/14, action for annulment under Art. 263 TFEU and Art. 106a(1) EA brought on 30 January 2014, European Parliament v. Council.

48 See Wolf, Sebastian (2011): EURATOM Before the Court: A Political Theory of Legal Non-Integration, European Integration online Papers (EIoP), Vol. 15, Art. 10 http://

possibilities for sanctions foreseen under Art. 83 EURATOM. Nonetheless, in contrast to the above, the European Court of Justice has intervened several times to deny a broad range of applications of EURATOM's basic articles.

7 Secondary legislation as timid opening towards a broader scope

The inability of the European Member States to produce a common European approach rather than individual national mandates can also be seen in the difficulties of the European Commission when proposing legislation under the EURATOM Treaty, especially concerning nuclear safety. Countries with a plan to phase out nuclear, such as Germany, are reluctant to cede too much control to the Community because they fear a dilution of their safety standards following a European compromise and therefore remain adamant that the key authority in respect to nuclear power is the nation and not the Community. On the other side are Member States seeming reluctant to any interference when calling a different security standard than their own. Jointly, this situation creates a very hesitant European policy approach to EURATOM-based secondary legislation.

Major secondary legislative work is carried out on the basis of Art. 30 EURATOM. Art. 30 and Art. 31 provide for the establishment in the Community of basic standards for the protection of the health of workers and the general public against the dangers arising from ionising radiations. Art. 30 provides a definition of the basic standards and Art. 31 describes the procedure for the adoption and enforcement of those standards.

Art. 33 EURATOM particularly states:

> "Each Member State shall lay down the appropriate provisions, whether by legislation, regulation or administrative action, to ensure compliance with the basic standards which have been established and shall take the necessary measures with regard to teaching, education and vocational training.
> The Commission shall make appropriate recommendations for harmonising the provisions applicable in this field in the Member States.

---

eiop.or.at/eiop/texte/2011-010a.htm, page 9: "The result of this research for data is a compilation of only 30 cases which mainly concern EAEC primary and/or secondary law (see Table 2 for the full list). Given the sheer mass of ECJ cases, this small number already can be interpreted as another indicator of the rather meagre legal development of the EAEC."

> To this end, the Member States shall communicate to the Commission the provisions applicable at the date of entry into force of this Treaty and any subsequent draft provisions of the same kind."

The following examples suggest a move towards a more European approach to regulation under EURATOM:

7.1 Directive 2009/71/EURATOM establishing a Community framework for the nuclear safety of nuclear installations and its amendment, Directive 2014/87/EURATOM

It should be underlined that the European Commission over the years has worked intensively to reach a high safety standard within EURATOM to improve safety in the respective Member States. This initiative is once again limited by the EURATOM provisions which state that the responsibility for the safety of nuclear installations is solely attributed to the Member States and the nuclear utilities and other respective license holders on their territory.

Subparagraph (b) of Art. 2 EURATOM provides for the establishment of uniform safety standards to protect the health of workers and of the general public. Art. 30 EURATOM defines "*basic standards*" for the protection of the health of workers and the general public against the dangers arising from ionising radiations.

Pursuant to Art. 30, Art. 31 and Art. 218 EURATOM, the Community initially, in 1959, issued basic safety standards with Council Directive of 2 February 1959 laying down rules and standards for the protection of the health of workers and the general public against the dangers arising from ionising radiations.[49]

These standard rules were revised fairly regularly: in 1962 by Directive of 5 March 1962, in 1966 by Council Directive 66/45/EURATOM, in 1976 by Council Directive 76/579/EURATOM, in 1979 by Council Directive 79/343/EURATOM, in 1980 by Council Directive 80/836/EURATOM and in 1984 by Council Directive 84/467/EURATOM.

In 1996 the basic standards Directives as revised were replaced by Council Directive 96/29/EURATOM of 13 May 1996 laying down basic safety standards for the protection of the health of workers and the general public against the dangers arising from ionising radiation ('the Basic Standards Directive')[50] which repealed the previously applicable rules as of 1 of May 2000.

---

49 OJ, English Special Edition, 1959–1962, (I), p. 7.

50 OJ 1996, L 158, p. 1.

Nuclear Safety Directive 2009/71/EC regulates that the Member States are required to submit certain practices involving a hazard from ionising radiation to a system of reporting and prior authorisation and to ensure protection from radiation for the population in normal circumstances.

The Directive comprises provisions for the establishment of a national legislative and regulatory framework for nuclear safety of nuclear installations, for the organisation, duties and responsibilities of the competent regulatory authorities, for the obligations of the licence-holders, for the education and training of all parties' staff and for the provision of information to the public. In terms of the organisation of the competent regulatory authorities, it constitutes the separation principle, which indicates that the competent regulatory authorities must be functionally separate from any other body or organisation concerned with the promotion or utilisation of nuclear energy. In addition, Member States shall arrange at least every ten years for periodic self-assessments of their national framework and competent regulatory authorities and invite an international peer review of relevant segments of their national framework and/or authorities. Outcomes of any peer review shall be reported to the Member States and the Commission.

Recital 9 of the Directive underlined that each Member State may decide on its energy mix in strict accordance with its relevant national policies and Recital 8 again put forward the principle of national responsibility as well as the principle of the prime responsibility of the respective license holder under national supervision.

7.2 A long and winding road towards the 2009/71/EURATOM Directive of the Council

Again, the road towards Council Directive 2009/71/EURATOM was littered with conflicts between the Commission and the Council, with Member States not wishing to lose any sovereignty over their nuclear energy and their power of oversight there over beyond what could not be denied under EURATOM.

The Commission started to work on a proposal for a joint Nuclear Safety regime in 2003[51] which it finally withdrew in 2010, having given up on finding a compromise with the Council.[52]

Council Directive 2009/71/EURATOM transfers major provisions from the International Convention on Nuclear Safety (CNS) of July 1994 and further safety

51 COM (2003) 32 – 1: Proposal for a Council (EURATOM) Directive Setting out basic obligations and general principles on the safety of nuclear installations.

52 Withdrawal of obsolete Commission Proposals (2010/C 252/04), O.J. C 252/7, 18.9.2010.

principles into European EURATOM legislation. Its basic structur mirrors the Convention's.[53]The CNS entered into force the year of and as result of the Chernobyl nuclear disaster and is designed to protect individuals, society and the environment from harm by establishing and maintaining effective defences against radiological hazards in nuclear installations. As sharply outlined by Dehousse, the CNS, despite the catastrophe in the middle of the European continent, does not contain any mandatory provisions for safety control.[54]

Directive 2009/71/EC could not have seen the light of day, especially as unanimous as it was actually voted in the Council, without a preceding permissive decision by the European Court of Justice in 2002. It was the ECJ that laid the ground for the Commission's legal initiatives in this field.

In the background of the specific conflict between the Commission and the Council before the ECJ was Council action of December 1998 that declared the accession of EURATOM to the CNS but also a reservation of the applicability of some provisions of the CNS for EURATOM. The Commission had proposed the original text of such restricting declaration but with far fewer reservations.

The possibility for the Community to derogate from certain provisions of the CNS follows from Art. 30,para. 4 CNS. Para. 4 regulates accession to the treaty for *inter alia* regional organisations. These organisations, reflecting their mandate, shall under Art. 30, para. 4 (iii) CNS *"communicate to the Depositary (…) a declaration indicating which States are members thereof, which articles of this Convention apply to it, and the extent of its competence in the field covered by those articles"*.

The Commission requested annulment of that specific part of the Council declaration detailing the reservation to the IAEA *"on the ground that, by limiting the scope of that paragraph, the Council sought to establish that the Community's competence in the fields covered by the Convention is limited to Articles 15 and 16(2) thereof and does not extend to the fields covered by Articles 1 to 5, 7, 14, 16(1) and (3) and 17 to 19 of the Convention."*[55]

The Commission argued that the third paragraph of the declaration infringes Community law in that it does not refer to all the competences of the EURATOM Community in the fields covered by the Convention and that that provision, and not

53 Dehousse Franklin, The Nuclear Safety Framework in the European Union after Fukushima, Egmont Paper 73 (2014), p. 17; Dehousse saw the following reason for the similarity of structure: "This was meant to distinguish clearly between the objectives and the obligations of the Member States. This, however, has not been fully achieved since there is an 'essential overlap between the scope of application, the definitions and the operational articles."

54 Dehousse, Franklin, ibid. p. 15.

55 ECJ Case C-29/99, para 2 (I-11283).

the whole decision of accession to the CNS, should therefore be annulled based on Art. 146 EURATOM. The European Commission accepted the limited competence of EURATOM but maintained that even if EURATOM would not have a specific right for regulation on the opening and operation of nuclear installations, it is competent concerning the risk resulting from the operation of such installations. The Council kept its positions of national sovereignty as principle where no specific rules were established in the EURATOM treaty and underlined its view that "*no article of the EURATOM Treaty bestows on the Community the competence to regulate the opening and operation of nuclear facilities. That competence was retained by the Member States. The Community would have competence only when it concerns the protection of the general public, and in consequence all the articles of the Convention which concern that protection were explicitly referred to in the above declaration*".[56]

Council and Commission were in agreement that the Community possesses shared competences to take, subject to Art. 15 CNS, the appropriate steps to ensure that in all operational states that the exposure of the workers and the public to radiation caused by a nuclear installation be kept as low as reasonably achievable and that no individual be exposed to radiation doses which exceeds prescribed national dose limits. The same agreement concerned Art. 16(2) of the Convention and its details on the appropriate steps to ensure that in an emergency where radiation could endanger the population, the competent authorities of the States in the vicinity of the nuclear installation are provided with appropriate information for emergency planning and response.

In essence, the Court had to judge whether the Community possesses other competences in the fields covered by the Convention on Nuclear Safety.

The Court underlined that "*the EURATOM Treaty does not contain a title relating to installations for the production of nuclear energy*", and needed to review if an interpretation of the provisions in Title II, Chapter 3 (Health and Safety) EURATOM could lead to a broader competence for the Community then authorised by the Council in its limiting decision.[57]

Interestingly, the Court upheld this definition and exercised of a broad interpretation of EURATOM.[58] The Court referred to its previous judgements under EURATOM. According to the Court, such interpretation had to be carried out in the light of the objective set out in the preamble to the EURATOM Treaty to "*create the conditions of safety necessary to eliminate hazards to the life and health of the public*". The Court took into account the fact that Title II, Chapter 3 of the EUR-

56 Case C 29/99, para 65 (I-11305).

57 Case C 29/99, para 74 (I-11221).

58 Case C 29/99, para 78 (I-11308).

ATOM Treaty implements Art. 2(b), which instructs the Community to establish uniform safety standards to protect the health of workers and of the general public and ensure that they are applied. The Court deduced from this that the objective for protection cannot be achieved without controlling the sources of harmful radiation. But it also outlined that on "*the other hand, the Community's activities in the field of health protection must observe the competences of the Member States defined, inter alia, in Title II, Chapter 3, of the EURATOM Treaty itself*".[59]

In the end, the Council correctly introduced a new reservation to the IAEA, reflecting the Court's judgment.[60]

In the end of 2008, the Commission started to rekindle a legislative process for a nuclear safety Directive.[61] For the Commission, this revised proposal aimed to build on: "*a) the technical work of the Western European Nuclear Regulators Association (WENRA) completed in 2006 for existing nuclear installations, with the participation of all European nuclear safety regulators; b) the principle that only strong and independent regulators can ensure the continued safe operation of the nuclear power plants in the EU; c) enshrining in the Community legislation the principles of the main international instruments available, namely the Convention on Nuclear Safety (CNS), concluded under the auspices of the International Atomic Energy Agency (IAEA), and the safety work carried out by the IAEA*".[62]

---

59 Case C-29/99, para 75 (I-11307).

60 See IAEA, "Declaration by the European Atomic Energy Community according to the provisions of Art. 30 (4)(iii) of the Convention on Nuclear Safety:
"The Community declares that Articles 15 and 16 (2) of the Convention apply to it. Articles 1 to 5,
Art. 7 (1), Art. 14 (ii) and Articles 20 to 35 also apply to it only in so far as the fields covered by
Articles 15 and 16 (2) are concerned. The Community possesses competence, shared with the above-mentioned Member States, in the fields covered by Articles 15 and 16 (2) of the Convention as provided for by the Treaty establishing the European Atomic Energy Community in Art. 2 (b) and the relevant Articles of Title II, Chapter 3 entitled "Health and Safety"."

61 COM(2008) 790 final Proposal for a COUNCIL DIRECTIVE (EURATOM) setting up a Community framework for nuclear safety: "The present draft Directive setting up a Community framework on Nuclear Safety aims at restarting the process of establishing a common EU framework on nuclear safety, by updatingand replacing the Commission proposal for a Council (EURATOM) Directive setting out basicobligations and general principles on the safety of nuclear installations, included in the initialNuclear Safety Package."(of 2003).

62 See EU Commission, Proposal for a Council Directive (Euratom) setting up a Community framework for nuclear safety, Brussels, 26.11.2008, COM(2008) 790 final, Explanatory memorandum, page 2

The final text of Directive 2009/71/EC ultimately adopted by the Council echoed the fact that Europe, with this Directive and despite following the structure of the CNS, remained below the rules on nuclear safety under the IAEA's Nuclear Safety Convention. The ball remained strictly in national courts.

The Directive entered into force on 22 July 2009 and all EU Member States had till 22 July 2011 to implement its contents in to their national laws. Member States' first reports on the implementation of the Directive were to be submitted to the Commission by 22 July 2014.

The Fukushima nuclear accident in March 2011 strengthened the initiatives of the European Commission to increase security and safety standards in the Union, but, again there was no change towards a unified European stance or interest on nuclear. The amending of Directive 2014/87/EURATOM reviewed the EU framework on nuclear safety in the light of the Fukushima accident in 2011 and the findings of the Commission triggered EU stress test exercises.

The amended Directive, which came into force in August 2014 and which had to be transposed into Member States' legislation by 2017, reinforces the provisions of the existing Directive.[63]

Five years after the Fukushima disaster, during a scientific workshop at Cambridge, Ludo Veuchelen, who had worked during his career at the Belgian Nuclear Research Centre and was Chairman of the Working Group on Safety and Regulation of the International Nuclear Law Association, criticised the entanglement of EURATOM with industry and underlined that the organisation had too much power for a single body. He added that a self-fulfilling interest of the civil servants working under EURATOM was propping up the system. He deplored e.g. a lack of democracy (control) and a lack of decisive power and control by the EU Parliament.[64]

---

63 The main objectives envisaged were :a stronger role and clearer independence of the national regulatory authority; the introduction of an EU-wide nuclear safety objective, focusing on accident prevention and risks of significant radioactive releases; a European system of regular topical peer reviews and regular safety reassessments of nuclear installations; more transparency on nuclear safety matters (information and cooperation obligations and involvement of the public) enhancing accident management and on-site emergency preparedness and response arrangements and procedures; promoting nuclear safety culture in the workplace.

64 Report (J. WEITZDÖRFER, Fukushima Five Years On – Legal Fallout in Japan, Lessons for the EU Workshop at the University of Cambridge on 4 and 5 March 2016, p 303, https://www.law.cam.ac.uk/press/events/2015/11/expert-workshop-fukushima-five-years-legal-fallout-japan-lessons-eu .

8 Accessing to the Union without access to EURATOM-Leaving EURATOM without leaving the EU Treaty-BREXIT as game changer?

On 1 January 1973 the United Kingdom (UK) became member of the EURATOM, and of the European Community.

As has been outlined above, EURATOM and the Communities shared some institutions since 1958. In 1967, the so-called Merger Treaty[65] brought together the separate Councils' and Commissions' institutions which the three Communities (the EEC, EURATOM and the European Coal and Steel Community) had kept separated until then.

Since that point, the provisions on the institutions in the EURATOM Treaty have been updated every time the corresponding rules in the EEC Treaty were amended. Those institutional rules are now split between the TFEU and the TEU. The constitutional link is now established in Art. 106a of the EURATOM Treaty, which was inserted by the Treaty of Lisbon.

Subparagraph 1 of this Article reads as follows:

> "1. Article 7, Articles 13 to 19, Article 48(2) to (5), and Articles 49 and 50 of the Treaty on European Union, and Article 15, Articles 223 to 236, Articles 237 to 244, Article 245, Articles 246 to 270, Article 272, 273 and 274, Articles 277 to 281, Articles 285 to 304, Articles 310 to 320, Articles 322 to 325 and Articles 336, 342 and 344 of the Treaty on the Functioning of the European Union, and the Protocol on Transitional Provisions, shall apply to this Treaty."

8.1 Exit à la carte- the process before Lisbon

Over the last decades, the European Commission clearly denied the possibility that any nation in Europe could access full membership of the Union if they do not also accede to the EURATOM Treaty.

In a request for a written answer, MEP Franz Obermayr (NI) in 2010 had asked the EU Commission inter alia as follows:

> "1. According to an Austrian report commissioned by the Austrian Greens before the 1994 referendum on membership of the EU, it would not be necessary for Austria to

65 Treaty establishing a Single Council and a Single Commission of the European Communities (8 April 1965), signed in Brussels on 8 April 1965 entered into force on 1 July 1967.

join EURATOM if it acceded to the European Union. Can the Commission endorse this report from the point of view of European law?(...)
3. Since the Lisbon Treaty entered into force, a new legal situation has arisen as regards the possibility of withdrawing from EURATOM: as Article 49a of the Lisbon Treaty[66] also applies to the EURATOM Treaty, it must be possible de jure for an EU Member State to withdraw unilaterally from it. Does the Commission anticipate that one or more Member States will withdraw de facto? What are the practical arrangements for implementing this clause?"[67]

In his answer, Commissioner Oettinger outlined on behalf of the Commission that no path would have been open for Austria e.g. to access the EU without acceding EURATOM in answering just with "*No.*" to the above question of the MEP. On selective withdrawal from EURATOM the Commission was a bit more detailed in its answer. Overall, the EU Commission is of the opinion that a Member State cannot withdraw just from the EURATOM Treaty under the new provisions of Art. 50 Lisbon Treaty.[68]

---

66 He seems to refer to Art. 50 Lisbon Treaty:
"1. Any Member State may decide to withdraw from the Union in accordance with its own constitutional requirements.
2. A Member State which decides to withdraw shall notify the European Council of its intention. In the light of the guidelines provided by the European Council, the Union shall negotiate and conclude an agreement with that State, setting out the arrangements for its withdrawal, taking account of the framework for its future relationship with the Union. That agreement shall be negotiated in accordance with Art. 218(3) of the Treaty on the Functioning of the European Union. It shall be concluded on behalf of the Union by the Council, acting by a qualified majority, after obtaining the consent of the European Parliament.
3. The Treaties shall cease to apply to the State in question from the date of entry into force of the withdrawal agreement or, failing that, two years after the notification referred to in paragraph 2, unless the European Council, in agreement with the Member State concerned, unanimously decides to extend this period.
4. For the purposes of paragraphs 2 and 3, the member of the European Council or of the Council representing the withdrawing Member State shall not participate in the discussions of the European Council or Council or in decisions concerning it.
A qualified majority shall be defined in accordance with Art. 238(3)(b) of the Treaty on the Functioning of the European Union.
5. If a State which has withdrawn from the Union asks to rejoin, its request shall be subject to the procedure referred to in Art. 49."

67 E-8740/2010, Parliamentary questions 26 October 2010 – Question for written answer to the Commission/Rule 117,Franz Obermayr (NI)

68 E-8740/2010 6 December 201; Answer given by Mr Oettinger on behalf of the Commission: "1. No.... 3. According to Article 50 of the Treaty on European Union, any Member State may decide to withdraw from the European Union in accordance with its own

It seems safe to say that the Union is of the opinion that one can only accede to all treaties, meaning that a departure from the Union membership invokes a dissolution from all treaties including the EURATOM treaty. A 'half divorce' does not appear to be an option.

In September 2002, the Secretariat of the Convention sent a discussion paper to its Praesidium (under former French President Giscard d'Estaing) for information concerning the beginning of the "*simplification procedure*". Some Member States did not want to see the occasion used to reopen discussion on matters that were firmly established: EURATOM was in this respect a particularly sensitive point.

In March 2003, the Praesidium published a paper '*Suggested approach for the EURATOM Treaty*'. This approach explicitly did not think it "*appropriate*" to become involved in an operation "*to amend the EURATOM Treaty substantially*".[69]The Praesidium instead favored amendment of the EURATOM Treaty allowing it to continue to exist independently.

8.2 The European Parliament calling for sunset

In 2002, the European Parliament passed a resolution that included a call for the EURATOM Treaty to be abandoned by 2007. If this proposal were adopted, it would have enabled the Convention and its subsequent Intergovernmental Conference to acknowledge that a fundamental reform of EURATOM was necessary but allowing more time for the process of assessing which parts of the Treaty should remain and in what framework.[70]

In 2003, several Convention Members called for more reform:

constitutional requirements. This Article also applies to the European Atomic Energy Community (Article 106a EURATOM Treaty). The EU and EURATOM share the same institutions, the same budget and staff, and are designed to function together with the same number of Member States. Hence, there appears to be no 'à la carte' withdrawal only from the EURATOM Treaty…".

69 See: Fouquet, Doerte, Froggatt, Antony" Options for the EURATOM Treaty in the framework of a New European Constitution" May 2003.

70 See: Barnes, Pamela, Going forward into the past: the resurrection of the EURATOM Treaty, EUSA 0507 EURATOM Treaty, Tenth Biennial International Conference, Montreal, Canada, May 17th-19th 2007

> "We wish to make the following recommendations to the Convention in relation to the EURATOM Treaty:
> The Convention has already achieved consensus on the following points: There should be a single constitution treaty. The Union should have a single legal personality and a single institutional structure.
> Therefore it is necessary to repeal the EURATOM Treaty. We argue here that it is now appropriate -to abolish the 'special economic zone' that the EURATOM created, and to respect the principles of fair competition and the creation of a level playing field for different energy sources, thereby ceasing to give nuclear energy undue advantages over its rivals. We offer an analysis of the present functions of EURATOM and make proposals concerning their transposition into the Part Two of the Constitution (see Praesidium preliminary draft Constitutional Treaty (CONV 369/02)), while proposing that others be simply repealed."[71]

8.3 The declaration of reform-minded Member States

The Treaty establishing a Constitution for Europe as signed in Rome on 29 October 2004 and published in the Official Journal of the European Union on 16 December 2004 contained an important formal declaration which unfortunately subsequently fell into oblivion, until today. Declaration No. 54 made by the Federal Republic of Germany, Ireland, the Republic of Hungary, the Republic of Austria and the Kingdom of Sweden and annexed to the Final Act of the Intergovernmental Conference which adopted the Treaty of Lisbon, signed on 13 December 2007; reads as follows:

> "Germany, Ireland, Hungary, Austria and Sweden note that the core provisions of the Treaty establishing the European Atomic Energy Community have not been substantially amended since its entry into force and need to be brought up to date. They therefore support the idea of a Conference of the Representatives of the Governments of the Member States, which should be convened as soon as possible".[72]

Brexit may be the right moment to rekindle this initiative.

---

71 Official statement by Convention Members: Marie Nagy, Renee Wagner, Neil MacCormick Contribution to the Convention; THE EUROPEAN CONVENTION –THE SECRETARIAT -Brussels, 18 February 2003-, CONV 563/03 – Contribution 250.

72 Consolidated Version of the Treaty on the Functioning of the European Union, 26.10.2012, Official Journal of the European Union, C 326/47.

8.4 The Reform of EURATOM debate- to be rekindled in the light of BREXIT

Art. 208 EURATOM stipulates that the treaty is concluded for an unlimited period.

The questions of whether EURATOM is an "eternal treaty" or if it can be phased out in view of a changed energy system, or if it should at least be adapted to the current reality of the energy market and be stripped of a certain allegiance to and promotion of nuclear technology may be rekindled by the current BREXIT debate. For such a move, Member States would finally need to commit to a new, specific EURATOM reform convention process.

As the Nuclear Monitor described in 2007: "*Obviously, EURATOM was meant to be for eternity. And its fathers were not even aware of nuclear's eternity problem since there are no explicit provisions for nuclear waste in the EURATOM Treaty!*"[73]

8.5 The Withdrawal option since Lisbon

Art. 50 of the Treaty on European Union sets out the procedure for a Member State to withdraw from the European Union should it wish to do so. It was first introduced by the Lisbon Treaty in 2007.

The corresponding Article, integrating e.g. Art. 50 TEU as also applicable for EURATOM, is Art. 106a EURATOM, introduced under Title III (Institutional and financial provisions)

In general, a Member State must notify the European Council of its intention to leave. The withdrawal agreement must be negotiated in accordance with Art. 218 (3) TFEU.

The UK government in its White Paper on BREXIT in February 2017 outlined very briefly that for the government, invoking Art. 50 TEU would also mean invoking the exit from EURATOM:

> "When we invoke Article 50, we will be leaving EURATOM as well as the EU. Although EURATOM was established in a treaty separate to EU agreements and treaties, it uses the same institutions as the EU including the Commission, Council of Ministers and the Court of Justice. The European Union (Amendment) Act 2008 makes clear that, in UK law, references to the EU include EURATOM. The EURATOM Treaty imports Article 50 into its provisions....As the Prime Minister has said, we want to collaborate with our EU partners on matters relating to science and research, and nuclear energy is a key part of this. So our precise relationship with EURATOM, and the means by

73 See EURATOM: Countries free to step out, Nuclear Monitor Issue: #658, 13/07/2007.

> which we cooperate on nuclear matters, will be a matter for the negotiations – but it is an important priority for us – the nuclear industry remains of key strategic importance to the UK and leaving EURATOM does not affect our clear aim of seeking to maintain close and effective arrangements for civil nuclear cooperation, safeguards, safety and trade with Europe and our international partners. Furthermore, the UK is a world leader in nuclear research and development and there is no intention to reduce our ambition in this important area. The UK fully recognises the importance of international collaboration in nuclear research and development and we will ensure this continues by seeking alternative arrangements."[74]

The United Kingdom's White Paper on the exit from and new partnership with the EU says that the European Union (Amendment) Act 2008 "*makes clear*" that, in UK law, references to the EU include EURATOM. The EURATOM Treaty "*imports Article 50 into its provisions*".

The tasks for the UK when leaving the Union and EURATOM are enormous: Falling out of all European funded or co-funded research is one issue. UK needs to set up a new national regulatory system and re-negotiate contracts to ensure supply of nuclear fuel, ores and fissile materials, not only for nuclear energy and indirectly military use,[75] but also for disrupting time-sensitive supply chains, which transit radioisotopes used in the diagnosis and treatment of cancer. Around 500,000 scans are performed in the UK every year using imported radioisotopes.

The UK does not have any reactors capable of producing these isotopes and at present must rely on a continuous supply from reactors in France, Belgium and the Netherlands.

This situation may force Member States to decide to use the task of separation from the UK in order to straighten out issues in EURATOM by e.g. increased democratisation of EURATOM, the need of a level playing field in the internal energy market, clarity on responsibility over the entire life cycle of an installation, full responsibility of the nuclear industry in case of accidents, phase out of old nuclear power plants in a coordinated and secure fashion, waste management and overall the urgent recognition that there is no common interest in promoting new nuclear energy in the Union.

---

74 HM Government, The United Kingdom's exit from and new partnership with the European Union, February 2017, page 44

75 At present, UK maintains a fleet of four nuclear-armed submarines in Scotland, each carrying 16 Trident missiles. The UK parliament voted in 2016 to overhaul its nuclear forces and for building four new nuclear-powered submarines to carry US Trident missiles armed with modernized nuclear warheads for the next decades. At present UK has approx. 215 warheads; see

9 The subsidy question – or how to shelter any public nuclear investment in a liberalised market?

9.1 The Hinkley Point C State Aid case

At present, Austria, supported by Luxemburg, pleaded to the European Court[76] to annul a positive state aid decision of the European Commission authorising substantial state aid for a new nuclear power plant at Hinkley Point in Somerset.[77] The case is now in appeal before the European Court of Justice (ECJ). There are several grounds on which Austria is fighting the decision of the European Commission. I will concentrate on Austria's argument on the notion that the promotion of nuclear power is an objective of Common Interest under EURATOM.

After the preliminary examination of the state aid package, the Commission had doubts as to the legality of the aid and opened a formal investigation.[78] But in October 2014, almost a year later, after intense back and forth between many stakeholders and the Commission (e.g. from EU independent power producers many using renewable energy as well as traders[79]) the, exiting, Commission under President Barroso gave a positive decision.[80]

The most pertinent argument of the European Commission to allow the UK subsidy regime, its Feed-in Tariff option via the so-called Contract for Difference for Nuclear, combined with state guarantees, was in its view due to the fact that, EURATOM with its technology promotion approach in Article 2, describes a common European interest in the promotion of building new nuclear power stations.

---

76 Austria v Commission Case T-356/15

77 State Aid procedure SA. 34947 (2013/C) (ex 2013/N) – United Kingdom- Investment Contract (early Contract for Difference) for the Hinkley Point C New Nuclear Power Station

78 Commission Decisions State aid SA. 34947 to initiate the formal investigation procedure, Brussels, 18.12.2013 C(2013) 9073 final

79 In view of EURATOM, their main arguments are summarised by the Commission as follows: "Several parties commented that the aid measures are incompatible with the Altmark criteria, whereby electricity generation would be a standard economic activity and thus nuclear energy should compete with other electricity sources in a liberalised internal electricity market; the measure lacks an objective of common interest; there appears to be no objective criterion for justifying the duration of 35 years; it treats differently nuclear power and renewable energy sources; it is based on unknown parameters and there is a lack of a cost-benefit analysis." See Decision Brussels, 08.10.2014C(2014) 7142 final cor

80 Press Release IP/14/1093, full text of decision published in 2015:O.J. L 109/44 of 28.04.2015

In view of the above analysis and in view of the arguments put forward by Austria and supported by Luxembourg before the Court, this Commission's view does not reflect the reality of the development under the EURATOM treaty and its limitations versus the development of an internal energy market. The European Commission takes the position that the EURATOM Treaty establishes in Art. 2(c) that the Community shall "*facilitate investment and ensure, particularly by encouraging ventures on the part of undertakings, the establishment of the basic installations necessary for the development of nuclear energy in the Community*" and points out to Art. 40 EURATOM which envisages the Community publishing of illustrative programs "*to stimulate investment, indicating production targets*".

The Commission correctly describes its obligation under Art.107 TFEU to investigate aid granted by Member States that distorts competition or threatens to do so. In addition, especially "*in the context of liberalised and increasingly competitive markets, the role of State aid control is increasingly important in EU electricity markets. The commitment of the European Union to promote investment into nuclear must be carried out in ways which do not distort competition.*"

In the case of the UK aid mechanisms for Hinkley Point C, the Commission concluded that no distortion exists.[81] The Commission seems to see itself bound by the objectives of EURATOM: "*The Commission however accepted that the measure was in line with the EURATOM Treaty. As recognised in past Commission decisions, the EURATOM Treaty aims at creating the "conditions necessary for the development of a powerful nuclear industry, which will provide extensive energy sources." This objective is further reiterated in Art 1 of the EURATOM Treaty, which establishes that "it shall be the task of the Community to contribute to the raising of the standard of living in the Member States (...) by creating the conditions necessary for the speedy establishment and growth of nuclear industries." On this basis, the EURATOM Treaty establishes the EURATOM Community, foreseeing the necessary instruments and attribution of responsibilities to achieve these objectives. The Commission must ensure that the provisions of this Treaty are applied*".[82]

The European General Court recently decided on the annulment plea by Austria. and with its judgment in effect underlying that there exists a compelling case to reform the EURATOM treaty and to clarify that there is no common interest to promote further nuclear power projects in the Union.

It was not only Austria and Luxembourg that addressed the Court over a decision by the EU Commission. Several German communal energy utilities as well as an Austrian and a German Green electricity producer and trader introduced an

81 See Decision Brussels, 08.10.2014C(2014) 7142 final cor

82 See Decision, Brussels, 08.10.2014C(2014) 7142 final cor, Rn 394 cons.

annulment procedure before the European General Court.[83] The Court declared their case as inadmissible, in line with its (in the majority of cases) restrictive view on access to justice for applicants concerning Commission decisions on state aid directed to a Member State granting aid to a competitor in the internal energy market. This adds to the effect of nuclear being a sheltered species within an internal EU energy market.

The General Court in its judgment gave full support to the Commission for its decision), especially on its points concerning EURATOM Treaty provisions as legal bases for the justification of a common European interest under Art. 107 (3) (c) TFEU. The General Court supported the assessment of the Commission that the state aid measure contributes to the long-term security of supply in particular "based on capacity forecasts and the role which Hinkley Point C's supply of electricity will play when it is expected to start operating".

The General Court decided that the Euratom Treaty would underline the promotion of building and operating new nuclear power plants in the general interest, and that the authorisation of State aid within this argumentation would also be applicable by a Member State even if that public interest of one Member State is not shared by all the Member States.[84]

The General Court followed and supported the views of the Commission and the United Kingdom that the Euratom Treaty gives the legal basis for state support for the construction and operation of nuclear power stations and this specific UK aid package considering that a specific nuclear market failure would allow for aid mechanisms in the common interest.

It remains unclear whether the General Court is trying to enlarge the current established definition of "common" or European Union interest under Art. 107 (3) (c) TFEU in suggesting that the Euratom Treaty, via a quite lenient interpretation of Art. 2 (c) in Recital 97, equates the building of nuclear power plants as in the common interest. This view of the General Court following the Commission seems a quite novel interpretation of Art. 2 (c) which from its wording does not at all tackle the promotion of investment into nuclear power plants but rather seeks to define the performance of the Euratom community as to "facilitate investment and ensure, particularly by encouraging ventures on the part of undertakings, the

---

83 See Ordinance/Beschluss, Rechtssache T-382/15 Greenpeace Energy eG mit Sitz in Hamburg (Deutschland) und die weiteren im Anhang namentlich aufgeführten Klägerinnen Prozessbevollmächtigte: Rechtsanwältinnen D. Fouquet und J. Nysten

84 For the following and an analysis in detail see: D. Fouquet, The Hinkley Point C Judgment of the General Court in view of a changing internal electricity market RELP Volume 9 Issue 1 2018, p. 35 cons..

establishment of the basic installations necessary for the development of nuclear energy in the Community". The Euratom Treaty itself then clarifies what is meant with the concrete execution of this investment policy under Art. 40. Specific instruments such as the illustrative programmes under Article 40 and the publication of projects under Article 44 are examples but there is no mention to the promotion of investment of nuclear power plants. The building of nuclear power plants remained, from the first day of the Euratom Treaty, solely in the competence of a sovereign national State. The Euratom Treaty, one might say, facilitates the groundwork for nuclear research and safety policies but certainly does not establish a Community project of investment facilitation into nuclear power plant builds.

9.2 The European Commission and the Hungarian Nuclear Build Case

In the same line as the Hinkley Point case, the Commission has also given a green light and recently accepted a state aid package by Hungary for a new Russian-built nuclear power reactor in Paks, Hungary. This is all the more astonishing as Russian involvement in a large part of European electricity could see, in principle, the same security concern arise that the Union has in the field of gas supply from Russia. A situation which has led to the adoption of serious safety regulations in the gas field in the Union. Moreover, the situation around Rosatom and waste export issues from Hungary to a country which seems set to not allow open and access to its waste processing and storage facilities adds significantly to security considerations. Concerns on security of supply lead the European Union to introduce a specific energy security strategy and one of the major reasons for such was its on Russia.[85]

The Russian Federation and Hungary signed in January 2014 a specific intergovernmental agreement (IGA) on a nuclear programme.[86] Based on the IGA,

---

85 The Commission underlines the security issues and needs for a strategy in view of " geopolitical events, i.e. the crisis in Ukraine. Temporary disruptions of gas supplies in the winters of 2006 and 2009 already provided a wake-up call for the EU, underlining the need of infrastructure development, increased cooperation and of a common European energy policy. Since then, the EU has done a lot to strengthen its energy security in terms of gas supply. However, the work is not completed yet and further steps are needed. ", see European Commission, Memo, Questions and answers on security of energy supply in the EU, 28 May 2014

86 Agreement between the Government of the Russian Federation and the Government of Hungary on cooperation on peaceful use of nuclear energy, concluded on 14 January 2014 and ratified in Hungary by Act II of 2014 of the Hungarian Parliament (2014. évi

both countries shall cooperate in the maintenance and further development of the current Paks nuclear power plant (Paks NPP). This includes the design, construction, commissioning and decommissioning of two new power units 5 and 6 with VVER (water-cooled water moderated) type reactors with a combined capacity of at least 1 000 MW in addition to the existing power units 1–4. The operation of units 5 and 6 is intended to compensate for the loss in capacity when units 1–4 (2 000 MW altogether) retire. Hungary submitted that units 1–4 will be in operation until the end of 2032, 2034, 2036 and 2037 respectively, without envisaged prospect of further lifetime extension.

However, the Commission seems doomed to repeat the Hinkley Point reasoning for the Paks State aid package:

The Commission again underlined that under the EU Treaties, Member States are free to determine their energy mix and have the choice to invest in nuclear technology. The Commission's role would only be to ensure that when public funds are used to support companies, this is done in line with EU state aid rules which aim to preserve competition in the Single Market.

The Commission's state aid investigation defined and concluded that in view of a waver by the Hungarian state to ask for a higher return of its investment than a private investor would ask for, the mechanisms constitutes State aid within the meaning of Article 107(1) TFEU. These rules require state aid to be limited and proportionate to the objectives pursued in order to be approved.

According to the Commission, Hungary had proven that the measure avoids undue distortions of the Hungarian energy market. In particular, it has made a number of substantial commitments to limit potential distortions of competition[87]:

II. törvény a Magyarország Kormánya és az Oroszországi Föderáció Kormánya közötti nukleáris energia békés célú felhasználása terén folytatandó együttműködésről szóló Egyezmény kihirdetéséről), quoted in EU Commission decision COMMISSION DECISION (EU) 2017/2112 of 6 March 2017, published in Official Journal L 317/45 on 1st of December 2017 on the measure/aid scheme/State aid SA.38454 — 2015/C (ex 2015/N) which Hungary is planning to implement for supporting the development of two new nuclear reactors at Paks II nuclear power station

87 The main observations and arguments for the decision of the Commission when it comes to proportionality were as follows:
To avoid overcompensation of the operator of Paks II, any potential profits earned by Paks II will either be used to pay back Hungary for its investment or to cover normal costs for the operation of Paks II. Profits cannot be used to reinvest in the construction or acquisition of additional generation capacity. To avoid market concentration, Paks II will be functionally and legally separated from the operator of the Paks nuclear power plant (the incumbent MVM Group) and any of its successors or other state-owned energy companies. To ensure market liquidity, Paks II will sell at least 30% of its total electricity

In its argumentation during the investigation procedure, Hungary even compared the EURATOM Treaty with the former European Coal and Steel (ECSC) Treaty on the basis that they both are of a sectoral nature and that the ECSC Treaty contains a far-reaching prohibition against State aid which was, in practice, aligned with Art. 107 TFEU by virtue of Art. 67 and Art.95 of the ECSC Treaty. Hungary stated that in applying the rules on State aid laid down in the TFEU the Commission would misconstrue the regulatory goal pursued by the drafters of the EURATOM Treaty, which lacks any specific State aid provisions.[88]

Many who gave comments during the full investigation procedure (Austria, IG Windkraft, Oekostrom AG, Greenpeace Energy and others) outlined once again, in this spectacular state aid case for nuclear power, that subsidising the construction and operation of new nuclear power plants is not provided for under the principles laid down in Article 107(3) TFEU as being compatible with the internal market. Nuclear power was clearly defined as not being a new, innovative or sustainable technology for electricity generation capable of contributing to achieving the EU goal: increasing the proportion of energy generated by renewable technologies. Moreover and once again, as especially underlined by Austria, neither Article 2(c) nor Article 40 EURATOM would allow for the promotion of new nuclear investments to be considered as an objective of common interest due to the fact that no common interest within the meaning of Article 107(3) TFEU could be derived or integrated from the EURATOM Treaty. In addition, it was stressed by Austria and others, also once again, that such an objective would be in conflict with other principles of the Union under TFEU, namely the precautionary principle under Article 191 TFEU and the sustainability principle of the Union.

Again, the Commission upheld the applicability of State aid rules in support of nuclear power; but remained somewhat true to its view that "*in fact, whilst Article 2(c) of the EURATOM Treaty creates an obligation on the Union to facilitate investments in the field of nuclear energy and Article 40 of the EURATOM Treaty obliges the Union to publish illustrative programmes in order to facilitate the development of nuclear investments, the EURATOM Treaty does not foresee any specific rules to control the financing, by a Member State, of such investments. According to Article 106a (3) of the EURATOM Treaty, the provisions of the TFEU shall not derogate from the provisions of the EURATOM Treaty*".[89]

---

output on the open power exchange. The rest of Paks II's total electricity output will be sold by Paks II on objective, transparent and non-discriminatory terms by way of auctions, see. European Commission, press release of 6 March 2017- IP/17/464

88 See published EU Commission decision, in Official Journal L 317/45, para 122.

89 See published EU Commission decision, in Official Journal L 317/45 para 277.

It seems that (and in case the European Court does not accept Austria's plea) the need to reform EURATOM remains a pressing necessity. Without a reform of EURATOM, there cannot be a level playing field in the Union for other energy technologies and modern energy services with renewable energy producers being especially concerned.

10 Conclusion

From the above, it is clear that EURATOM never was a harmonising treaty for a joint common approach and objective. It was, from the beginning hampered, by interferences from the nuclear weapon state France and later the UK, when joining EURATOM, in order not to hinder their own national interests in nuclear weapon planning and development.

The EURATOM Treaty and its original objective of promoting and guaranteeing nuclear energy development no longer corresponds to modern reality and is now completely outdated.

The EURATOM Treaty does not fit in the actual internal energy market driven by consumers' interests. The technology it was established to support is no longer economically competitive in electricity generation. There is now a multitude of players to guarantee security of supply without the risks and internalised burdens associated with nuclear energy production, storage and radioactive waste.

Without heavy state aid and guarantees, new nuclear power has no leg to stand on.

After more than 60 years of industrial production of nuclear power plants, now is the time to say goodbye and to ensure the safe dismantling of all obsolete nuclear power stations as well as safe final storage of all waste in and within the European Union.

European Nuclear policy will, in the coming years, have to manage nuclear risks, decommissioning of reactors and nuclear waste management. This is not covered by the treaty.

As it does not appear legally feasible for EU Member States to exit EURATOM without leaving the EU, there is an urgent need to revise the treaty in accordance with the modern environmental, social and economic objectives of the European Union.

A paramount task for Europe!

In an EU with a nuclear legacy of hundreds of old nuclear power plants, an understanding between those Member States that still want nuclear energy as part of their energy mix and those who are phasing out or never had nuclear energy as a source may be appropriate. This agreement, in light of a reform of the EURATOM treaty, should see the creation of a European dismantling and safe storage support mechanism which integrates the whole life cycle approach and full responsibility of the nuclear power producers. It should establish a progressive European liability regime and the reform needs to remove any mention from the archaic preamble of the EURATOM treaty of a promotional objective while fully opening the treaty to democratic scrutiny and a legislative process that are totally commonplace in a parliamentary democracy. After such reform of EURATOM, any new nuclear power will need to face the full market conditions and would not be able to hide behind its ancient subterfuge. After more than 60 years, there is no time better than the present.

Economic Management of Future Nuclear Accidents

Tomas Kåberger[1]

Abstract

Nuclear core melts with large emissions of radioactive substances are not paid for by nuclear power companies but by the victims and by taxpayers. This subsidy is often the result of legislation with that purpose.

Experience shows that the relative frequency of such accidents is several orders of magnitude larger that the risk estimates publicised by the nuclear industry and nuclear proponents.

This chapter describes the how the problem was created in order to make the nuclear development economically possible. In the end, it is described how a market may be created based on compulsory paying capacity, possibly provided via catastrophe bonds that would internalise many costs of accidents. At the same time, such regulations would provide a market evaluation, by responsible actors, of the nuclear risk costs.

1 Tomas Kåberger, Chalmers University of Technology, Göteborg, Sweden, tomas.kaberger@chalmers.se

R. Haas et al. (Eds.), *The Technological and Economic Future of Nuclear Power*, Energiepolitik und Klimaschutz. Energy Policy and Climate Protection, https://doi.org/10.1007/978-3-658-25987-7_9

1 Creating the problem

As private cars were introduced and became more powerful during the 20th century they created a social problem: People who could readily afford to buy a car, were sometimes unable to pay the costs of the damages they caused when crashing into other cars, people or houses.

The frequency and cost of car accidents, however, were small enough for car drivers to manage the costs by sharing them. Legislation was introduced making traffic insurance compulsory, ensuring that compensation could and would be paid even for very unlikely large accidents. Today, this compulsory payment capacity may be in the order of 1 000–100 000 times the price of a car in many countries.

Regarding nuclear reactors, accidents pose a similar problem: Companies may well afford to pay for the construction of a nuclear reactor. Still, they are not able to pay for the damages after a nuclear core has melted and radioactive materials escaped from a broken containment.

This was understood already in the mid 20th century, after the US report on consequences of a major reactor accident was published in 1957. [Beck et al 1957]

In retrospect, this report appears naive as the understanding of the risks of ionising radiation was not well developed. Still, the economic impact of an accident was understood as a so large liability that no private investor would be interested in investing in a nuclear power plant. At the same time there was a common notion that nuclear power was desirable. Therefore, according to dominant actors, economic market conditions should be created to make nuclear investments appear profitable.

Jasper [1989] describes how the nuclear industry asked a group of legal experts to provide a proposal for a legislation that would make nuclear power profitable by socialising the costs of potential accidents.

For car accidents a solution was to share the costs among all car users via mandated insurance. In the nuclear case, a tiny part of the liabilities were shared, but it was necessary to relocate most of the costs outside the nuclear industry.

The report by Murphy et. al [1957] and the resulting Price-Anderson Act have served as blueprints for nuclear accident legislation in most countries with privately owned nuclear power reactors, as well as for international conventions[2]. The key elements are:

1. In case of a nuclear reactor accident the operator of the plant is the only one actor that can be held liable.

2 Most notably the Paris convention on third party liability in the field of nuclear energy. See IEA 1989.

This reduces the cost of building a reactor, as all suppliers are relieved of the risk that faulty equipment or mistakes during construction may imply economic costs to them, were they to cause a reactor accident. Without this component in the legislation, large companies would refuse to deliver because they either would have to accept high insurance costs, or they would run the risk of bankruptcy if they had delivered something that later was declared the cause of an accident.
2. The economic liability of the operator is strictly limited to an amount far below the potential costs of a major accident.

This component in the legislation has two important aspects. Without a limited accident liability the operator would face bankruptcy in case of a major accident. To banks and others, considering lending money to an operating company, that risk for bankruptcy would be a reason to increase interest rates to compensate for their risk of losing their money in case of an accident. Increased interest rates would lead to higher costs of nuclear power.

An important consequence of this legislation is that the victims of large accidents are without the right to compensation. It is sometimes assumed that the governments, i.e. all people paying taxes in the country where the reactor is operating, take over the full liability. However, this is rarely explicitly stated, as the size of potential accidents is large also in relation to government budgets. It follows that the actual victims of nuclear accidents will carry a large part of the burdens, for example as loss of their habitat.

Deliberately externalising environmental costs is an extraordinary measure in economic policy. Arguments in support of this kind of legislation sometimes illustrate why. In Sweden, the law was based on a government commission that justified this subsidy by arguing "As seen from the statements above, it is necessary to utilise nuclear power – at any cost – if we do not want to accept a lowered standard of living". [SOU 1959:34 p.25]. (In Swedish: *"Såsom framgår av det ovan sagda blir vi nödgade att ta atomkraften i anspråk – kosta vad det kosta vill – om vi inte vill acceptera en standardsänkning."*)

We may speculate if this blunt expression of illogical economics is a result of the commission secretary intentionally leaving a clear signal to coming generations, or an expression of enthusiastic nuclear zeitgeist beyond rational economic thinking.

Many would claim that this was a decision driven by nuclear weapon ambitions and therefore outside the realm of economic logic.

Regardless of the motives, legislation limiting the liabilities of nuclear actors appears as decisive energy policy. The law had immediate and significant economic implications for the competitiveness of this particular energy technology. It was designed to, and served as, a substantial subsidy.

In the half century since this legislation was spreading over the world, evidence has accumulated regarding both the frequency of large accidents and the health impact of ionising radiation.

2 Experience

The frequency of core melts and of accidents with emissions far exceeds earlier expectations. We now count around[3] 10 nuclear power reactors whose operations were closed after full or partial core melts[4]. Out of these core melt events, Chernobyl and three Fukushima reactors have resulted in significant emissions of radioactive materials.

This experience has come during a period where IAEA estimate the accumulated number of reactor operation years is around 17 000. [IAEA 2017]. So far, the relative frequency of core melts has been one in a couple of thousand reactor-years, while core melts with significant radioactive contamination has been in the order of one in 5 000 reactor-years.

Regarding ionising radiation, the series summary reports provided by the International Commission on Radiation Protection, ICRP, now expects roughly 10 times as many cancer cases from a certain collective dose of ionising radiation compared to the estimate in the 1950s.

The scientific issue of health effects of ionising radiation is one of the most intensive scientific controversies as the economic stakes involved are high. One reason is the industrial interests to defend the privilege of legally limited liability. Another, the potential liabilities and political implications of the exposure of people to radiation from nuclear weapons testing[5]. The latter became an issue in

3 The words "around" is used as there are other reactors closed after core damages, while not immediately after, and there are reactors whose status as commercial power reactors are unclear.

4 While definition of core melts as well as which reactors to include among power reactors may be challenged, a rough list may include St Lucens GHHWR in 1966, Bohunice A1 in 1977, TMI 2 in 1979, Chernobyl 4 in 1986, Greifswald 5, and finally three reactors at Fukushima Dai-ichi in 2011. [World Nuclear Association, 2016]. Cochran [2011] also include The Sodium Reactor Experiment, Sankt Laurent A-1 and A-2 as well as Chapelcross-2, but not the Bohunice A1, and ends up with 11 power reactor core melts. There are also non-power producing reactors where core melting has occurred.

5 This is the same kind of scientific controversy as those previously experienced with tobacco smoking and asbestosis.

1981 when a UN committee estimated nuclear weapon testing being responsible for 150 000 premature deaths[6].

The costs of severe reactor accidents may be in the order of several hundred or even several thousand billions EUR or USD. Costs resulting from such accidents include evacuation of people, health effects among people despite evacuations, costs of limiting emissions of radioactivity, decontamination as well as some sort of decommissioning of the remains of reactors and waste management.

Chernobyl and Fukushima experiences show these costs are not within the capacity of even the largest nuclear operators in the world.

In the Chernobyl case, most of the consequences occurred in Belarus and the Ukraine. As the costs were initially paid out of the Soviet Union budget, most of the resources came from the Soviet republic of Russia. When Russia withdraw from the Soviet Union, they could stop paying, which provided an economic incentive for Russia to dissolve the union. A compilation of incomplete cost assessments are provided by Samets & Seo 2016. Health effect due to low-level exposure is not included in the cited studies.

Ukraine had difficulties managing the costs of safeguarding the Chernobyl site. Hence, taxpayers in other European countries have been forced to assist in financing the construction of a shelter over the damaged reactor, which would reduce the distribution of radioactivity over the coming 100 years. In order to protect their own interests, the potential victims have been made to pay for prevention.

Fukushima is still in a state where it is impossible to estimate a limit to the costs as the long term strategies to control the radioactivity are not yet fixed. Suzuki et al (2016) estimate costs excluding health effects at almost 100 trillion yen, or 1 trillion US dollar.

Using experience to get a possible order of magnitude of the real risk cost of nuclear reactors, we may combine 100 billion -1 trillion US dollar per accident with a relative frequency of one in 5 000 reactor-years which would give an average risk cost of 20–200 million dollar per reactor-year. With reactors producing 5–10 TWh/year this would provide a range from 0.2-4 cent/kWh. Despite enormous uncertainties the accumulated experience suffices to dismiss the idea that the risk costs are "clearly negligible".

The argument that externalising the costs of accident risks is acceptable because the risk costs are negligible is not consistent with experience.

The question is instead how to manage this risk costs in a rational way.

6 United Nations Comprehensive Study on Nuclear Weapons 1981, p 86, § 260

3 Managing nuclear risk

Often, a risk of lost property that is large for the actor exposed, is managed by an insurance system. When there is a large economic risk for third parties, such as with car accidents, the insurance system is made compulsory[7].

In an insurance market, voluntary actors use experience and other relevant information to reach a price for the risk, and the actors offering to share the risk, normally insurance companies, become economically liable for the outcome.

For nuclear accident risks, the analogy with the compulsory car insurance system is, however, not immediately applicable. The reason is that insurance companies are not able to cover damages of the magnitude of nuclear accidents. With less than 500 nuclear power reactors in the world, the owners of reactors cannot finance an insurance system by themselves. This argument has been used to claim that a compulsory insurance is not possible.

However, it is possible to find insurance systems that do not rely on traditional insurance companies. Such systems have been developed to manage natural disasters and other rare catastrophes. Radetzki & Radetzki [2000] describe how catastrophe bonds[8] could be used for nuclear accident liabilities.

Operators could be forced to pay for catastrophe bonds to collect capital enough to compensate victims of large accidents. Such bonds would be normal interest bearing bonds, were those who provide capital are offered an extra premium on the condition that the bond value is used to cover the cost of the catastrophe in question if realised. Using the catastrophe bond market opens access to relevant magnitude of capital to cover also large nuclear accidents, and the premium paid by the reactor operators would provide a measure of the accident risk.

Creating the market would include specifying the kind of costs to be covered caused by an accident. Property costs, compensations for evacuations are directly payable to the victims. What may be significant costs for health effects will not be individually identifiable but may occur as large economic burdens on health, or medicare, insurance systems. This liability may then be payable to the health insurance systems rather than to individuals, even when individual suffering may be a significant part of the economic impact.

7 The reason for a compulsory insurance is that the responsible actor may lack paying capacity. Electric Power Companies may be very large, but no EPC is large enough to manage a large reactor accident. Even more important is that EPCs often organise nuclear power plants in separate limited liability companies to avoid liability for other risks including future waste and decommissioning costs.

8 For a definition of catastrophe bond, see Investopedia 2017.

Another relevant issue in the regulation is how large the required paying capacity shall be. This may prove an important issue if the paying affects every provider of capital for any accident. As described by Radetzki & Radetzki I have also earlier suggested that a solution that reduces the importance of this capacity decision I have presented is to make the catastrophe bonds consecutive rather than exchangeable. Consecutive bonds would be specified so that one bond may be used to cover cost between 1 000 M USD and 1 001 M USD, while another bond would only be used if the accident cost more than 1 001 M USD to cover the next million. The former may be expected to cost more than the latter, as the probability of the larger accident is lower than the probability of the smaller.

Under such a regime of consecutive bonds the importance of the level of required capacity may be reduced as the cost of marginal bonds would be small when reaching trillions of USD.

By introducing compulsory arrangements of this kind, a market value would be created for nuclear accident risks. Political debates on probabilities and effects would be replaced by professional assessments made by parties who would have to assume economic responsibility for their assessments. This is likely to result in more scientific assessments of risks and their economic value.

4 Possible consequences, and imperfections

The obvious consequence of a compulsory insured paying capacity for third party liabilities, is that the operators would pay a market price of the risk costs. This would reduce their competitiveness in relation to other energy technologies.

Still, this proposal is embraced by many pro-nuclear spokespersons as they expect the assessment in the market will be that most nuclear reactors are so safe that the risk costs are small. Paying a small amount to cover the full accident liability would be a low price to do away with the political arguments of accident risks.

At the same time nuclear critics find the proposed market solution palatable as they are convinced the price for nuclear catastrophe bonds would be so high reactors would be out-competed by modern renewable energy alternatives.

There have been some critical arguments. Swedish nuclear opponent Eija Liljegren-Palmær has argued that the nuclear industry has such an advantage regarding access to information that they would be able to mislead representatives of financial institutions, such as pension funds, to accept theoretical calculations with too small premiums. Instead of misleading politicians representing tax payers, they would mislead those managing the same tax-payers pension funds.

Such a risk cannot be disregarded. When Compulsory, government controlled pension funds are involved the risk is similar.

However, often pension funds are managed without pro-nuclear policy objectives. Further, markets would allow individuals or groups to withdraw their money from funds they see exposing them to risks. Thus there would at least be more of voluntary risk taking then after parliamentary majority decisions.

However, there are proponents who would say that the hysteria around radiation is making the assessments biased so as to over-estimate risk costs. Here, the response is that the nuclear industry only has to find a sufficient number of financial actors who see the realistic risks to buy the required paying capacity via the bond market. The required capacity is likely to be small compared to potential global capital markets.

There is a risk in both directions that lack of information or ignorance may result in decisions resulting in misleading pricing of the accident risks. However, some may lose a small potential income as they over estimate risks. To others entering the marked via misinformed decisions may lose their assets. Still, the victims and taxpayers are not affected, as those who voluntarily took on the risks will carry both the economic liability and guarantied paying capacity.

For electricity consumers a demand on full paying capacity would not have a significant effect on electricity prices. In the short term this is true as nuclear power rarely provide the marginal, price setting generation. In the longer term, the set of different generation technologies now available at similar costs is providing many alternatives, and in most parts of the world nuclear is clearly not an option when considering new capacity.

If the market evaluation of the risk turns out low enough, reactors would continue to operate and the insurance would not affect short term marginal cost and therefore not the electricity price.

If, however, the risk cost was high and reactors forced out of the market, the cost of renewable energy has lately proven so low that the price effect would also be small.

As described elsewhere, [Kåberger 2001], this reform may come together with the internalisation of fossil fuel externalities. Making fossil fuelled generators pay for local air-pollution and greenhouse gas emissions would initially have a greater effect on electricity prices as fossil fuelled power often constitute price-setting marginal generation. An important reason to do both is that nuclear power under a carbon pricing regime may be ably to continue operating with higher risk costs than in a situation without pricing emissions from fossil fuelled generators in the short term.

5 Conclusion

Experience shows that currently prevailing regulation of nuclear accident liabilities is a significant subsidy to nuclear power. As new, renewable energy supply is economically competing with existing nuclear power, re-regulating accident liability will have importance for the rate of decommissioning nuclear reactors and therefore also for the number of reactor accidents in the future.

Experience of large reactor accidents with significant emission of radioactivity has shown that even the most endowed reactor owners are unable to compensate the victims. Governments concerned also have had great difficulties in meeting the need for compensation. In the Chernobyl case tax payers of other countries have even stepped in to subsidise the mitigation of the consequences.

The experience of reactor owners creating significant costs they are unable to compensate, has proved to be a problem similar to when the owners of cars caused damages beyond their paying capacity.

In the current era of competitive electricity markets this opportunity appears all the more urgent for markets to find the efficient rate of change of power supply.

As the catastrophe bond concept and market was developed by the end of the 21st century it has become possible to address the problem of large nuclear accidents by making it compulsory for anyone operating a nuclear reactor to be able to pay for the consequences of large accidents. As with cars, the required paying capacity may be orders of magnitude higher then the cost of building a reactor, but the societal problem of rolling-off external costs on citizens and communities around nuclear accident sites can be addressed to some degree.

References

Beck, C.K. et al., 1957: Theoretical Possibilities and consequences of major accidents in large nuclear power plants – a study of possible but highly improbable, were to occur in large nuclear plants. United States Atomic Energy Commission. Washington.

Cochran, Thomas B., 2011: Statement before the US senate on the Fukushima Nuclear Disaster and its Implications for 'us Nuclear Power Reactors. https://www.energypolicyblog.com/2011/04/27/reassessing-the-frequency-of-partial-core-melt-accidents/

Investopedia, 2017: Catastrophe Bond. http://www.investopedia.com/terms/c/catastrophe-bond.asp

Jasper, J., 1990: Energy and the State: Nuclear Politics in the United States, Sweden and France. Princeton University Press.

IAEA, 2017 https://www.iaea.org/pris/ viewed 2017–03-11.

Kåberger, T., 2002: Swedish Nuclear Power and Economic Rationalities. Energy and Environment, Vol. 13, No. 2 pp 185–200.
Murphy, A.W. et al., 1957: Financial protection against atomic hazards. Atomic Industrial Forum, Inc. New York.
OECD, 1994: Liability and Compensation for Nuclear Damage – An International Overview. OECD, Paris.
Radetzki M. & Radetzki M., 1997: The Liability of Nuclear and Other Industrial Corporations for Large- Scale Industrial Damages Journal of Energy & Natural Resources Law, Vol. 15, No. 4.
Radetzki M. & Radetzki M., 2000: Private Arrangements to Cover Large-Scale Liabilities Caused by Nuclear and Other Industrial Catastrophes. Geneva Papers on Risk and Insurance, Vol. 25 No 2, April 2000.
Samet, J.M. & Seo, J., 2016: The Financial Costs of the Chernobyl Nuclear Power Plant Disaster: A Review of the Literature. http://www.greencross.ch/wp-content/uploads/uploads/media/2016_chernobyl_costs_report.pdf
Suzuki, T et.al., 2016: Aiming at a Low Carbon Society in Japan by 2050: Impact of the Fukushima Nuclear Accident and COReduction Target Economics of Energy & Environmental Policy, Vol. 5, No. 1.
IEA, 1989: Paris convention on third party liability in the field of nuclear energy. https://www.oecd-nea.org/law/pubs/1989/154-paris_third_party_brussels_conv.pdf
SOU, 1959:34 Om atomansvarighet. (On nuclear liability)
SOU, 1994:107/108 Kärnbränslefondsutredningen
United Nations, 1981: Comprehensive Study on Nuclear Weapons. Department of Political and Security Council Affairs. United Nations Centre for Disarmament. Report of the Secretary-General. https://unoda-web.s3-accelerate.amazonaws.com/wp-content/uploads/assets/publications/studyseries/en/SS-1.pdf
WorldEnergyAssociation, 2016: Decommissioning Nuclear Facilities. http://www.world-nuclear.org/information-library/nuclear-fuel-cycle/nuclear-wastes/decommissioning-nuclear-facilities.aspx, as of November 2016.

Corporate Policies of the Nuclear Vendors

Stephen D. Thomas[1]

Abstract

The nuclear reactor supply industry, once seen as an essential component of diversified companies with an electrical engineering capability, is now seen in Europe, USA and Japan as a risky niche business for specialist companies. Vendors from Russia and China now appear likely to win the vast majority of new reactor orders although in both cases, their technologies have not been reviewed by experienced Western regulators. Russia may not have the financial strength to back up its large order book while China has yet to win orders in open export markets.

1 Stephen Thomas, University of Greenwich, United Kingdom, stephen.thomas@greenwich.ac.uk

R. Haas et al. (Eds.), *The Technological and Economic Future of Nuclear Power*, Energiepolitik und Klimaschutz. Energy Policy and Climate Protection, https://doi.org/10.1007/978-3-658-25987-7_10

1 Introduction

When nuclear power began to be commercially exploited in the 1950s, it was seen by the major 'national champion' diversified engineering companies as the technology of the future and therefore a key capability to acquire. In the USA, such companies included Westinghouse and GE, in Germany Siemens and AEG, in Japan Hitachi, Mitsubishi and Toshiba and in France Compagnie General d'Electricité (CGE) and Empain Schneider. Russia also developed its own technologies but it had little impact outside the Soviet Republics and the Comecon countries until the collapse of the Soviet Union in 1990. China has only become a force in nuclear markets since 2008, while, for various reasons, India, despite pursuing nuclear power since the 1960s has not built a competitive nuclear reactor supply industry.

By 2018, the picture was very different to that of the 1950s. Nuclear power had consistently failed to meet expectations and was increasingly seen as a technology that was not central to corporate ambitions. Westinghouse had sold its nuclear division to the British government in 1999 who sold it on to Toshiba in 2006; GE effectively exited the business in 2006 leaving it largely in the hands of its Japanese collaborator, Hitachi; Siemens exited the sector in 2009; AEG collapsed in the mid-70s; Mitsubishi had made little impact outside Japan; the French companies had been reorganised several times. The sector has generally not been profitable and by 2016, Areva, the latest incarnation of the French nuclear industry was effectively bankrupt after six consecutive years of losses and was split up with a majority stake in its reactor division being bought by the French utility, EDF, reverting to its previous name, Framatome. Westinghouse declared bankruptcy in March 2017 and was sold by Toshiba to a Canadian company, Brookfield Business Partners. However, there were major players both integrally connected to their national government. Russia's champion was the massive Rosatom group which, through a huge number of subsidiaries, contained the whole range of nuclear activities including fuel cycle, reactor sales, and reactor ownership and operation. China had three vendors all with a wide range of nuclear activities, attempting to make an impression on the international market: China General Nuclear (CGN), China National Nuclear Power Corporation (CNNC) and the SNPTC division of State Power Investment Corporation (SPI). Korea emerged as a potential reactor exporter in 2010 through Korea Hydro Nuclear Power Co (KHNPC) winning an order for four reactors to the United Arab Emirates. By 2018, it had not won further export orders and it remains to be seen whether it can compete with the stronger, better established vendors.

Other vendors include, Atomic Energy of Canada Limited (AECL), Mitsubishi and The Nuclear Power Corporation of India Ltd (NPCIL). However, their prospects for exports appear limited and they are not examined further

In this chapter, we examine the strategies of the seven established reactor vendors: Framatome, Rosatom, Hitachi-GE, Toshiba/Westinghouse, CGN, CNNC and SPI. We look at their strengths and weaknesses including:

- Experience with their current technologies;
- Their home market;
- Their target export markets;
- Their ability to offer financing to their prospective customers.

2 The historical structure

The main pioneers of the technology were two US companies, Westinghouse and GE offering light water cooled and moderated reactors developed from submarine propulsion reactors. Westinghouse developed the Pressurised Water Reactor (PWR) while GE developed the Boiling Water Reactor (BWR). Japan and Germany tried to develop their own technologies but with limited success and their main effort went into licensing and indigenising these US technologies. The companies took licenses with their long-term partners in the heavy electrical industry, Mitsubishi and Siemens with Westinghouse and Hitachi, Toshiba and AEG with GE. France followed its own technological route until 1969 when it too adopted US technology, Empain Schneider licensing the PWR from Westinghouse and CGE, the BWR from GE.

The exceptions to this pattern of licensing US were the UK and Canada. The UK tried to develop its own technologies (gas-cooled, graphite moderated reactors) until 1977 when it adopted the Westinghouse PWR. Unlike the other countries, the major companies in the UK, like GEC never put their weight behind the nuclear programme preferring to take stakes in weak consortia that frequently failed. By the time the PWR was adopted, it was too late for UK companies to develop a distinctive capability. Canada developed heavy water cooled and moderated reactors, Candu, for its home market and it has exported these to Argentina, China, Korea, Romania, India and Pakistan. A particular attraction for some buyers was that unlike PWRs and BWRs, this design did not need enriched Uranium. Enrichment is a technology with military sensitivity and is controlled mainly by USA, Russia, UK and France and using PWRs or BWRs inevitably involves some degree of dependence on these countries. However, the technology AECL is offering is old and appears to have very few prospects of new orders.

3 The vendors

Toshiba

Toshiba's roots in the nuclear business go back to the 1960s and its participation in Japan's Boiling Water Reactor (BWR) programme. Japan's electricity system is split into two parts with all equipment for the northern half, including Tokyo Electric, supplied by Toshiba and Hitachi using designs licensed from GE. This pattern continued with nuclear power. Japan's first BWRs were imported, then orders were split reasonably evenly between Hitachi and Toshiba with, typically, one of the companies the primary contractor for the nuclear steam supply system (NSSS) and the other supplying the other high value item, the turbine generator. The US market collapsed in the mid-70s and Hitachi and Toshiba took a more central role in new reactor design development, notably with the Advance Boiling Water Reactor (ABWR) announced in 1980 and with its first order in 1986. An important exception was the development of the Economic Simplified Boiling Water Reactor (ESBWR) announced in 2000, which was very much a product for the US market and has not been seriously considered for Japan.

Toshiba and Hitachi never played a lead role in exports of the BWR with, for example, the most recent BWR export, for Taiwan, going to GE. Whether this was a conscious decision by the Japanese companies not to pursue exports or due to restrictions imposed by their licensor, GE, is not easy to determine. However, other technology licensees such as Siemens and Framatome chose to end their license agreements, in 1975 and 1982 respectively, when they felt they did not need the support of their licensor.

This arrangement between Hitachi, Toshiba and GE ended in 2006 when Toshiba chose to buy the Westinghouse reactor division, outbidding its rivals Mitsubishi and Hitachi. The Westinghouse nuclear division had been bought by the UK government in 1999 via its fuel cycle company, British Nuclear Fuels Limited (BNFL) for $1.1bn. BNFL subsequently bought ABB's and Combustion Engineering's nuclear business for $485m[2]. The government planned to privatise BNFL and presumably create a reactor vendor business. These plans were destroyed by the effective financial collapse of BNFL in 2002 and the subsequent break-up of BNFL's assets including the sale of a 77 per cent stake in the Westinghouse/ABB/Combustion Engineering group to Toshiba in 2006 for $4.16bn[3]. Inevitably, this led to the breaking of the

2 The Times 'BNFL makes Pounds 300m US acquisition' December 30, 1999

3 The Shaw Group took a 20 per cent stake and Ishikawajima-Harima Heavy Industries Co. purchased the remaining 3 per cent stake. Associated Press 'Toshiba completes acquisition of Westinghouse, U.S. atomic power plant company' October 17, 2006

relationship between Toshiba and GE and Hitachi. Toshiba chose to continue to offer the ABWR design independently. This design had received generic regulatory approval from the US safety regulator, the Nuclear Regulatory Commission (NRC) for 15 years from 1997[4]. Toshiba applied to renew this approval in 2010, but in July 2016, the application was withdrawn because of the lack of potential US customers[5]. One utility, NRG, had said it would buy two ABWRs in 2009 for its South Texas prospect, but this project made little progress and is unlikely to go ahead.

Toshiba's main option is the Westinghouse AP1000 PWR, which in 2006 appeared close to completing regulatory review by the NRC, although this was only finally achieved in 2011. Four AP1000s were ordered for the US market, two each for the Summer and Vogtle sites and construction started on these in 2013/14 but a combination of serious cost overruns and delays, and the bankruptcy of Westinghouse led to the abandonment of the Summer project in 2017 with the Vogtle project also at risk of abandonment

The AP1000 was submitted to the UK safety regulator, Office of Nuclear Regulation (ONR), in 2007 in the Generic Design Assessment (GDA) process but the process was suspended in December 2011 due to the lack of UK customers. At that time, there were still 51 design issues to be resolved[6]. Subsequently, Toshiba bought a 60 per cent stake in a UK-based consortium, NuGen, which plans to build three AP1000s. The GDA process was re-opened in 2014 and was completed in 2017[7]. Toshiba sold its bankrupt Westinghouse nuclear division in 2018 and it is not clear whether it expects to continue to try to sell reactors. The NuGen project seems unlikely to proceed, at least using AP1000 technology. The Combustion Engineering capability gave Toshiba System 80+ PWR technology, which had received regulatory approval from the NRC in 1997, while the ABB stake gave them a 50 per cent stake in a company, HTR, offering high temperature gas-cooled reactor technology, HTR-Modul. The System 80+ and HTR-Modul capabilities were not directly marketed but these technologies had already been licensed, the System 80+ to Korea and HTR-Modul to South Africa and China.

---

4 For details of NRC reactor design reviews, see http://www.nrc.gov/reactors/new-reactors/design-cert.html (Accessed March 1, 2016)

5 https://www.mynewsdesk.com/toshiba-global/pressreleases/toshiba-withdraws-application-to-u-s-nuclear-regulatory-commission-to-renew-abwr-design-certification-1462359 (Accessed July 15, 2016)

6 http://www.onr.org.uk/new-reactors/ap1000/index.htm (Accessed July 15, 2016)

7 http://www.onr.org.uk/new-reactors/ap1000/index.htm (Accessed September 2, 2018)

Hitachi

The history of Hitachi in the nuclear business was intimately connected with that of GE and Toshiba until 2006. Under the reorganisation resulting from Toshiba's acquisition of the Westinghouse nuclear division, two new joint ventures were set up: GE-Hitachi, 80 per cent owned by GE which was to operate primarily in the USA and Hitachi-GE, 80 per cent owned by Hitachi, which was to operate in all other markets[8]. GE-Hitachi applied to renew the license for their version of the ABWR in 2010 despite it having no immediate US customer, but by February 2016, little progress appeared to have been made on this. The ESBWR did receive US regulatory approval in 2015. However, the prospects for sales of reactors in the USA, either of the ABWR or ESBWR designs, appear poor and GE-Hitachi is not considered further. Hitachi-GE purchased a UK-based consortium set up to build and operate nuclear power plants in Britain, Horizon, in 2012 for $1.2bn[9]. Horizon owns two sites at each of which it plans to build two ABWRs. The ABWR design was submitted to the UK safety regulator, ONR, for Generic Design Assessment in 2014 with expected completion of the review in 2018[10]. The Horizon projects can only proceed if Hitachi-GE finds a strong investor, either a utility or the UK government.

Framatome

Framatome achieved a dominant position in reactor supply for France in 1975 when its version of the PWR, licensed from Westinghouse, was chosen for the large programme of reactors that followed. By 1990, 58 PWRs had been sold to the French market. Framatome was initially privately owned but in 1975, the Commisariat Energie Atomique, the French national nuclear R&D organisation took a 30 per cent stake, expanding to 34 per cent in 1982 and in 1984, when the parent of the other shareholder, Creusot Loire went bankrupt, the government re-organised the shareholding with CEA taking 35 per cent, the large diversified French engineering company, CGE, taking 40 per cent and EDF with 10 per cent taking most of the rest.

Framatome had progressively moved away from the Westinghouse licensed design terminating its technology license with them in 1981 and the last four of the 58 reactors ordered were for a design designated N4, the first design that was claimed to be wholly French. However, the Chernobyl disaster led to a perceived need for greater safety and in 1992, a new design, the European Pressurised Water Reactor

8 An exception to this is GE-Hitachi's PRISM fast reactor design which is being marketed in export markets, for example, the UK. See http://gehitachiprism.com/ (Accessed July 22, 2016)

9 Associated Press 'Japan's Hitachi to buy Horizon Nuclear' October 30, 2012

10 http://www.onr.org.uk/new-reactors/uk-abwr/index.htm (Accessed March 1, 2016)

(EPR) was announced developed by a joint venture of Framatome and Siemens, Nuclear Power International[11]. The aim was that the design would be licensable in both France and Germany. In 2000, Framatome (66 per cent) and Siemens (34 per cent) merged their nuclear businesses as the Areva NP part of the Areva group[12]. In January 2009, Siemens announced it would exercise its right to withdraw from the joint venture and this was completed in 2011.

The EPR was submitted to the UK's GDA process in 2007 and was given a design acceptance certificate, valid for 10 years in December 2012[13]. In 2008, Areva NP also submitted its EPR design to the US NRC for generic appraisal. However, in 2015, with no US orders likely, it froze the process[14].

The financial collapse of Areva led to it being split into its component parts, a reactor vendor division and a fuel cycle company. A majority stake in the reactor divisions was bought by the French utility, EDF, in 2018 and the company was renamed Framatome. By mid-2018, it was not clear what plans EDF had for Framatome.

Rosatom

Rosatom, the Russian national nuclear company, is a vast diversified company involved in every aspect of nuclear technology including fuel cycle activities, ownership and operation of reactors as well as reactor sales.

Russia was the first country in the world to operate a power reactor with the 6MW Obninsk reactor opened in 1954. Russia commercialised two types of reactor, the VVER, the Russian version of the PWR, and the RBMK, the design used at the Chernobyl site. The VVER has been built in two sizes, 440MW and 1000MW. For this analysis, we focus on the period after the 1986 Chernobyl disaster and we do not look at the RBMK which ceased to be an option for new orders after 1986. We focus particularly on the period from 2007 onwards when Russia began again for the first time since 1986 to market reactors in large numbers for Russia and for export. For an overview of Russia's nuclear history up to the Chernobyl disaster, see Schmid (2015) and IAEA (1997).

At the time of the Chernobyl disaster, Russia was building reactors in Russia, Ukraine, Lithuania Czechoslovakia, the German Democratic Republic (GDR), Poland, Bulgaria, Cuba and Hungary. Some of these were completed without interruption, such as Hungary and Lithuania, for some construction was halted for

11 Nuclear News 'Joint Franco-German design partly unveiled' August 1992

12 Nuclear Engineering International 'Framatome and Siemens to merge' February 2000

13 http://www.onr.org.uk/new-reactors/uk-epr/design-acceptance.htm (Accessed July 15 2016)

14 Nuclear News 'Areva suspends work on U.S. EPR certification' April 2015

a significant period but later restarted, such as Russia, Ukraine, Czech Republic, Slovak Republic while other programmes were abandoned, such as the GDR, Cuba and Poland. By that time, for the VVERs, Russia was concentrating on the 1000MW design, the V-320.

The first new activity of the Russian nuclear industry after Chernobyl was with the order in 1997 of two reactors using the AES-91 design for the Tianwan site in China[15]. This was followed by the order in 2002, after several years of negotiation, of two reactors using the AES-92 design for the Kudankulam site in India[16]. These reactors were essentially the V-320 design with additional safety systems, greater 'passive' safety and a core-catcher for the AES-91. The AES-91 was developed by the St Petersburg design studio of Rosatom, Saint-Petersburg Atomenergoproekt, while the AES-92 was developed by the Moscow Office, Moscow Atomenergoproekt[17].

However, it was the announcement of a new design, AES-2006, in 2006 along with ambitious targets for new reactor orders for the Russian home market that saw a sudden change of gear for the Russian nuclear industry (Mukhatzhanova, 2007). Although this time there was a single designation, AES-2006, as with AES-91 and AES-92, there was a Moscow (V-392M) and a Saint-Petersburg (V-491) version with the first four orders, all for Russia, split between the two versions[18]. Forecasts of a steady flow of three orders a year for the home market were quickly proved wrong. Only one further order beyond the first four, on which construction started in 2008–10, was placed (in Kaliningrad in 2012) and this was effectively abandoned within a year of construction start. The focus switched to exports with an order for four reactors won with Turkey in 2010 and by 2017, Rosatom was claiming an export order book of about 30 reactors, although construction had started on only one project, in Belarus.

Some of these markets are for countries with no experience of commercial nuclear power plants, including Vietnam, Jordan, Bangladesh and Egypt and, historically, attempts to build nuclear plants in such countries often come to nothing. The potential orders for Finland and Hungary are most strategically important. The perception will be that safety regulation in EU countries is rigorous and if Russia

15 Xinhua News Agency 'Work on nuclear power station begins' October 20, 1999

16 Nuclear Engineering International 'Koodankulam deal signed' January 2002

17 For details of the history of the VVER design, see http://www.rosatom.ru/en/resources/b6724a80447c36958cfface920d36ab1/brochure_the_vver_today.pdf (Accessed March 2, 2016)

18 For a detailed specification of the differences, see http://www.rosatom.ru/en/resources/b6724a80447c36958cfface920d36ab1/brochure_the_vver_today.pdf (Accessed March 2, 2016).

can satisfy the safety regulators in these countries, this will be seen as a strong endorsement of the design. In 2018, construction on the first reactors for orders to Turkey and Bangladesh belatedly started

In 2010, a new design, VVER-TOI (V-510), developed by the Moscow office was announced. It was claimed it would have lower design, construction, operation, maintenance and decommissioning costs, would be up to 38 per cent more efficient than the AES-2006 VVER design and would have a slightly higher rated capacity of 1300MW[19]. When it was announced, it was expected to be available for ordering in 2012, but it was not till 2018 that construction on the first unit began.

China

China has three nuclear reactor vendors but with very different backgrounds. China National Nuclear Corporation (CNNC) was the original company set up in the 1960s, China General Nuclear (CGN)[20] set up in 1994 and State Power Investment Corporation (SPI) set up in 2007[21].

CNNC and CGN

Because of their common technology roots with technology licensed from Framatome, it is logical to deal with CNNC and CGN together. CNNC has its roots in the Second Ministry of Machine Building, from which the China Ministry of Nuclear Industry was created and was renamed CNNC in 1988. It makes no secret of its military roots and its continued military connections and its web-site states[22]: 'Historically, CNNC successfully developed the atomic bomb, hydrogen bomb and nuclear submarines and built the first nuclear plant in the main land of China. CNNC is the main body of the national nuclear technology industry, the core of the national strategic nuclear deterrence and the main force of the national nuclear power development and nuclear power construction and shoulders the duel historical responsibilities for building of national defence force, increasing the value of state assets and developing the society.'

---

19 Nuclear Engineering International 'Atomenergoproekt on track to market VVER-TOI in 2013' February 2012

20 Until 2013, CGN was China Guangdong Nuclear (CGN).

21 The State Nuclear Power Technology Company was created in 2007 and merged with the smallest of the big four Chinese generation companies, China Power Investment Corporation (CPIC) to form the SPI

22 http://www.cnnc.com.cn/tabid/643/Default.aspx (Accessed December 18, 2015)

It was in 1985 that construction work began on the first reactor for the Chinese market, a small (300MW) indigenous design of Pressurised Water Reactor (PWR)[23] designed by the Shanghai Nuclear Engineering Research and Design Institute (SNERDI), established in 1970 and supplied by CNNC. Four reactors of this design were exported to Pakistan between 1985 and 2011, China's only export market to date. In 1987, construction started on the first large reactors (Daya Bay), two 1000MW units imported from the French vendor Framatome. However, despite it being the only Chinese reactor vendor, CNNC was not a major player in this project. The Chinese partners in Daya Bay, primarily Guangdong Electric Power Company were consolidated into a new state-owned company, China Guangdong Nuclear (CGN)[24] in 1994. In 1992, CNNC signed a technology transfer agreement with Framatome for the technology ordered at Daya Bay, the M310. In 1995, CGN signed a similar agreement with Framatome[25]. In 2008, construction work in China took off with six reactors beginning construction (four from CGN and two from CNNC) all based on the imported M310 design, CNP-1000. In 2009 and 2010, ten more reactors of this design started construction.

The M310 design dates back to the 1960s, having been licensed by Framatome in the early 1970s from Westinghouse. The Chinese authorities had already acknowledged more modern designs were needed. Its plan was to select one of the advanced designs on offer from foreign vendors importing a few reactors initially but progressively transferring the technology to Chinese companies. Two designs were considered, the French EPR supplied by Framatome's successor company Areva and the AP1000 supplied by Westinghouse, which, by then was owned by Toshiba. The AP1000 was chosen in 2006 with four reactors ordered, reportedly because of greater willingness of Westinghouse to transfer technology. However, a year later, an order for two EPRs was placed with CGN partnering Areva and EDF for this project.

It became clear that the AP1000 and the EPR were too expensive in their existing form and all three vendors began work to develop advanced reactor designs, using the designs of their licensee as the basis. These designs would meet the requirements of European and US regulators and would be Chinese Intellectual Property. CNNC and CGN developed smaller reactor designs (1000MW) than the EPR (1600MW) based on the M310, ACPR-1000 and ACP-1000 respectively. Four reactors in China using the CGN ACPR-1000 design and two reactor exports to Pakistan using the

23 The PWR is the most widely used type of reactor worldwide accounting for about two thirds of the world's operating reactors

24 http://www.cgnpc.com.cn/n1500/index.html (Accessed December 15, 2015)

25 Nuclear Engineering International 'Growth in China' November 2001

CNNC ACP-1000 design had started construction by 2018. However, in 2013, the Chinese government required CGN and CNNC to 'merge' these designs to create the Hualong One or HPR-1000 design. It appears that CGN and CNNC have their own versions of the HPR-1000 which may have more in common with ACPR-1000 and ACP-1000 than with each other. By 2016, CNNC had started construction of two reactors in China using their version of the HPR-1000 and CGN had started construction of one.

Following the Fukushima disaster in March 2011, there was a sharp reduction in ordering and from 2011–14, construction started on only six reactors, two using the old design based on M310, two imported from Russia and two ACPR-1000s. In 2015 six construction starts took place, three for the HPR-1000, two for the ACPR-1000 and one for the CNP-1000. However, by September 2018, there had been only one construction start in China since the beginning of 2016. It is not clear what has led to this new pause in construction starts. Possible factors include the slowdown of electricity demand growth, which has led to serious overcapacity in some regions, concerns about the new technologies and shortage of skills and technological capacities.

By 2013, CGN and CNNC were beginning to look to export markets and an order with CNNC was agreed with Pakistan for two ACPR-1000s[26] (construction on the first unit started in August 2015 and the second in May 2016). The export markets are coordinated by the Chinese Atomic Energy Authority (CAEA) and the National Development and Reform Commission and the three Chinese vendors do not appear to compete in the same market. In December 2015, a CNNC/CGN joint venture company, the Hualong International, was announced to export Hualong One technology[27]. In May 2016, CGN and CNNC signed an agreement that they would not compete with each other in export markets[28].

CNNC would focus outside Europe with South America, including Argentina (building one Canadian design reactor then an HPR-1000) and Africa, including Sudan, its most likely markets. CGN would focus on Europe where its best opportunities appear to be in UK and Romania (building a Canadian supplied reactor). The UK would be a particular prize for CGN bringing prestige that would enhance CGN's prospects in other markets.

---

26 Most reports now state the design is the HPR-1000.

27 Nucleonics Week 'CNNC and CGN set up joint venture to export Hualong One reactors' January 7, 2016

28 Nuclear Engineering International 'China's CGN and CNNC agree not to compete' July 2016, p 5

SPI

The AP1000 was chosen in 2007 over the EPR with four reactors ordered and a new company created, State Nuclear Power Technology Company[29] (SNPTC), which merged with a utility to form State Power Investment Corporation (SPI) in 2015, to indigenise the technology. SNERDI was made a subsidiary of SNPTC giving it experience and substance. SNPTC's advanced design was the CAP1400, a scaled up version of the AP1000. By September 2018, construction on the two CAP1400 units firmly planned had not begun despite press reports forecasting an imminent start. In May 2016, it was reported that the design was only complete enough for one year of construction work to be carried out and there was said to be discussion whether the CAP1400 should be for export only[30]. Whether China can credibly offer a design for export that has not been built for the home market is questionable. There are reports that China is delaying start of construction on a CAP1400 until the first AP1000 is operating successfully and the first units only went critical in mid-2018[31].

In November 2014, Turkey announced it was in exclusive talks with SPI and Toshiba/Westinghouse to buy four reactors, two using the Toshiba AP1000 design and two using the CAP1400 with construction start forecast for 2018/19. However, by 2016, it was reported that the talks were not going well and were no longer exclusive[32]. At best, the timetable is likely to slip and at worst, not to go ahead. SPI is one of five vendors competing in South Africa for an order for 6–8 reactors but it did not appear to be a front-runner and in August 2018, South Africa effectively abandoned its nuclear programme.

4 A comparison of the competitive positions of the vendors

For this analysis (see Table 1), we look at the current design being offered by each of the vendors: for Westinghouse, this is the AP1000, for Framatome, the EPR, for Hitachi-GE the ABWR, for Rosatom the AES-2006, for CNNC and CGN the HPR-1000 and for SPI CAP1400.

29 http://www.snptc.com.cn/en/ (Accessed December 18, 2015)

30 Nuclear Intelligence Weekly 'Weekly Round-up' May 20, 2016, p 1

31 Nuclear Intelligence Weekly 'Nine Projects Top Priority List' May 6, 2016, p 5

32 Nuclear Intelligence Weekly 'Akkuyu's Prospects Pull Past Sinop' July 22, 2016, p 3–4.

Home market

A strong home market provides a vendor with a market that may be less cost sensitive and competitive than export markets giving it a more assured flow of orders and a profitable base. An assured flow of orders will allow the build-up of a strong, efficient supply chain and reactor importers will see regulatory approval in the home market as a demonstration of the licensability of the design, especially where the home market has a long history of building and operating reactors and where the regulatory body is open and accountable.

On this criterion, the Chinese vendors appear strongest with the likelihood that all three vendors will receive at least two orders per year, although doubts about the role of the CAP1400 may weaken the position of SPI. Ten years ago, Russia and France were forecasting a strong market to replace existing reactors and to meet electricity demand growth. For France, the economics of life-extension appear far more attractive than new-build and there is little prospect of many orders being placed. Russia is still talking about new projects but these have been constantly delayed. Prospects for the Japanese home market appear for Hitachi and Toshiba appear poor with the priority likely to be getting existing plants back on line rather than trying to build new ones. Westinghouse is often seen as a US company because of its US base, but the prospects in the USA for new orders, particularly following the problems at the Summer and Vogtle projects also appear equally bleak there.

Regulatory approval

The USA and the UK are both carrying out rigorous generic reviews of designs to establish for all sites the licensability of the design, a process that has taken 5–10 years. Design approval means that for a period of 10 (UK) or 15 (USA) years the design is approved leaving only site-specific issues to be reviewed in any application to construct or operate a plant. The AP1000 has completed the US and the UK process. EPR has completed the UK process but the US process was abandoned in 2015 largely because there were no immediate prospects of reactor sales to the USA. The GE/Toshiba/Hitachi ABWR was licensed in the US in 1997 but this expired in 2012. Renewal was applied for independently by both GE-Hitachi and Toshiba but by 2016, little progress appeared to have been made and Toshiba effectively abandoned its renewal application. Hitachi-GE submitted their updated version of the ABWR to the UK authorities in 2014 and was completed in December 2017. The Russian regulatory process is opaque and documentation is not available so buyers would have to trust in the rigour of the Russian process. Equally, the Chinese process is not transparent. The reviews that do take place seem to take about a year, suggesting they are not in the same depth as the US/UK equivalent.

Experience with their current designs

Only the ABWR has actual operating experience with four reactors completed in Japan, the first in 1992. These have not operated very little since the Fukushima disaster but up to that point, their lifetime load factor was poor, only about 60 per cent. Four further ABWRs have started construction, two in Japan and two in Taiwan but none of them were actively being built in 2016 and they are unlikely to be completed. The EPR has four reactors under construction (two in China and one each in Finland and France) and by 2016 these were 4–10 years late and well over-budget. The two reactors under construction in China were started last but the first reactor in China went critical in June 2018. The eight AP1000s that started construction are also very late and over-budget. The Summer project was about 4 years late when it was abandoned and the Vogtle project is also at least 4 years late. The AP1000s nearest to completion are the four reactors under construction in China, the first two of which went critical in mid-2018. Six AES-2006s have started construction with the four in Russia are all at least four years late. The first two reactors were completed in 2017 and 2018. In July 2016, one of the reactor vessels for the Belarus project was dropped while it was being manoeuvred into position. Belarus has demanded that the vessel be replaced and Rosatom has agreed. It is not clear how far this will delay the project[33]. Two each of the CNNC and the CGN versions of HPR-1000 were under construction by September 2018. There is no construction experience yet with CAP1400.

Government support

The support of the vendor's national government is increasingly key for winning orders, particularly providing finance and coordinating other companies to participate. China appears to have particular advantages in this respect because of the strength of China's economy and its ownership of the vendors. These advantages remain unproven in export markets and if economic growth in China is not sustained, there may be less scope for China to support its vendors. The Russian government is also intimately connected with its vendor, Rosatom, and most of the export orders it claims are reliant on Russia providing the finance. However, international sanctions resulting from the Ukraine issues, the collapse of the oil price and the money spent by the Kremlin trying to defend the rouble mean the capacity of Russia to provide finance for exports must be in doubt. One of the main

33 TASS 'Rosatom to observe Minsk's demand for replacement of Belarus NPP reactor vessel – company' August 12, 2016

reasons given for delays completing nuclear power plants in Russia is lack of funds (Thomas, 2015).

Ironically, the Japanese government was in the process of setting up government mechanisms to support Japanese vendors Hitachi and Toshiba, including provision of loan guarantees in 2010 at the time of the Fukushima disaster. The Japanese government still seems determined to support nuclear power but it remains to be seen whether it can get the political support to do this and by 2018, Japanese vendors had won no orders with this government support and Toshiba's future as a reactor vendor was in doubt.

Areva claimed that it could supply sovereign loan guarantees for reactor exports[34] for example to China and South Africa. It did provide €610m in loan guarantees for the Finnish Olkiluoto project in 2003[35], but this project has gone badly and it may be that the guarantees will have to be paid. However, this sum, 20 per cent of the expected construction cost, would appear not to be adequate coverage for a project now and the expected cost of nuclear has increased markedly. For example, the expected cost of the two reactor Hinkley Point project is about £30bn and this is expected to require 70 per cent coverage by loan guarantees for the deal to be viable. This would result in guarantees worth about £21bn. Whether the French government is willing to guarantee such sums must be in doubt. Whether Framatome, under the control of EDF, would continue to try to sell reactors was not clear by mid-2018.

Vendor's financial position

The decline of markets for reactors has left several of the vendors in financial difficulties. Most serious of these are the collapses of Areva and Westinghouse. In March 2015, Areva announced annual losses of €4.8bn, the fifth consecutive year of losses[36]. The public stake in Areva had continued to increase to around 87 per cent by then. It became clear that the company could not continue in its existing state and the French government launched a rescue attempt. The company was split into two main parts, the fuel cycle business, Areva NC, renamed Orano, and the reactor vendor and servicing business, Areva NP, renamed Framatome. EDF, itself 83 per cent state-owned, was required to take over up to 80 per cent of Areva NP for about €2bn. The plan is that EDF would sell up to a 29 per cent stake to a

---

34 Nucleonics Week 'French export credit agency to insure loans for Cgnpc, Eskom' August 21, 2008

35 Nucleonics Week 'European 'green power' generators challenge EPR's competitiveness' December 16, 2004

36 http://www.areva.com/finance/liblocal/docs/doc-ref-2014/DDR_EN_310315.pdf (Accessed March 1, 2016)

third party leaving it with a 51 per cent stake. In July 2016, Mitsubishi and EDF signed a deal increasing cooperation and in 2017, Mitsubishi took a 19.5 per cent stake in Framatome[37].

The French government had to recapitalise Areva with about €4bn of public money[38] and it also had to assume liabilities for failings with existing orders, for example, the cost overrun at Olkiluoto.

In July 2015, a report by an independent panel of accountants and lawyers from within the company showed that Toshiba had been overstating its profits for seven years. This led to mass resignations at board level including the CEO[39]. This resulted in all the credit rating agencies reducing Toshiba's credit rating and in December 2015 both Standard & Poors and Moody's reduced the rating to 'junk' (non-investment grade)[40]. In May 2016, Toshiba wrote down the value of the Westinghouse nuclear business, which it had acquired in 2006 for $5.4bn by $2.4bn[41]. In May 2016, Toshiba announced its largest ever loss of Yen460bn ($4.2bn) for FY 2015[42] and in March 2017, Westinghouse declared bankruptcy. Toshiba effectively cut the Westinghouse division adrift and it was sold to the Canadian company, Brookfield Business Partners. By mid-2018, it was not clear whether Westinghouse's new owners would pursue new reactor orders or whether it would concentrate on the less risky and more lucrative reactor servicing and maintenance market. By mid-2018, it was not clear whether Toshiba would attempt to rebuild a reactor vendor division based on its BWR capability.

The Hitachi group does not appear to be in serious difficulties although its nuclear division has not sold a reactor for nearly 20 years. The three Chinese vendors appear profitable. It is difficult to determine the strength of the Rosatom group but the weakness of the Russian economy in general is likely to restrict its scope.

---

37 Nuclear Intelligence Weekly 'EDF's Balancing Act Between MHI and CGN' July 8, 2016, p 4

38 Nucleonics Week 'EDF taking over Areva reactor business: government' June 4, 2015

39 Japan Times 'Heads roll at Toshiba as scandal claims top brass' July 22, 2015

40 Agence France Presse 'Moody's, S&P cut scandal-hit Toshiba's credit rating to junk' December 22, 2015

41 Nuclear Intelligence Weekly 'Toshiba Warns of $2.4 Billion Westinghouse Impairment' April 29, 2016

42 http://www.toshiba.co.jp/about/ir/en/pr/pdf/tpr2015q4e_ca.pdf (Accessed July 20, 2016)

5 Conclusions

The nuclear reactor supply industry, once seen as an essential component of diversified companies with an electrical engineering capability, is now seen in Europe and USA as a risky niche business for specialist companies. In Japan, the capability is still in the hands of national champion companies but their commitment to the sector must be in doubt following the Fukushima disaster.

If the nuclear reactor vendor business is to have a future, it appears increasingly likely that it will be driven by Russia and China. For both countries, their nuclear industry appears to be part of national policy to expand their political and economic influence. However, neither of these has significant experience in developed countries with well-resourced critical customers, with experienced, independent safety regulators and with well-developed public participation channels. For some developing country markets that have less capability to be critical customers, this may not be a restriction, but such markets are usually of limited scale, require significant financial support and, historically, nuclear programmes in such countries often do not materialise.

The volume of nuclear orders being placed for China are often seen as China 'going nuclear' but the reality is that nuclear only makes up about 3 per cent of its electricity supplies and because of its rapid electricity demand growth, even if its most ambitious plans are realised, it will still get less than 10 per cent of its electricity from nuclear power. If these plans are to be realised, China needs to get over a major hurdle of siting plants inland rather than on the coast where all the existing plants are. There is considerable resistance to inland siting (King and Ramana, 2015) and if this is not overcome, the scope for further nuclear capacity will be heavily restricted. So while the world nuclear industry may well need China for its survival, China does not necessarily need nuclear power. So if exports of nuclear plants are not bringing it the political influence and economic influence it is hoping for, it may not pursue the export market, even if it does continue to build in China.

For Russia, the dynamics are rather different. Electricity demand in Russia is falling, and, in the short-term, it probably lacks the financial resources and the supply chain to build many reactors either for export or the home market. Its economy is nowhere near as flexible and competitive as China's so it cannot as easily as China switch the focus of its export efforts to other sectors.

It is hard to see the vendors from Japan and Europe – Hitachi-GE, Toshiba and Framatome – being competitive in nuclear export markets. Their technologies are problematic, they lack the comprehensive government backing that Russia and China give, and their home markets are minimal. The question then becomes do they need to sell new reactors to survive and perhaps profit. From 1991, when it started building the last completed reactor in France, Framatome/Areva has started building only four reactors, two in China, and one each in France and Finland. These orders have been highly problematic are unlikely to be profitable and have seriously damaged their reputation. The flow of work servicing and providing replacement parts for operating plants is much more predictable and probably more profitable. With utilities under more financial pressure than in the past, this work is perhaps less profitable than it was in the past and the original vendor is no longer so sure of getting the servicing work, but it still appears a better route than very risky new reactor projects. With about 160 reactors in USA and France beginning to reach the end of their design life and with their owners generally anxious to run them a further 20 years or more, this appears a market that would provide a continued flow of work for Hitachi-GE, Toshiba and Framatome. It would also give them the opportunity to think afresh on new reactor designs, taking into account the lessons from Fukushima, which are only now beginning to emerge.

References

IAEA, 1997. Performance analysis of WWER-440/230 nuclear power plants, IAEA-TEC-DOC-922, Vienna. http://www-pub.iaea.org/books/IAEABooks/5573/Performance-Analysis-of-WWER-440-230-Nuclear-Power-Plants (Accessed March 3, 2016)

Mukhatzhanova, G., 2007. Russian Nuclear Industry Reforms: Consolidation and Expansion, Centre for Non-Proliferation Studies, Monterey. http://www.nonproliferation.org/russian-nuclear-industry-reforms-consolidation-and-expansion/ (Accessed March 3, 2016)

Schmid, S., 2015. Producing Power: The pre-Chernobyl history of the Soviet nuclear industry, MIT Press.

King, A. & Ramana, M.V., 2015, The China Syndrome? Nuclear Power Growth and Safety After Fukushima. Asian Perspective: October-December 2015, Vol. 39, No. 4, pp. 607-636.

Thomas, S., 2015. Nuclear construction problems, PSIRU, Greenwich http://www.psiru.org/reports/nuclear-construction-problems (Accessed March 10, 2016)

Tab. 1 A Comparison of the Competitive Position of the Major Nuclear Reactor Vendors

Vendor	Government backing	Home market	Technology experience	Regulatory approval	Main expected markets	Financial condition
Toshiba	Untested	Weak	AP1000. 8 orders (China & USA), none in operation & all going badly. ABWR, participation in 3 complete units (Japan) & 3 under construction (Japan & Taiwan). All in partnership with Hitachi & GE	AP1000 approved by US NRC in 2011 & under review by UK ONR. ABWR approved by US NRC 1997, expired 2012, applied for renewal but withdrawn in 2016	AP1000: China, UK, India	Weak
Areva	Limited	Weak	EPR. 4 orders (Finland, France, China) none in operation & all going badly.	EPR approved by UK ONR 2012. Began review by US NRC but abandoned 2014	India, UK, China	Effectively bankrupt
Hitachi-GE	Untested	Weak	ABWR, participation in 3 complete units and 3 under construction. All in partnership with Hitachi & GE	ABWR approved by US NRC 1997, expired 2012, applied for renewal but no progress. Under review by UK ONR	UK, USA	OK
Rosatom	Apparently strong	Weak	AES-2006, 6 under construction (Russia & Belarus), none in operation & 4 at least 4 years late. Many orders claimed with no construction yet	Not reviewed outside Russia	Claimed orders: Egypt, Iran, Nigeria, India, Bangladesh, Jordan, Finland & Vietnam. Likely orders: Hungary, South Africa, China, Jordan	OK
CGN	Strong	Strong	CGN HPR1000, 1 under construction (China) since Dec 2015	Not reviewed outside China, proposal to submit it to UK ONR in 2016	UK, Romania	OK
CNNC	Strong	Strong	CNNC HPR1000, 2 under construction (China) since Dec 2015	Not reviewed outside China	Pakistan, Argentina	OK
SPI	Strong	Strong	CAP1400 not yet under construction	Not reviewed outside China	Turkey, South Africa	OK

Technical

The Technological Development of Different Generations and Reactor Concepts

David Reinberger, Amela Ajanovic, and Reinhard Haas[1]

Abstract

This chapter documents the technical development of different generations of nuclear power plants and provides an outlook for possible future concepts and their market prospects. The objective is to assess whether there is really significant technological progress on the horizon and whether these "new" concepts have prospects to become cost-effective. A major conclusion is that most of the so-called Generation IV concepts have already been discussed in the 1950s. At that time, they have not been pursued further due to problems such as costs, limiting factors in material properties and problems in appropriately controlling the fission processes. Yet, since about 2000 a modest revival of the discussion on these concepts is observed, obviously mainly motivated by securing the flow of public money for nuclear research and the broad recognition that with present reactor concepts the nuclear industry will not succeed.

---

1 Reinhard Haas, Technische Universität Wien, Austria, haas@eeg.tuwien.ac.at; Amela Ajanovic, Technische Universität Wien, Austria, ajanovic@eeg.tuwien.ac.at; David Reinberger, Magistrat Stadt Wien, Austria, david.reinberger@wien.gv.at

R. Haas et al. (Eds.), *The Technological and Economic Future of Nuclear Power*, Energiepolitik und Klimaschutz. Energy Policy and Climate Protection, https://doi.org/10.1007/978-3-658-25987-7_11

1 Introduction

Since the first nuclear power plants (NPP) were built in the 1950s, several technical developments have taken place. This chapter documents the changes in various generations of nuclear power plant designs, their major features and differences and to analyze technological problems arising mainly with respect to safety. In addition, we focus on the prospects of so-called Generation IV (GEN IV) reactors.

According to the WNISR (2018), as of mid-2018, world-wide 413 nuclear power reactors were operating in 31 countries. The design, size and age of these reactors vary widely. Some of them are from the 1960's, directly derived from designs for military purposes. Over two thirds (69 %) of the reactors in operation are Pressurized Water Reactors (PWR), 13 % Boiling Water Reactors (BWR), 11 % Pressurized Heavy Water Reactor (PHWR). The majority of them is older than 30 years.

Apart from the basic features (i.e. type of coolant, moderator material, working temperature, thermal/fast neutrons), the designs are categorized by "generations" (GEN); i.e. Generation I, II, III, III+, and IV. For GEN III and GEN III+, there are no clear definitions of which reactor design fits to which generation. It can be seen as an indicative classification. The major motivation for the nuclear industry for developing new reactor concepts are: (i) high current costs; (ii) low current fuel use efficiency; (iii) limited available uranium-resources; (iv) problems with refueling schedule; (v) problems concerning safety and waste management.

2 Historical development of reactor concepts

This sub-chapter provides a concise overview and critical review of the history of reactor concepts from GEN-I to GEN-III+, their advantages and their weaknesses. In addition, an outlook on GEN-IV reactor types is given. It is claimed that they are radically different designs compared to those in use today. If ever, they are expected to enter markets in 20 to 40 years.

Today three generations of nuclear reactors are operating worldwide. (Goldberg et. al (2011)). Figure 1 shows when specific concepts entered the market or are expected to do so. Today, both GEN II and GEN III plants are still being planned, built and used. Even GEN I plants are still in operation despite their lack of safety features.

Generation I

GEN I plants are the early prototypes and power reactors from the 1950s and 1960s that were launched to generate nuclear power for commercial (civilian) purposes. They were deployed directly from military applications for commercial purposes without any additional safety devices. At least some of them were clearly built for dual-use (civil and military) purpose. GEN-I plants did not have any (active or passive) dedicated safety devices.

Examples of GEN I are the Soviet AMB in Belojarsk or the AM-1 in Obninsk (both are graphite moderated types which have advantageous properties for the plutonium production) , the UK's "Magnox" gas-cooled reactor – e.g. Calder Hall-1 (1956–2003), Wylfa (1959-2012) (also Graphite moderated) – and the US reactors of the first generation such as Shippingport (1957–1982) in Pennsylvania (PWR) and Dresden-1 (BWR) (1960–1978) in Illinois. Most of them have already been shut down.

The last European remaining commercial Gen I plant, the Wylfa Nuclear Power Station in Wales, has been scheduled for closure in 2012, and finally was shut down in December 2015. In Pakistan the last GEN-I plant – a CANDU-137 – is still operational at Karachi.

Generation II

GEN II systems began operation in the late 1960s and comprise the vast majority of today's operating reactors ordered from the mid-60s to the 2010s. They are designed for a typical average operational lifetime of 40 years. The major difference to the GEN-I reactors are dedicated active safety designs and that they were in principle designed for civil use only.

GEN-II reactors are typically light water reactors (LWRs) even though there are heavy water designs too. They use safety features involving electrical or mechanical operations that are initiated automatically and, in many cases, can be initiated by the operators.

Figure 1 depicts the evolution of different generations of nuclear power reactors: GEN I, II and III in operation, GEN III+ as near-term deployment, and finally the GEN IV expected to be deployed not before 2030.

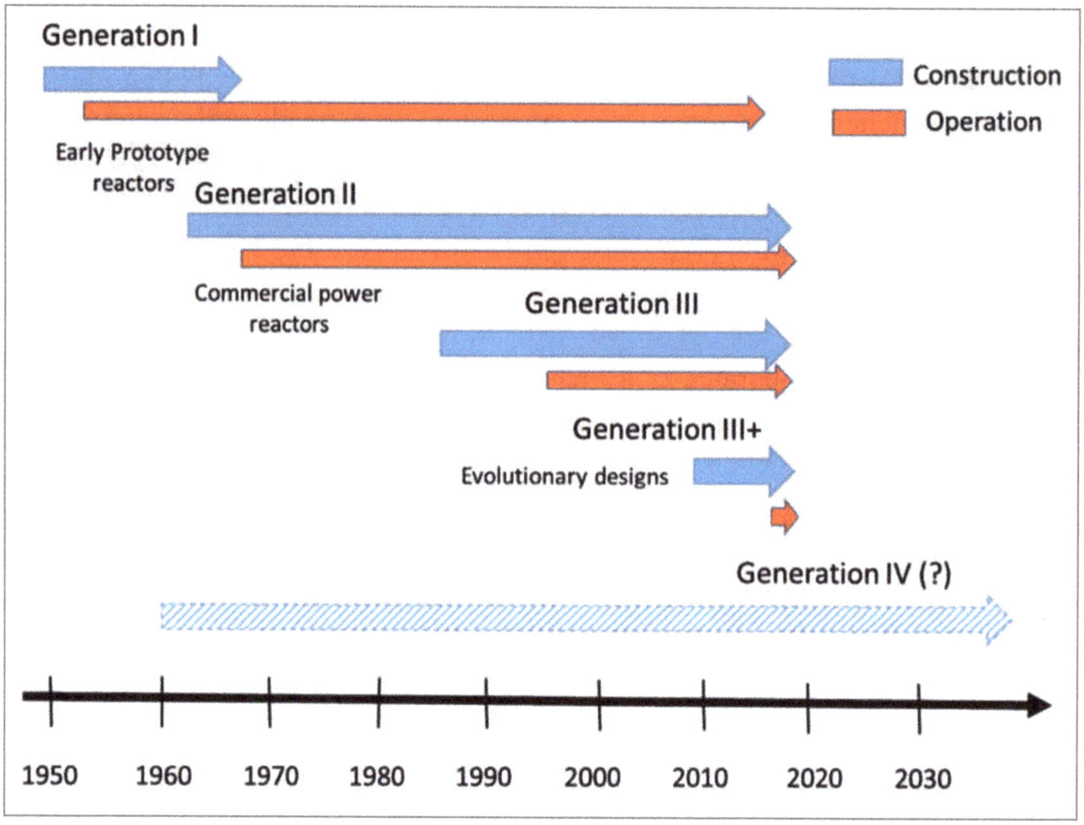

Fig. 1 The evolution of different generations of nuclear power reactors

Most of the GEN II plants still in operation in the Western countries were manufactured by Westinghouse, Framatome, and General Electric (GE). In the following section, the most important GEN-II systems are described in more detail.

Pressurized water reactors

The pressurized water reactor (PWR) was developed from the reactors used to propel submarines. In contrast to submarine reactors, which use high-enriched uranium, PWRs employ low-enriched uranium. They are water cooled and moderated. The power density of PWRs can be relatively high (100 MW/m^3) due to the effectiveness of the heat removal.

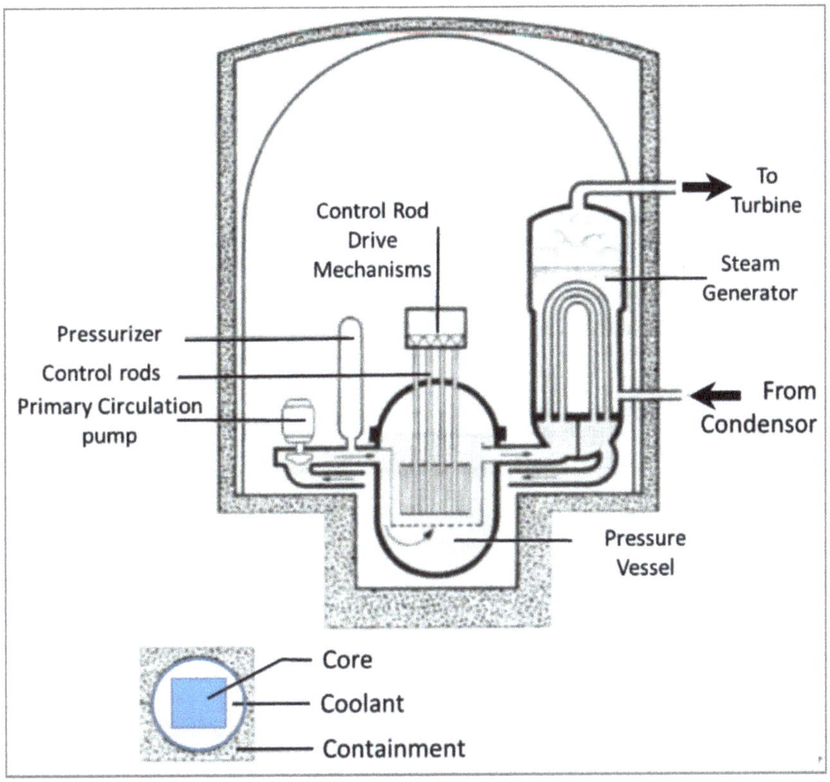

Fig. 2 Description of a pressurized water reactor (based on Greenpeace (2005))

Their primary circuit is characterized by high pressure (~15M Pa) and high temperature (~600°F/ ~300 °C). Safety shall be guaranteed by different systems to control the power output and to cool the core and the barrier system, consisting subsequently of the material structure of the uranium-pellets, the cladding of the fuel rods, the reactor pressure vessel (RPV) itself, the concrete structure around the RPV and the containment, which encloses the primary circuit.

The steam generators, the link between the primary and secondary circuits, are a potential source for radioactivity leaking to the outside. Due to the high power density and the correspondingly high decay heat generated after shutdown, PWRs depend heavily on the availability of the cooling system. The active systems depend on a continuous electricity supply. Even when emergency diesels, redundant grid connection and batteries are installed, station blackouts represent a serious risk, which i. e. in Fukushima led to the destruction of all safety barriers.

Safety systems are usually redundant (i.e. more components are provided for a task than needed). However, redundancy becomes useless if a so-called 'common cause-failure' like flooding or fire disables all parallel trains of a safety system. In principle, there is a continuous trend worldwide towards increasing automation in nuclear power plants' safety, which potentially can reduce the hazards of human error. There is also a trend to replace original analog I&C with digital systems. The implications on safety of this shift are discussed still controversial.

Boiling Water Reactors

The boiling water reactor (BWR) was developed from the PWR, in an attempt to achieve greater simplicity of design and higher thermal efficiency by using a single circuit and by generating steam within the reactor core. As for PWRs, water acts as moderator and coolant. The result is a reactor that still exhibits most of the hazardous features of the PWR, while introducing a number of new problems.

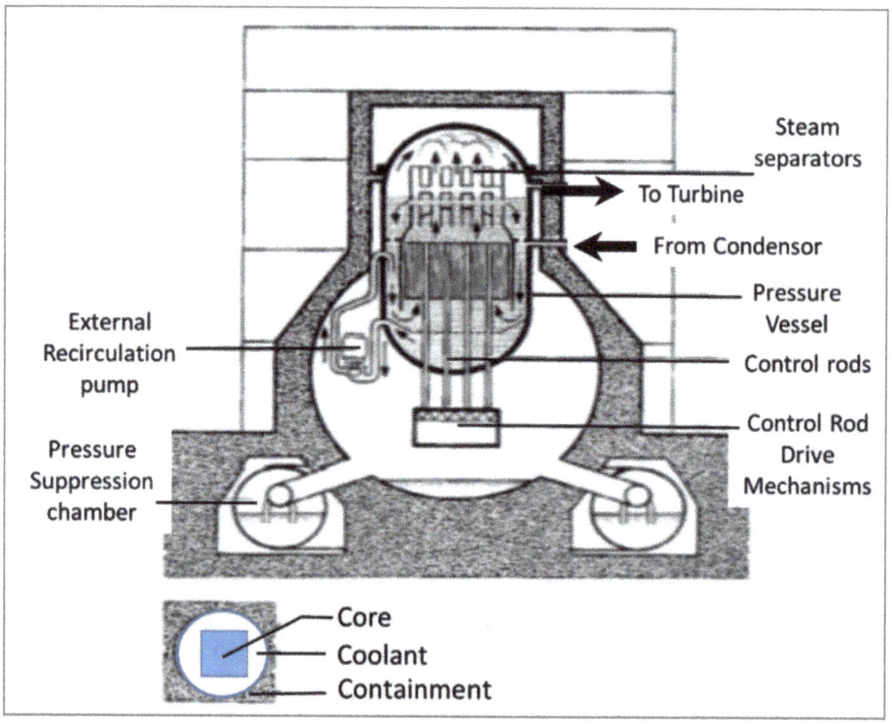

Fig. 3 Description of a boiling water reactor (based on Greenpeace (2005))

BWRs have lower power density (~50 MW/m^3) in the core as well as lower pressure and lower temperature in their cooling circuit than a PWR. The uranium inventory in the core is higher than in PWRs.

The primary circuit of a BWR passes outside the reactor containment. The radioactive steam is directly used to power the turbine(s). Neutron fluxes are considerably lower (by a factor of 10) than in a PWR, leading to significantly less neutron induced aging of the reactor materials. On the other hand, the vessel is much larger; longitudinal welds may be required, whereas there are only circumferential welds in a PWR vessel. There is also a much more complicated inner structure, as well as many penetrations at the bottom. Like a PWR, a BWR depends heavily on fast and reliable active safety systems, while the plumbing of the BWR's emergency core cooling system is much more complex. Control rod injection is – against gravity – from underneath the pressure vessel, Regulating the operation of a BWR is generally more complex than in a PWR. Under certain circumstances, the collapse of so-called steam voids in the core can lead to increasing reactivity and thus increasing power during an accident.

3 Generation III and Generation III +

GEN III designs began to emerge in the mid-1980s onwards, based on learning from the Three Mile Island and Chernobyl accidents. A number of 'evolutionary' GEN III designs were developed from GEN II reactor types without any drastic changes.

According to the World Nuclear Association (WNA), GEN III reactors are characterized by improvements in the following areas (WNA 2004):

- a more standardised design for each type to expedite licensing, reduce capital cost and reduce construction time,
- simpler and more rugged design, making them easier to operate and less vulnerable to operational upsets,
- longer operating life – typically 60 years,
- reduced possibility of core melt accidents,
- better fuel technology, higher thermal efficiency, higher burn-up to reduce fuel use and amount of waste,
- modularized construction.

The most significant improvement of GEN III systems is the incorporation of passive safety features in some designs, which do not require active controls or operator intervention but instead rely on gravity or natural convection to mitigate the impact of abnormal events. Passive systems do not work under any circumstances. They may not relay on functioning of some technical systems but they relay on certain external (uncontrollable) parameters under which the desired physical/chemical process works.

Different concepts bearing the labels GEN III and GEN III+ are in various stages of development and implementation today. In the following, the most important examples as mentioned by WNA (2004) and the International Atomic Energy Agency (IAEA, 2004) are provided. Usually, the concepts are classified into two categories: Large designs >700 MWe and medium designs <700 MWe.

1. Pressurized Water Reactors

The principal large designs are APWR (Mitsubishi Heavy Industries (MHI)/ Westinghouse), APWR+ (MHI), EPR (AREVA), AP-1000 (Westinghouse), KSNP+ and APR-1400 (Korean Industry) and the CNP-1000 (China National Nuclear Corporation). Regarding the Russian VVERs, an advanced VVER-1000 has been developed by Atomenergoproject and Gidropress.

The main small- and medium-size advanced PWR designs are the AP-600 (Westinghouse) and the VVER-640 (Atomenergoproject and Gidropress).

2. Boiling Water Reactors

The main large concepts are the ABWR, (Hitachi, Toshiba, GE), the BWR 90+ (Westinghouse Atom of Sweden), the SWR-1000 (Framatome ANP) and the ESBWR (GE). The HSBWR and HABWR (Hitachi) are small- and medium-sized advanced BWR concepts.

Three ABWRs were already operating in Japan: Two at Kashiwazaki-Kariwa since 1996, a third started operating in 2004. As of the end of 2018, none of them is operational due to shutdown following earthquakes in July 2007 and March 2011.

Generation III+

Generation III+ reactors are slightly modified GEN III designs initiated in the late 1990s, when the nuclear industry began to promote a 'nuclear renaissance'. These should solve three key problems: safety, cost and buildability. The central claim was based on the premise that the existing designs have become too complex and expensive due to new safety systems.

Construction costs of US$1,000/kW were forecast, a level that would make nuclear competitive with gas, and construction times of four years or less were expected (see previous WNISRs). The promise that nuclear power could be the cheapest option reignited interest in nuclear power in a number of key markets.

What are the lessons learned today? Regarding the claim that designs would be easier to build, 20 years since a 'nuclear renaissance' has been announced, only a handful of GEN III+ reactors – four AP-1000, one EPR – have started operating, in 2018, in China. Standardization did not take place, and the introduction of modularized design seems to have simply shifted the quality issues from construction sites to module factories.

By end of 2018, an additional 11 reactors claiming to meet GEN III+ criteria – whatever they are – were under construction: six AES-2006, three EPRs and two AP-1000, all of them years behind schedule and significantly over budget. The construction of two AP-1000s in the U.S., at the VC Summer site in South Carolina, was abandoned in 2017 after builder Westinghouse went bankrupt.

4 Small Modular Reactors (SMRs)

As nuclear power generation has been established in the 1950s, the capacity of reactor units has grown from 60 MWe to more than 1400 MWe, with corresponding economies of scale in operation. However, these large plants operating today set a high and expensive standard for safety.

Several firms are working on designs that are smaller in scale than the current GEN III designs. A basic feature of these plants is that they would make use of modular construction techniques. Small components shall be assembled in a factory environment (offsite or onsite) into structural modules weighing up to 1,000 tonnes. The idea is that these new reactors—at a 300 MW scalable, smaller than a rail car and one-tenth the cost of a big plant—could be built quickly.

The driving forces for SMRs are the reduction of investment per unit and the need for integration into smaller grids in developing countries. However, there are serious doubts concerning these prospects. Furthermore, the opinion that the only way to make nuclear power cost competitive is the use of small modules is not shared by all nuclear industry experts. The reason is – see above – the achieved economies-of-scale with respect to LWR. Capital construction costs (price per unit of electric capacity, or $/kWe) of a nuclear reactor decrease with size, but the economy of scale applies only if reactors are of a very similar design, as has historically been the case.

The design characteristics of SMRs, however, are significantly different from those of large reactors. SMRs approach the economies of scale problem by achieving significant cost savings elsewhere leading to significant learning-by-doing efficiencies. For example, SMR designs seek to streamline safety and safeguard requirements by replacing (at least some) security guards with concrete security barriers and/or by building underground, streamlining the requirements for operators and the emergency planning zone requirements.

The new designs do stretch out refuelling schedules, from 18 months to possibly 3–5 years and potentially to as long as 10 years.

5 Generation IV

The above-mentioned problems of the industry have led to the development of a so-called GEN IV reactors. They are described as radically different designs to reactors operating today, some involving a closed fuel cycle. Moreover, GEN IV reactors are said to be highly economical, to incorporate enhanced safety and reliability features and embed physical protection. They should produce minimal amounts of waste and be proliferation resistant, as these designs include advanced actinide management.

In addition, GEN IV reactors should have all of the features of Gen III+ units, as well as the ability, when operating at high temperature, to support economical hydrogen production, thermal energy off-taking, and perhaps even water desalination.

Yet, today, in 2018, GEN IV reactors are considered to be decades away from commercialization, as they have been when the discussion on these concepts started again in the early 2000s. The few attempts in the past based on concepts like fast neutron reactors nowadays called GEN IV ended with severe technical issues (e.g. Phénix and Superphénix in France) or accidents (e.g. Monju in Japan), which forced a closure of the respective project. Only two are currently in operation (Beloyarsk-3 and -4).

The start of the GEN IV initiative can be pointed to the year 2000, when the U.S. DOE launched the "Generation IV International Forum" (GIF) with the aim of developing a technically new generation of reactors. Today, ten member-countries participate in this initiative (Argentina, Brazil, Canada, France, Japan, Republic of Korea, South Africa, Switzerland, U.K. and the USA), as does EURATOM. The initial goal was to develop innovative nuclear systems (reactors and fuel cycles) likely to enter markets by as early of 2020.

The concepts selected for Generation IV were discussed within the GIF groups of international experts from industry, universities and national laboratories. They were organized to undertake the identification and evaluation of candidate systems, and to define research and development (R&D) activities to support them. Initially, some 100 different designs were identified as candidates and evaluated.

These designs ranged from concepts that rather belonged to Generation III+ to a few that were radically different from all known technologies. At the end of the process, six concepts were recommended for further development in the framework of GIF. A roadmap describes the R&D required to develop each of the six systems as well as the approximate time and cost for completion. Many of the technological gaps were common to more than one system and the roadmap identifies several areas where crosscutting R&D would be required. The necessary R&D would be very expensive, and no single country had the necessary facilities and expertise to carry it out alone (DOE, 2002).

5.1 Sodium-Cooled Fast Reactor System (SFR)

The SFR system consists of a fast-neutron reactor and a closed fuel cycle system. There are two major options: One is a medium size (150 to 500 MWe) reactor with metal alloy fuel, supported by a fuel cycle based on pyrometallurgical reprocessing in collocated facilities. The second is a medium to large (500 to 1,500 MWe) reactor with MOX fuel, supported by a fuel cycle based upon advanced aqueous reprocessing at a centralized location serving a number of reactors. The primary coolant system can either be arranged in a pool layout or in a compact loop layout. (DOE, 2002; Lineberry, 2002).

According to GIF, the SFR has the broadest development base of all the GEN IV concepts. The existing know-how, however, is based mainly on old reactors, which have already been shut down for various reasons (safety, economics, resistance from the population), e.g Superphenix in France (1988) and Kalkar in Germany (1989).

Considering its history, as well as the significant hazards of this reactor line it is difficult to understand, why the SFR has been selected by GIF. According to GIF, research on both, the fuel cycle and the reactor system, is necessary to bring the SFR to deployment. Furthermore, there is important work to be done regarding safety. It has been argued that the SFR system is top ranked in sustainability because of its closed fuel cycle and potential for actinide management. It has been rated good in safety, economics, proliferation resistance and physical protection. In 2002 the SFR system was estimated to be deployable by 2015 (DOE, 2002). It did not happen.

5.2 Very-High-Temperature Reactor System (VHTR)

The VHTR is considered as a next step in the evolutionary development of high-temperature gas-cooled reactors (HTGR). This reactor line has been pursued until the late 80s in several countries; however, only prototype and demonstration plants were ever operated, all of which were decommissioned after rather short and unsuccessful overall operating times about twelve years of operation at most such as the small Dragon reactor experiment (20 MWth, 1966 -1975, U.K.), the THTR Hamm-Uentrop, (308 MWe, 1986–1988, Germany) as well as the U.S. plants at Peach Bottom (42 MWe, 1967–1974) and Fort St. Vrain (342 MWe, 1976–1989).

The VHTR system uses a thermal neutron spectrum and a once-through uranium fuel cycle. The reference reactor concept has a 600-MWth graphite-moderated helium-cooled core based on either the prismatic block fuel of the GT-MHR or the pebble bed of the PBMR. It is regarded as the most promising and efficient system for hydrogen production, either using the thermochemical iodine-sulphur process, or from heat, water, and natural gas by applying the steam reformer technology at core outlet temperatures greater than 1,000°C. The VHTR is also intended to generate electricity with high efficiency (over 50%).

Furthermore, it is hoped that the concept could benefit from the experience gained with the Japanese HTTR research reactor and the Chinese HTR-PM still under construction, as well as from the GT-MHR and the PBMR projects at present in the planning phase.

5.3 Supercritical-Water-Cooled Reactor System (SCWR)

The SCWRs are high-temperature, high-pressure water-cooled reactors that operate above the thermo-dynamic critical point of water ($T > 374{,}12$ °C and $p > 22{,}06$ GPa). The reference plant has a 1700 MWe power level, an operating pressure of 25 MPa, and a reactor outlet temperature of 550°C. Fuel is uranium oxide. SCWRs could be designed as thermal or as fast-spectrum reactors, but current worldwide efforts focus on the thermal design.

The thermal efficiency of a SCWR can approach 44%, compared to 33–35% for LWRs. Because no change of phase occurs in the core and the system utilizes a direct cycle (like the BWR), steam separators, dryers, pressurizes and recirculation pumps are not required, resulting in a considerably simpler and more compact system than traditional LWRs. SCWRs are hoped to be more economical than LWRs, due to plant simplification and high thermal efficiency. The Governments of Japan, the U.S. and Canada are developing the SCWR. There have been no prototypes built so far.

The technology for the SCWR is based on the existing LWRs and supercritical-water-cooled fossil-fired power plants. However, there are important SCWR technology gaps in the areas of materials and structures, including corrosion and stress corrosion cracking (SCC), safety and plant design. The main feasibility issues are the development of suitable in-core materials and the demonstration of adequate safety and stability.

5.4 Lead-Cooled Fast Reactor System (LFR)

LFR systems are reactors cooled by liquid metal (lead or lead/bismuth) with a fast-neutron spectrum and closed fuel cycle system. A full actinide-recycle fuel cycle with central or regional facilities is envisaged. A wide range of unit sizes is planned, from 'batteries' of 50–150 MWe, and modular units of 300–400 MWe to large single plants of 1200 MWe. The LFR battery option is a small factory-built turnkey plant with very long core life (10 to 30 years). It is designed for small grids, and for developing countries that may not wish to deploy a fuel cycle infrastructure.

Among the LFR concepts, this battery option is regarded as the most promising, concerning fulfilment of Generation IV goals. However, it also has the largest research needs and longest development time.

Experience with the technology is restricted to seven Russian Alpha class submarines, which stopped operation in 1995, and on the advanced liquid-metal fast breeder reactor (ALMR), the design of which was withdrawn from the U.S. NRC review at an early stage (WANO, 2004).

The LFR system is top-ranked in sustainability because a closed fuel cycle is aimed at, and in proliferation resistance and physical protection because it employs a long-life core. It is rated good in safety and economics. The LFR system was estimated in 2002 to be deployable by 2025 (DOE, 2002). That is out of reach today.

5.5 Gas-Cooled Fast Reactor System (GFR)

The GFR system is a helium-cooled reactor with fast-neutron spectrum and closed fuel cycle. It is primarily envisioned for electricity production and actinide management. The GFR reference assumes an integrated, on-site spent fuel treatment and re-fabrication plant, but the viability of the planned technology has yet to be demonstrated. Fuel cycle technology is the most comprehensive technology gap of the GFR.

In spite of large technology gaps, according to GIF, the GFR system is top-ranked in sustainability because of its closed fuel cycle and excellent theoretical performance in actinide management. It is rated good in safety, economics, as well as proliferation resistance and physical protection. The GFR was estimated in 2002 to be deployable by 2025 (DOE, 2002). That will not happen.

Several GIF members have a specific interest for a sequenced development of gas-cooled system: The first step of the 'Gas Technology Path' aims to develop a modular HTGR, the second step would be the VHTR, and the third step the GFR

(Carré, 2004). The gas-cooled systems VHTR and GFR are seen as the top priorities of GIF members in Europe and the USA.

5.6 Molten Salt Reactor System (MSR)

During the 1960s the USA developed the molten salt breeder reactor as the primary back-up option for the conventional fast breeder reactor (cooled by liquid metal). A small prototype (8 MWth), the Molten Salt Reactor Experiment (MSRE), was operated for only four years. The next project planned, the Molten Salt Breeder Reactor (MSBR), was never built. The present work rests only on these projects. Detailed designs of an MSR have not been produced since the 1970s (Forsberg, 2002).

The MSR system is based on a thermal neutron spectrum and a closed fuel cycle. The uranium fuel is dissolved in the sodium fluoride salt coolant that circulates through graphite core channels. The heat, directly generated in the molten salt, is transferred to a secondary coolant system, and then through a tertiary heat exchanger to the power conversion system.

5.7 Discussion of GEN IV concepts

GIF considers a closed fuel cycle as a major advantage of Generation IV concepts. A system with a closed fuel cycle is regarded as more effective, and sustainable. However, not all of the six concepts selected for development employ a closed fuel cycle. The VHTR, most favoured, relies on an open cycle; and for the SCWR, once-through constitutes the nearer-term option. Furthermore, it is questionable, whether it will actually be possible to successfully develop and implement a closed fuel cycle. In addition, the costs of such closed fuel cycle concepts would be very high. According to the study "The Future of Nuclear" of the U.S. Massachusetts Institute of Technology (MIT, 2003), a convincing case has not yet been made that the long-term waste management benefits of advanced closed fuel cycles involving reprocessing of spent fuel are not indeed outweighed by the short-term risks and costs, including proliferation risks. The waste problem of nuclear industry can only be reduced even in an optimistic closed cycle scenario but remains far from being solved. (e.g. Gutachten Transmutation, Gerald Kirchner et al., 2015) Also, the MIT study found that the fuel cost with a closed cycle, including waste storage and disposal charges, to be about 4.5 times the cost of a once-through cycle. Therefore, it is not realistic to expect that there will be new reactor and fuel cycle technologies that simultaneously overcome the problems of cost, safe waste disposal and proliferation. As a result, the MIT study concludes that the once-through fuel cycle best meets the criteria of low costs and proliferation-resistance (NEI, 2003).

The basic concepts of the "new generation" have been around as long as nuclear power, but they were forced out of the market in the early years by the LWR – not without reason, considering the experiences so far, which are dominated by technical and economic problems, and safety deficits.

In order to overcome these problems, materials, processes and operating regimes that are significantly different from those of currently operating systems or previous systems would have to be developed. So far, none of the six reactor concepts selected for development fulfils all GEN IV aims.

6 Conclusions

Practically all of the 413 world-wide operating reactors (as of mid-2018) are GEN-II reactors carrying the well-known risks with respect to safety issues and proliferation. By the end of 2018, only a hand full of GEN III+ reactors – attributed with higher safety levels – are in operation (all in China). There are only 11 are under construction world-wide. They should be safer due to additional passive safety features. The lessons learned so far are that these generations' plants are much more expensive and can experience huge delays in construction times.

Different international, government sponsored organisations such as GIF are selling the idea of so-called GEN IV reactors. They pretend that under that umbrella term a completely new generation of reactors is being developed. The envisaged concepts are surely different from the light water concepts used so far. The message for the media, politicians and the public is: GEN IV is a safe, economical competitive, proliferation-resistant power source without the problem of increasing greenhouse gas emissions.

However, these expectations are very high, and their realization is at least 20 years away. In addition, the underlying basic concepts of "GEN IV" are decades old and encompass a range of ideas, which were already discussed in the 1950s. They were not pursued further due to severe problems such as high costs, limiting factors in material properties and difficulties in controlling the fission processes. They were forced out of the market already in the early years by the LWR.

It is of core interest to understand what the real motivation behind the GEN IV initiative is. To some extent it seems the revival of the discussion on GEN IV is mainly driven by the wish to secure the flow of public money for nuclear research. Whether it will really lead to the development of new reactors remains highly doubtful. The sobering final conclusion is that substantial amounts of money are invested in efforts which are not likely to solve any of the problems of nuclear power, climate change and energy supply security.

References

Carré, 2004: R&D Program on Generation IV Nuclear Energy System: The High Temperature Gas-cooled Reactors; Annual Meeting on Nuclear Technology, 2004, May 25-27, Düsseldorf

DOE, 2002: United States Department of Energy's the Generation IV International Forum (GIF): A Technology Roadmap for Generation IV Nuclear Energy System; December 2002, (http://energy.inel.gov/gen-iv/docs/gen_iv_roadmap.pdf)

Forsberg, 2002: Molten Salt Reactors (MSRs); Americas Nuclear Energy Symposium (ANES 2002), Miami, Florida; October 16–18, 2002

Goldberg S. M., Rosner R., 2011. Nuclear reactors: Generation to Generation, American Academy of Arts and Sciences.

Greenpeace, 2005. Nuclear Reactor Hazards: Ongoing Dangers of Operating Nuclear Technology in the 21st Century. April 2005.

IAEA, 2004: International Atomic Energy Agency: Nuclear Technology Review 2004; Vienna, August 2004

Lineberry, 2002: Lineberry, M. J.; Allen T. R:. The Sodium-Cooled Fast Reactor (SFR); Americas Nuclear Energy Symposium (ANES 2002), Miami, Florida; October 16–18, 2002

MIT, 2003:An Interdisciplinary MIT Study: John Deutch (Co-Chair), Ernest J. Moniz (Co-Chair), Stephen Ansolabehere, Michael Driscoll, Paul E. Gray, John P. Holdren, Paul L. Joskow, Richard K. Lester, and Neil E. Todreas; The future of nuclear power, January 2003, Massachusetts Institute of Technology

NEI,2003: Nuclear Engineering International: The future lies in the past; October 2003, 42-45

WNA, 2004: World Nuclear Association: Advanced Nuclear Power Reactors; November 2004, www.world-nuclear.org)

WNISR, 2018. The World Nuclear Industry Status Report 2018.

Nuclear Waste, Proliferation

Decommissioning of Nuclear Power Plants and Storage of Nuclear Waste

Experiences from Germany, France, and the U.K.

Ben Wealer, Jan Paul Seidel, and Christian von Hirschhausen[1]

Abstract

The decommissioning of nuclear power plants and the storage of nuclear waste are major challenges for all nuclear countries. Both processes are technologically and financially challenging. We provide an analysis of the status quo of both processes in three major nuclear countries: Germany, France, and the U.K. Germany was able to gain some decommissioning experiences but not one large-scale reactor has been released from regulatory control. EDF was forced to cancel its target to immediately dismantle all GCRs by 2036 due to underestimated technological challenges, while in the U.K., decommissioning of the legacy fleet lasts well into the 22nd century. Until now, no scale effects could be observed, if EDF can reap scale effects due to the standardization of its fleet remains to be seen. The search for a deep geological disposal facility is the most advanced in France, where the start date is fixed by law to 2025. There are many uncertainties related to estimated future costs. In the three countries, three different funding schemes are implemented: Germany switched from internal non-segregated funds to an external segregated fund for waste management. In the U.K. the decommissioning of the

1 Ben Wealer, TU Berlin, Germany, bw@wip.tu-berlin.de; Jan Paul Seidel, TU Berlin, Germany, janpaulseidel@gmail.com; Christian von Hirschhausen, TU Berlin, Germany, cvh@wip.tu-berlin.de

R. Haas et al. (Eds.), *The Technological and Economic Future of Nuclear Power*, Energiepolitik und Klimaschutz. Energy Policy and Climate Protection, https://doi.org/10.1007/978-3-658-25987-7_12

legacy fleet will be paid by the taxpayer for the next 100 years while an external segregated fund was established for the current operational fleet. In France, an internal segregated fund finances the future liabilities of EDF.

1 Introduction

The decommissioning of nuclear power plants (NPP) and the long-term storage of nuclear waste are important elements of the life cycle of nuclear power plants, but both processes have been underestimated for a long time, both in terms of the technological challenges as well as the financial implications. Traditional economic analysis has discounted the future costs for decommissioning and storage so that these never appeared in the financial calculations. Furthermore, the little available experience leads to a high uncertainty about future pathways. In addition, technical and financial data are difficult to compare between countries.

In this chapter we provide an in-depth analysis of decommissioning and waste storage processes in three major European nuclear countries: Germany, France, and the United Kingdom. To do so, we are going to compare the different national strategies of organizing and financing the decommissioning of NPPs and the storing of high-level nuclear waste. We distinguish between the two main elements of the strategy: production and financing. First, someone has to manage and organize the decommissioning and storage process (the production): this can either be private or public companies, or a mixture of both. Second, both processes need to be financed: This can be done by the federal budget, an external segregated fund, or in-house financing by the companies, usually done in internal funds—segregated or non-segregated (OECD/NEA 2016a).[2]

The shutdown reactors in the three observed countries France, Germany, and the U.K. account for more than 77% of the reactors in shutdown state in the European Union (European Parliament 2013). In addition, these countries represent interesting features that reveal the different experiences and challenges. Germany is just exploring ways for large-scale decommissioning, having decided to phase out

2 External segregated fund: The operators pay their financial obligation in a publicly controlled fund managed by private or state owned external independent bodies. Internal non-segregated fund: The operator of a facility is obliged to form and manage funds autonomous, which are held within their accounts as reserves. Internal segregated fund: The operator feeds a self-administrated fund, which is separated from the other businesses.

nuclear power by 2022. Both, France and the U.K., are facing specific technological challenges with the graphite-moderated reactors of the 1960s. There are concerns in France that the financial challenges of the future decommissioning process—with 58 still operational reactors France will face the largest decommissioning project in Europe—are not yet well understood and underestimated. The case study of the U.K. reveals organizational and financial challenges of having to clean up after a few decades of operating the nuclear sector, and the long time frames to be expected.

The focus of this chapter is on the decommissioning of large-scale NPPs that were operated commercially for electricity production and the management of high-level radioactive waste (HLW). The two light water reactor (LWR) concepts—the pressurized water reactor (PWR) and the boiling water reactor (BWR)—are the most widely installed reactor designs in the world. However, in this case study, the only BWRs are located in Germany while France only operates PWRs. The concepts for gas-cooled and graphite-moderated reactors (GCR) were developed

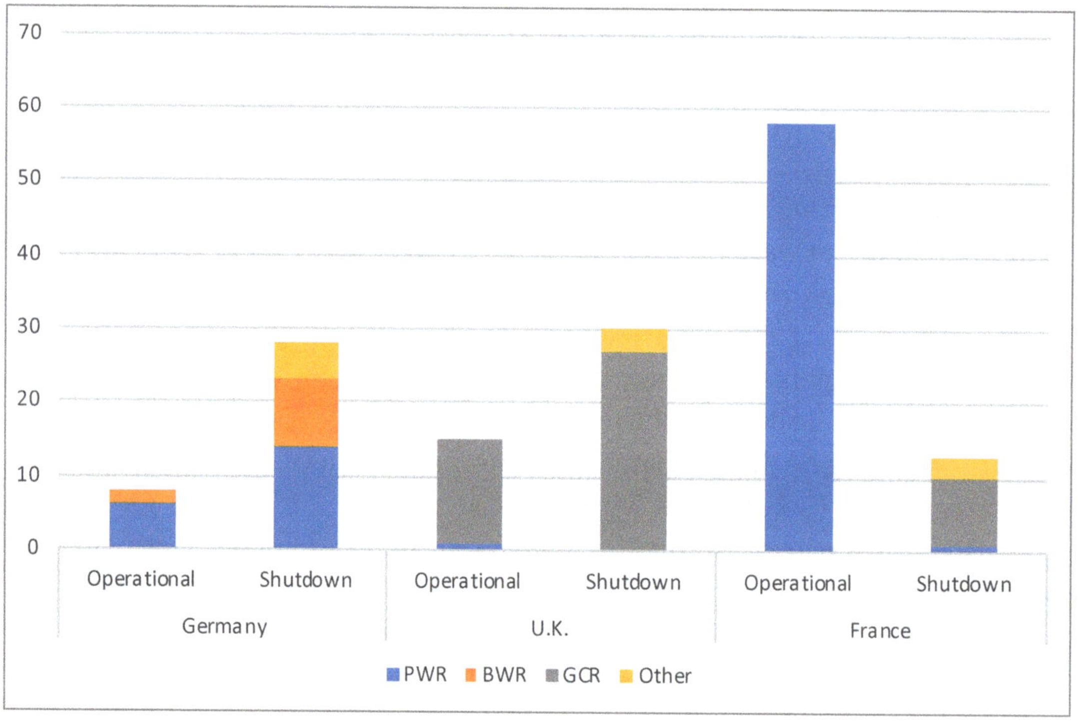

Fig. 1 Operational and shutdown reactors by reactor type in the observed countries, own depiction based on IAEA (2017).[3]

3 The cluster "other" comprises for France: two fast breeding reactors (FBR) and one heavy-water moderated gas-cooled reactor (HWGCR); for Germany: one high-tem-

and installed in France and the U.K. While France ceased operations of its entire GCR fleet, they are still operational in the U.K. The number of shutdown reactors in the United Kingdom and Germany outweigh the operational reactors by a large amount, as can be seen in Fig. 1.

The decommissioning and clean-up of the global civil nuclear legacy and the long-term storage of nuclear waste represents technological, safety and environmental challenges for all nuclear countries. During the period when nuclear energy was established, the focus of planners and operators was predominantly on designing, building and operating a safe plant and only limited on the eventual decommissioning of the facility and the management of resulting radioactive wastes (McIntyre 2012). In general, the decommissioning process of an NPP moves from the outer to the inner area of the reactor. Simultaneously, the degree of contamination of the handled parts increases. The eventual process of decommissioning of the nuclear power plant can be divided into five main stages.[4] A geological disposal facility for high-level waste is still missing worldwide. As spent nuclear fuel (SNF) has to be stored for 40–50 years in order to cool down, before it can be permanently stored in an underground disposal repository, the time was not that pressing to tackle this problem for the responsible actors, which led to the construction of interim storage facilities for high-level wastes.

As the goal of this chapter is an analysis of the technical, organizational and financial status quo of both processes in Germany, France, and the U.K., the different national strategies and approaches of organizing the production and the financing of the decommissioning process, and storing the nuclear waste will firstly be analysed and then compared; the final section concludes.

---

perature gas-cooled reactor (HTGR), one FBR, two pressurized heavy-water reactors, one HWGCR, and one FBR; for the U.K.: two FBRs and one SGHWR

4 Stage 1: Dismantling of systems that are not needed for the decommissioning process and installation of the logistic in the hot zone. Stage 2: Dismantling of higher contaminated larger system parts. Stage 3: Dismantling in the hot zone, e.g. deconstruction of the activated reactor pressure vessel (RPV) and its internals (RVI) and of the biological shield. Stage 4: Deconstruction of contaminated system parts, removal of operating systems and decontamination of buildings. Stage 5: Demolition of the building or further nuclear or non-nuclear use (Wealer et al. 2015).

2 France

With a nuclear share in the net electricity production of 72.28% in 2016 (IAEA 2017) France still relies heavily on nuclear power. The Energy Transition for Green Growth bill, approved by the National Assembly in 2015, foresees a reduction of this share to 50% by 2025. In 2017, Electricité de France (EDF) operates 58 NPPs on 19 sites across the country with a total installed capacity of 63.2 GW. The oldest running reactors are the two Fessenheim units which became critical in 1978, and the most recently installed NPPs are the two Civeaux units, commissioned in 2002. The French nuclear fleet has an average age of around 30 years and has with its three PWR designs[5] the highest degree of standardization in the world (World Nuclear Association 2017). All the nuclear steam supply systems of the operational NPPs were designed by Framatome (IAEA 2017). In 2010, EDF announced that it was assessing the prospect of raising the 40-year lifetimes to 60 years for all its existing reactors. This strategy would involve replacement of all steam generators in the 900 and 1,300 MW reactors and other refurbishments costing 400–600 million EUR per unit.[6] As the focus of this chapter lies on the decommissioning and waste management of commercial NPPs, the activities of EDF will be the focus in the following.

2.1 Production

2.1.1 Decommissioning of nuclear power plants

Currently, there are 13 NPPs in a stage of permanent shutdown; of which nine NPPs are from the first-generation GCRs, similar to the U.K. Magnox reactors. Chooz-A is the only PWR currently being decommissioned by EDF. The decommissioning of the former military installations in Marcoule G-1, G-2, and G-3 is taken over by the public nuclear research agency Commissariat à l'énergie atomique et aux énergies alternatives (CEA). French regulation states that NPPs have to be immediately dismantled and the process has to be carried out as fast as possible; depending on the complexity of the plant this could mean several years up to several decades (Autorité de sûreté nucléaire 2016b). Initially EDF planned to decommission its shutdown NPPs in two waves within 25 years with an estimated end in 2036 (Au-

5 The three-loop 900 MW reactors (CP0, CP1, CP2), the four-loop 1,300 MW reactors (P4), and the four-loop 1,450 MW reactors (N4).

6 So far EDF has replaced the three steam generators in 22 of their 900 MW reactors, ordered 44 steam generators for eleven of the 1,300 MW class and will proceed with the other nine (World Nuclear Association 2017).

torité de sûreté nucléaire 2016a). The first wave consisted of the FBR Super-Phénix, the HWGCR Brennilis, the PWR Chooz-A and the GCR Bugey-1; in the second wave the last five GCRs Chinon A 1–3 and Saint Laurent A 1–2 were planned to be decommissioned. This ambitious plan was changed in 2016.

Chooz-A (operational from 1967 to 1991) was the first European commercial PWR and built under a Westinghouse license. By December 1995 the reactor was defueled and the SNF dispatched to the reprocessing centre in La Hague. The policy change in the French decommissioning strategy in 2011 from Long-Term Enclosure to Immediate Dismantling accelerated the decommissioning plans and reduced the enclosure period from 50 to only a few years (European Parliament 2013). Since 2014, the reactor pressure vessel of Chooz-A is being decommissioned under water. The work was contracted to a Westinghouse-Nuvia France consortium (Hitchin 2010). The experiences with the PWR dismantling are seen as an important feedback source for the further dismantling of the still operational PWR fleet, but this is questionable (Seidel and Wealer 2016) as Chooz-A is an early reactor design and quite unique with its embedment inside a bedrock of a hill in the Ardennes. Chooz-A is therefore not really comparable to the other PWRs. EDF missed the ambitious target of completing the decommissioning process by 2016, the process is now expected to be completed by 2020–2025 (Martelet 2016).

As disposal routes for irradiated graphite waste are still missing, the Long-Term Enclosure of a GCR is the worldwide preferred decommissioning strategy. Nonetheless the change of EDF's strategy also affected its GCR decommissioning strategy, which was considered to be more global and interconnected as it had to cope with a considerable amount of irradiated graphite. To cope with this waste, EDF initially started with the construction of an interim storage facility on the Bugey site, but the construction was stopped in 2012. In 2016, EDF announced a change of its GCR strategy to the regulatory authority ASN (Autorité de sûreté nucléaire): the focus for the next 15 years would lie on dismantling nuclear installations except for the reactors and its buildings. The plans foresee that the first reactor (Chinon A-1) will start the dry dismantling in 2031; during the estimated dismantling duration of 25 years, the five remaining reactors will be enclosed. This new strategy will lead to a possible release of the GCRs from regulatory control in the beginning of the 22nd century. The major motivations for a switch to a dry dismantling strategy were constrains due to the long immersion times of the reactors, i.e. corrosion and leak tightness (Martelet 2016). The initial plan with the continuous flow of graphite waste and the very tight focus on the reactor core could not be implemented because the actual dismantling is technologically complex and needs more preliminary tests than expected.[7] ASN recognized the proposal and expects by the end of 2017 a detailed decommissioning plan for the next 15 years and EDF to take a position regarding this sudden change in strategy. In France, EDF officially uses the term "Safe Configuration" instead of Long-Term Enclosure (or Safe Storage), while in the U.K., where EDF Energy is responsible for the decommissioning of the operational GCR fleet, the company opts officially for the Long-Term Enclosure strategy. It is also unsure, if the new strategy is compatible with the implication of the French regulatory authority to carry out the decommissioning as fast as possible (Seidel and Wealer 2016).

2.1.2 Storage of high-level wastes

As France operates a closed fuel cycle, SNF is not declared as waste but as a resource and is reprocessed in La Hague by Areva. The glass canisters containing vitrified HLW are stored at the production sites Marcoule, Cadarache, and La Hague (Lehtonen 2015). The final forecasts for the generated waste of the operational nuclear fleet—assuming an average life of 50 years—is expected to be around 10,000 m^3 (OECD/NEA 2016b). The Waste Management Act established the way to treat radioactive waste and set the direction of research undertaken by the government agency ANDRA. Research for a final storage is mainly undertaken at the 500-metre deep underground rock laboratory Cigéo in Bure situated in clays. ANDRA expects to present its master plan to the government for operation and disposal at Cigéo site in 2017 and expects a construction permit in 2018. Construction should start in 2020 and the start of the pilot phase is set by law to 2025. Contrary to other countries, research is also undertaken in the field of partitioning and transmutation, and long-term surface storage of wastes following conditioning. A major part of the low-level long-lived waste is the graphite from the GCRs, which is probably going to be stored 200 meters underground in a layer of clay as ("Intact Cove Disposal") or going to be stored with the HLW in the Cigéo disposal (Ministry for Ecology, Sustainable Development and Energy 2014).

7 According to EDF's previous time schedule, the critical path of the former GCR initial decommissioning project consisted of the graphite removal from the reactor core and the decommissioning of the reactors was already well behind in schedule in 2011 (Laurent 2011).

2.2 Financing

Applying the polluter-pays-principle, the operators of nuclear power plants are responsible to bear all costs related to decommissioning and waste management. The financing scheme is based on two different kinds of funds. The first is characterized by a segregated internal fund set up by EDF and Areva[8] and managed under separate accountability. Besides that, there are two more internal restricted funds related to ANDRA: one for research for future storage facilities and one for the construction and operation of a future storage facility for medium and long lived high-level waste. The majority of future costs is related to the facilities of EDF. **Table 1** provides this cost estimation and the provisions set aside by EDF.

Tab. 1 Estimated costs of EDF end of 2016 in million EUR (EDF 2017).

Purpose	Estimated costs end of 2016 (Mio. EUR)	Provisions set aside end of 2016 (Mio. EUR)
Spent Fuel Management	18,460	10,658
Long-term Radioactive waste management	29,631	8,966
Nuclear plant decommissioning	26,616	14,122
Last cores	4,344	2,287
Total	79,051	36,033

In total, the estimated costs are more than 79 billion EUR, with the main parts being for the geological disposal facility Cigéo (~ 30 billion EUR) and decommissioning (~ 27 billion EUR). Critical reports about the cost estimations mention that, especially the decommissioning costs of 27 billion EUR for 58 reactors are underestimated. Extrapolations of costs estimations of other countries show that EDF is expecting comparatively low costs per unit (Cour des Comptes 2014). EDF argues that the costs will be lower due to the high standardization degree of their fleet and because multiple reactors are situated on the same site (Assemblée Nationale 2017). Nonetheless, the cost estimations are increasing continuously every year. In only three years, since the end of 2012, the estimated costs increased by nearly 10 billion EUR. The current provisions are discounted with an interest rate of 4.2% and an assumed inflation rate of 1.5%. As always, little changes of the estimated interest or inflation

8 Areva has no NPPs and operates nuclear facilities like the reprocessing center in La Hague.

rate for provisions and cost can have large influences on the calculations resulting in an underestimation of the needed financial resources. In its assessment of the EDF decommissioning and waste strategy, the French National Assembly could not the share "the excessive optimism" of EDF for its future decommissioning projects. The report concluded that the decommissioning and the clean-up will take more time, the technical feasibility is not fully assured, and will cost more money than EDF currently anticipates (Assemblée Nationale 2017).

The two funds set up by the waste management agency ANDRA are fed by payments from the operator's internal funds at the time they are needed. The only fund fed right now is the research fund, receiving payments through a tax paid by the operators. As there is not yet a construction license, the construction fund is currently not fed but the operators make payments from their internal funds to ANDRA's general budget to finance operations related to the storage facilities for short-lived, medium-level wastes. AREVA and EDF were forced to advance their back-end provisions and accountancy practice because of partial privatizations. Both have now set up restricted internal segregated funds for the financing of the nuclear back-end. EDF feeds its fund by a charge of 0.14 Eurocent/kWh included in the price of electricity. Due to the Waste Law of 2006, the assets in the funds of EDF and Areva have to be accounted separately and the market value has to be at least as high as the provisions to be covered. In cases of insolvency or bankruptcy of an operator, the state can claim right over the assets. The internal funds are supervised by an administrative authority, who is authorized to impose corrective measures. This also includes the right to impose payments to ANDRA's budget. A detailed report about the estimated costs, the timing and the value of the provisions has to be presented at least every three years (European Commission 2013).

2.3 Conclusion for France

The operators EDF and CEA are responsible for the decommissioning of their power plants. While the latter is clearly a public agency, this classification is not so clear for EDF. The major shareholder of the private company is the French state (over 85%), making EDF de facto a public enterprise. The gained decommissioning experiences are not sufficient and the strategic impact of Chooz-A for the future decommissioning of the operational PWRs is questionable. If Chooz-A finishes in 2025, the process will have taken 34 years to complete. The decommissioning process of the GCRs has not really started yet and an end is not in sight. A long time frame is to be expected, reaching well into the 22nd century. The management of high-level radioactive waste is in the hands of the public agency ANDRA. The anticipated start of operations of the HLW disposal facility to begin in 2025 is ambitious. The financial aspects of the nuclear back-end in France are dominated by questionable cost estimations and hence set aside financial resources which are likely to be too low. The internal segregated funds are managed by the operators and subject to an administrative control and oversight by national authorities. If this control will be able to prevent a shortfall of financial resources in the future is uncertain.

3 United Kingdom

The U.K. currently operates 15 NPPs on 8 sites—all operated by EDF Energy, a subsidiary of EDF—and has an installed nuclear capacity of 8.8 GW representing a nuclear share of 20.4% of the British electricity production in 2016 (IAEA 2017). The latest shutdown was Wylfa-1 in 2015. A particularity of the British nuclear feet is that, with the exception of Sizewell B (PWR) only GCRs are operational. At the moment, EDF Energy is considering lifetime extensions for its nuclear fleet until 2023[9], and is therefore investing about 600 Mio. GBP in plant upgrades (EDF 2017). Currently, there is a controversial discussion about plans to commission new reactors with a total capacity of 16 GW starting in 2030, including the well discussed NPP Hinkley Point C.

3.1 Production

3.1.1 Decommissioning of nuclear power plants

The first generation of British NPPs—the so-called Magnox line—was operated by the publically owned British Nuclear Fuels Limited and U.K. Atomic Energy Authority (UKAE) and is in a state of shutdown now. The public body Nuclear Decommissioning Authority (NDA) is responsible for the decommissioning of this legacy fleet. The NDA estate comprises besides the Magnox NPPs, research centres, fuel-related facilities, and Sellafield, the most hazardous site in Europe. Here the site operations include fuel reprocessing, fuel fabrication, and storage of nuclear materials and radioactive wastes. With the exception of the latter, all sites are managed through private-sector consortia, while Sellafield is managed by the NDA itself. The NDA has employed more than 3,500 contractors all over the U.K.

9 Sizewell B will probably be extended until 2055.

and already spent around 12 billion GBP; contracting is critical to the NDA, as 95% of the NDA's funding is spent externally (NDA 2016c).

The 17 sites of the estate are grouped into 6 Site Licence Companies (SLC). The NDA owns these sites and takes the role as the supervising and contracting authority and is turning the management over to the contractors, the so-called SLCs under European public procurement law. The latter are the long-term shareholders of the sites but the management is periodically open to competition. The winner of these contracts acts as the Parent Body Organization (PBO). The PBO receives the shares of the SLC and organizes the strategic management for the duration of the contract. This mechanism was introduced with the idea to increase the efficiency of the procedure by opening the work to private contractors (MacKerron 2015). The NDA is responsible for defining both the target and the timing of decommissioning and remediation, allowing the SLCs to determine how best to deliver this outcome. The current plans of the NDA indicate that it will take around 110 years to complete the core-mission of nuclear clean-up and waste management (NDA 2016c).

With the exception of Calder Hall 1–4, part of the Sellafield complex, all the sites with Magnox reactors are operated by Magnox Limited. Since 2014 Cavendish Fluor Partnership[10] is the current PBO and hence the long-term owner of Magnox Ltd. and supplier of the strategic management and additional resources. The Sellafield complex is operated by the SLC Sellafield Limited. The organization of this site was changed in April 2016 and Sellafield Ltd is now a wholly owned subsidiary of the NDA. A detailed review concluded that the complex, technical uncertainties at the Sellafield site were less suited to the PBO model (NDA 2016a). Its mission to retrieve nuclear waste from some of the world's oldest nuclear facilities extends well into the 22nd century and the sums of money involved are much greater than on other NDA sites.

Since 1977, 30 reactors were shut down and 26 of these are currently in the dismantling process. The current strategy for the Magnox fleet is the Long-Term Enclosure approach. With the exception of the Wylfa reactors, the reactors are defueled and most of the systems external to the biological shield have been removed. According to NDA's strategy, the biological shield, the reactor pressure vessel, the external pressure circuit, and steam generators would be sealed and stored. The dismantling of the reactors will begin 85 years after the shutdown of the plant. However, the NDA and Magnox Ltd are currently reviewing their strategy as there have been advances in remote decommissioning techniques and considerable experience gained in remote handling, packaging, and storage of highly activated waste at Magnox sites. In addition, an improved understanding of

10 A consortium of Cavendish Nuclear—a subsidiary of Babcock International—and Fluor.

the implications of radioactive decay has shown that after the long period of Long-Term Enclosure, the larger amount of the reactor waste will still not be suitable for management as low-level waste. A last development is the realization that the reduction in decommissioning costs with the increase in deferral time is largely offset by the increased cost of preparing and managing the Long-Term Enclosure of the reactor. The waste has been conditioned on-site and interim storages have been built to store the waste until the final disposal route is available. Some site decommissioning and remediation work has been undertaken at most sites with a major focus on the preparation of the ponds for the Long-Term Enclosure state. Since 2011 the focus has been on the plants in Bradwell and Trawsfynydd. Magnox is working towards a target of placing all the reactors into the Long-Term Enclosure state by 2028. The ultimate goal for NDA's mission is to achieve the end state of all sites by 2125 (NDA 2016c).

3.1.2 Storage of high-level wastes

The NDA advocates an approach where wastes are managed according to the radiological, physical, and chemical properties and divides its strategy of radioactive waste management in two topics: Higher Activity Waste (HAW) and Lower Activity Waste (LAW).[11] According to the NDA (2015), the radioactive waste stocks and future arising sum up to around 4,720,000 m^3 in terms of final packaged volume. About 90% of this volume can be attributed to LAW and about 10% consists of HAW. 75% of the NDA-owned HAW is from the Sellafield Site and about 22% from Magnox sites (NDA 2015). Within the U.K., there are large quantities of graphite present with approximately 60,000 tonnes on the Magnox sites alone (NDA 2016b). As the dismantling of the reactors is deferred, the biggest amount of graphite will arise from 2070 onwards. High-level vitrified waste stemming from reprocessing SNF is stored in stainless steel canisters in silos at Sellafield. In 2016, a dry cask storage facility for SNF was commissioned at the Sizewell B station. U.K. policy states that SNF management is a matter for the commercial judgement of its owners, subject to meeting the necessary regulatory requirements. SNF is not considered as waste and the U.K. has a closed fuel cycle in place, i.e. used fuel was reprocessed. If the U.K. government gives up reprocessing and declares SNF as a waste, SNF would be

11 The term HAW refers to all radioactive material that has no further use and is either LLW, ILW or HLW and deemed not suitable for the disposal at the LLW repository in Cumbria or the LLW repository in Dounreay. The strategy of the NDA consists in converting HAW into a form that can be safely stored and managed for many decades awaiting the opening of the geological disposal facility (NDA 2016b).

consigned to a geological disposal facility. The aim of the NDA is to reprocess at least all of the Magnox fuel, which should be achieved by the year 2020 (NDA 2016c).

The long-term management policy for HAW is to package and store wastes in interim storages until they can be transferred to a geological disposal facility. The inventory to be disposed of is currently being stored by the waste owners.[12] In 2014, the Department of Energy and Climate Change (DECC) published a White Paper on implementing a geological facility and the Radioactive Waste Management Limited (RMW) was established as an NDA subsidiary. As a legal entity, RWM will be able to apply for and hold the regulatory permits and licenses required for the siting, construction and operation of a geological disposal facility (DECC 2014). A public agency is therefore responsible for the preparatory work to plan the geological disposal of HAW and delivering the disposal facility. Until today no possible suitable sites have been identified. The detailed layout and design of the facility will depend on the waste inventory and the specific geological characteristics of the site. The underground facilities are expected to comprise a system of vaults for the disposal of ILW, and an array of engineered tunnels, for the disposal of HLW and SNF (DECC 2014). During the construction and operational stage, which will last around 100 years, wastes that have been placed in the facility could be retrieved, which is not an option after the closing of the facility. The siting is still based on a voluntarist approach, i.e. the willingness of local communities to participate in the process, although this approach has already failed once. Current plans predict the deep geological facility being available around 2060 (NDA 2016c).

3.2 Financing

Following the reorganization of the nuclear sector in the United Kingdom there, are three different financing systems for the nuclear back-end in place: one for the NDA facilities, one for the reactors owned by EDF Energy, and one for possible newbuild power plants—the decommissioning costs for the nuclear facilities in the U.K. have to be considered separately for the NDA and the EDF Energy sites. The NDA expects in their annual report for 2016/17 discounted costs of 116 billion GBP for more than 120 years; the majority (around 75%) of the costs are attributed to the Sellafield site alone (NDA 2017).[13]

---

12 NDA and its SLCs, EDF Energy, Urenco U.K. Ltd, Ministry of Defence, GE Healthcare and other non-nuclear users of radioactive material.

13 The uncertainty of this estimation was mentioned in the NDA annual of 2012/13 report as follows: "*Given the very long timescale involved and the complexity of the plants and*

The decommissioning process is managed by the NDA and undertaken by contractors, which are primarily financed through public funds. The annual budget for the NDA is set by the U.K. Department for Energy and Climate and HM Treasury (OECD/NEA 2016a). In addition to the governmental funding, the NDA generates income with commercial activities. In the commercial year 2016/17, the NDA earned around 1 billion GBP, with 612 million GBP coming from reprocessing and waste management activities (NDA 2017). With the shutdown of Wylfa, income through selling of electricity production, which used to decrease the payments from the taxpayers, came to an end. So the funding of the NDA will become even more dependent on the British taxpayers in the future.

The decommissioning of the EDF Energy NPPs will be primarily payed by the Nuclear Liabilities Fund (NLF), an external segregated fund established by the U.K. government in 1996. The only function of the fund is to provide funding to meet specific waste management costs and the decommissioning liabilities for the NPPs originally owned by public utility British Energy (OECD/NEA 2016a). The NLF had assets with a market value of around 8,935 million GBP at the end of the financial year 2015 (NLF 2015). In 2005, when British Energy was restructured and became EDF Energy, the U.K. government announced that it would fund the qualifying liabilities for the case that they exceed all the assets of the fund (OECD/NEA 2016a). The owner of the NLF is the Nuclear Trust, a public trust established under Scottish law.[14] When the NLF was established it received an initial endowment of 228 million GBP from the U.K. government. Today, the fund is fed by two sources: one source is a small quarterly payment by EDF Energy, the second and predominant sources are the revenues of the investments of the fund (OECD/NEA 2016a). In the financial year 2014/15, EDF Energy made contributions of around 26.5 million GBP to the fund and the operating profit before tax was around 155 million GBP (NLF 2015). If EDF Energy wants to receive payments from the fund to meet liabilities, it has to apply to the NDA, which acts as an agent of the U.K. government. The NDA as the administrator of the Liabilities Management Agreements approves the NLF payments for decommissioning and waste management. Table 2 provides the cost estimations and the provisions set aside by EDF Energy. The provisions for the eight operational NPPs of EDF Energy are calculated with a discount rate of 2.7% and an implicit inflation rate based on long-term forecast of adjusted retail prices.

---

material being handled, considerable uncertainty remains in the cost estimate particularly in later years."

14 The five trustees—three are appointed by the U.K. government and two by EDF Energy—also act as the directors of the NLF.

Tab. 2 Estimated costs and related provisions for EDF Energy nuclear backend in million GBP (EDF 2017).

Purpose	Costs based on year-end economic conditions of 2016 (Mio. GBP)	Amounts in provisions at present value in 2016 (Mio. GBP)
Spent Fuel Management	3,101	1.771
Long-term Radioactive waste management	5,326	888
Nuclear plant decommissioning	15,808	6,190
Last cores		1,373
Total	24,230	10,222

According to EDF's financial statement, the provisions for decommissioning and waste management are reported in the assets as "receivables" (EDF 2017, 93). The current value of the fund exceeds the discounted cost estimates of EDF Energy. But already in 2012, the NLF expressed the view, that the fund may not be large enough in the end. In addition, the U.K. government insists that the fund is deposited almost entirely in the National Loans Fund to earn an annual rate of interest used to reduce the overall U.K.'s public debt (MacKerron 2015). EDF Energy is responsible for all operational aspects of the decommissioning of the existing NPPs, but the U.K. government has the power to decide to transfer the decommissioning responsibility to the NDA at any point after the electricity generation at the power stations ended (OECD/NEA 2016a).

According to the Energy Act of 2008 operators that want to construct new nuclear power plants have to establish secure financing arrangements and exact plans for decommissioning and disposal before they get the application to build a new plant. The financing has to be realized by an independent external fund that will be fed during the operational time of the plant with levy of a certain amount per generated kWh.

3.3 Conclusion for the U.K.

The public agency NDA is responsible for the decommissioning of the legacy fleet. The licenses remain with the NDA, but the work is tendered to a PBO, a consortium of private enterprises. This was a changed in 2016 for the Sellafield site, where a public organizational model is seen as more suitable for the complex and long-lasting clean-up of the site. Nearly all the shutdown reactors have been defueled and are currently being prepared for the Long-Term Enclosure. A long time frame is to be expected, reaching well into the 22nd century. For the current operational NPPs the operator EDF Energy—a private enterprise and subsidiary of EDF—is responsible for the decommissioning. Concerning the high-level waste management, the construction of a disposal facility is the scope of a NDA subsidiary and thus a public matter. Up to now all site selection activities failed and have to restart again. The financing of decommissioning and waste management will be challenging too, especially for Sellafield. The costly decommissioning and site remediation of the legacy fleet have to be financed by the taxpayers over more than 100 years. The lessons learned from the shortfall of former provisions led to the establishment of an external segregated fund for the operational NPP fleet, which should prevent public payments in the future but it remains questionable if the financial resources set aside in the fund will cover these costs.

4 Germany

With a nuclear share of 13.1 percent (80.07 TWh) in 2016 (IAEA 2017), Germany has the smallest share of nuclear energy among the observed countries, but the 28 shutdown NPPs (in 2017) constitute the most diverse NPP fleet to dismantle of the observed countries (see Fig. 1). The 13th amendment of the Atomic Energy Act in August 2011 withdrew the operating licenses of the seven oldest NPPs and Krümmel. The remaining eight operational plants will be gradually shutdown by the year 2022. The current fleet is operated by the German utilities EnBW, PreussenElektra (E.ON subsidiary) and RWE as well as the Swedish utility Vattenfall Europe Nuclear Energy. Siemens Kraftwerke Union AG (KWU), later to be dissolved with the French firm Framatome into Areva NP, built all the nuclear steam supply systems—the reactor and the reactor coolant pumps and associated piping in an NPP—in the operational power plants. The operational Pre-Convoi PWRs (Vor-Konvois) will be shut down in 2021. The Convoi reactors (Konvoi)—the latest PWR-design commissioned in the late 1980s—are the last NPPs to be shut down in 2022.

4.1 Production

4.1.1 Decommissioning of nuclear power plants

The different shutdown reactors are in different stages of their decommissioning process: two are in a stage of Long-Term Enclosure (LTE), three plants have been successfully dismantled and released from regulatory control, two plants have been dismantled but await regulatory release, while the remaining NPPs are currently in different phases of the decommissioning process.

The BWR Lingen was put into Long-Term Enclosure[15] in 1998, the request for decommissioning was submitted to the regulatory authority by RWE in 2008 and approved in 2016. The three NPPs—two BWRs and one HWGCR—that have been successfully decommissioned and released from regulatory control were rather small prototype reactors. Of the three, VAK Kahl (BWR, 25 MW), was the only reactor that operated for a longer period of time (24 years) and was after its shutdown immediately dismantled and released as a greenfield in 1998.[16] The BWR Würgassen was the first larger commercial NPP to be dismantled. The reactor was of the first generation BWRs and had a capacity of 640 MW. After 19 years of operating time, the NPP was shut down in 1994 and de facto decommissioned by 2014. During the dismantling in the hot zone (stage three) PreussenElektra tendered the dismantling and conditioning of the reactor vessel internals to Areva NP GmbH. The site has not yet been released as a greenfield as parts of the buildings are used as an interim storage for low and medium level wastes awaiting the opening of the disposal site Konrad. Decommissioning of Gundremmingen-A, another first generation BWR (237 MW), started in 1983 and in 2016 the majority shareholder RWE finished the actual decommissioning process with the decontamination of the buildings (Bredberg et al. 2017), but as it is the case with Würgassen the site cannot be released from regulatory control as parts of the building are used for future decommissioning works for the still operational units B and C.

There are currently 12 power plants in the process of being dismantled, the major part of them are PWRs.[17] The NPP Stade has nearly finished decommissioning, here PreussenElektra again tendered the removal and conditioning of the reactor vessel

15 The other NPP in Long-term Enclosure is the pebble bed HTGR THTR-300.

16 HDR Großwelzheim (Superheated BWR, 25 MW, 1969–1971) was decommissioned from 1988 until 1998. Niederaichbach (HWGCR, 100 MW, 1973–1974) was decommissioned from 1987 until 1995.

17 Other NPPs being decommissioned are AVR Jülich (HTGR, 15 MW), KNK II (FBR, 17 MW), and MZFR Karlsruhe (PHWR, 1,219 MW); all three reactors are decommissioned by the public company EWN respectively EWN subsidiaries.

internals to Areva. The legacy fleet of the former German Democratic Republic (GDR) Rheinsberg and the five units of Greifswald are being decommissioned by EWN Entsorgungswerk für Nuklearanlagen (EWN), a public company under control of the Federal Ministry of Finances. For both sites, the deferred dismantling strategy was chosen. The Rheinsberg reactor pressure vessel was transported to the centralized on-site interim storage facility (Zwischenlager Nord), also operated by EWN. In Greifswald the reactor vessel internals of reactor one and two were immediately dismantled and conditioned. For the internals of reactor three and four as well as all the reactor pressure vessels of the five reactors deferred dismantling was the strategic choice; also in storage and planned to be dismantled later on are the 17 steam generators and parts of the primary cooling system. Decommissioning of Obrigheim, operated by EnBW, should be completed sometime between 2020 and 2025. Mülheim-Kärlich, the only RWE power plant and the only Konvoi reactor currently being dismantled, is entering the reactor decommissioning phase and is planned to be released from regulatory control in 2021.

All NPPs shut down in 2011 and Grafenrheinfeld shut down in 2015 have submitted their decommissioning proposal at the regulatory authority; the proposal was also submitted for Gundremmingen B, which shut down in December 2017. Of these, Brunsbüttel, Isar 1, Biblis A and B as well as Neckarwestheim 1 have been granted their decommissioning requests in 2017. The major part of the NPPs are still in the post-operational phase or are just starting with the decommissioning process. The German operators currently face several obstacles in order to be able to conclude the decommissioning process in a timely manner without escalating costs. At the moment there is still not a sufficient number of transport and storage casks being produced in order to defuel the reactors. The quick shutdown of the NPPs after the Fukushima incident caused a high number of special fuel rods—not completely burnt-down fuel—in the reactor cores. For these fuel rods no casks for the safe storage have been approved by the regulatory authority at this point. The defueling of the reactors cores and subsequently storing in an interim storage cannot be achieved until the required casks are available.

4.1.2 Storage of high-level wastes

The high-level radioactive waste consists of SNF and vitrified structures from the reprocessing process. The political decision to stop German reprocessing was final in 1989, after this the German operators invested in the French reprocessing facility in La Hague. Until 2005, nearly half of all the SNF was sent to France and the U.K. for reprocessing. From 2005 on, the policy was direct geological disposal—which meant interim storage of SNF and no more reprocessing (Hocke and

Kallenbach-Herbert 2015). For this, the utilities operated through Gesellschaft für Nuklearservice (GNS) two centralized interim storage facilities in Gorleben and Ahaus; a third facility is Zwischenlager Nord operated by the public company EWN. But the major part of the SNF is currently still stored in the storage pools or in one of the twelve de-centralized on-site storage facilities.

In 2016, the institutional framework of the waste management process was changed with the introduction of the law aimed to restructure the responsibilities in the nuclear waste management process.[18] The ownership of the centralized interim storage facilities was transferred to the newly created public company Gesellschaft für Zwischenlagerung (BGZ, "company for interim storage"), which will also take over the decentralized interim-storage facilities and the low-level waste repositories. According to the final report[19] of the high-level waste management commission set up by the Repository Site Selection Act in July 2013 –the site for the deep geological facility with "the best safety" for the 30,000 m^3 of high-level waste[20] is to be found in a three-phase process, accompanied by extensive public participation. For the up 200,000 m^3 of low- and intermediate waste and salt mixture to be retrieved from the Asse II geological facility currently no disposal solution exists. The goal of the Repository Site Selection act is a "deep geological repository with reversibility" in either clay, salt or granite. The plans foresee a start of operation of the disposal site after 2050, but more realistic estimates expect the start after 2080 (Thomauske 2015). After 50 years of operation time the disposal facility is planned to be sealed off.

4.2 Financing

The funding system in Germany differs between purely public-owned facilities, facilities with mixed-ownership and the facilities in private ownership. The costs for the decommissioning of the former owned nuclear facilities are financed from the current public budget; the Federal Government covers the majority of the costs, while some are covered by State Governments. The most common examples for public funding are the former GDR NPPs Greifswald and Rheinsberg, the decommissioning of which is totally funded by the Ministry of Finance. For the facilities in mixed-ownership, there is a proportional split of the costs between the public and the private utilities clarified by special arrangements (European Commission

18 Gesetz zur Neuordnung der Verantwortung in der kerntechnischen Entsorgung (BT 768/16).

19 See Kommission Lagerung hoch radioaktiver Abfallstoffe (2016).

20 This includes the high-level waste until the shut-down of the last power plant on 31.12.2022.

2013). However, the majority of the costs are related to the nuclear back-end of the privately-owned NPPs. In 2015, the auditing company Warth & Klein Grant Thornton AG provided on behalf of the German government an estimation of the whole costs for the nuclear back-end of 23 commercial NPPs: 47.5 billion in 2014 Euros. The several undiscounted cost categories are presented in Table 3.

Tab. 3 Estimated Nuclear Back-End Costs in Germany (Warth & Klein Grant Thornton AG Wirtschaftsprüfungsgesellschaft 2015).

Cost categories	Undiscounted costs 2015–99 in prizes of 2014 (Mio. EUR)	Discounted costs 2015–99 with nuclear specific inflation rate of 1.97% (Mio. EUR)
Decommissioning and dismantling	19,719	30,214
Casks, Transport, Operational Wastes	9,915	52,840
Interim Storage	5,823	26,770
Low and Medium Waste Disposal (Schacht Konrad)	3,750	9,016
High Level Waste Disposal	8,321	50,966
Total costs	47,527	169,808

In addition, there are costs the for the public funded decommissioning of Greifswald and Rheinsberg and for research facilities: The initial decommissioning costs for Greifswald were estimated to be about 4 billion EUR and for Rheinsberg 600 million EUR; the latest cost estimate in 2016 was around 6.5 billion for both facilities. As always, all cost estimations are subject to many uncertainties related to expectations about future inflation rates, cost increases, and time delays. The estimation of Warth & Klein Grant Thornton AG considered this by a computation of the estimated costs with a nuclear specific inflation rate of 1.97% until 2099, which resulted in total discounted costs of around 169.8 billion EUR. The audit concluded that the effect of changing the estimated nuclear-specific inflation rate on future costs is strong and causes the most uncertainties.

In the old financing system, the financial resources to cover decommissioning and waste disposal were managed by the private companies in internal non-segregated funds with no public authority controlling them. The companies set up the provisions according to international accounting standards and were free to choose where to invest it. The OECD/NEA (2016a) highlighted the unregulated and uncontrolled system of internal non-segregated funds itself as the most critical aspect of the German system. In the case of a bankruptcy of the operator, the financial resources to cover future costs would probably have been lost. The financial situation of the utilities was and still is not secured to exclude the risk of bankruptcy in the future. In the case of the loss of the funded provisions, the public budget would have been obliged to cover the costs. Table 4 presents the provisions of the companies as mentioned in their annual financial statements at the end of 2014. The calculations of the private companies were based on an average interest rate of 4.58% and the before mentioned nuclear specific inflation rate of 1.97%; both are highly uncertain. A lower real interest rate on the provisions set aside would have had a crucial effect. With an average interest rate of 2.03 %, the present value of the set provisions would have to be today around 77 billion EUR to cover the future costs (Warth & Klein Grant Thornton AG Wirtschaftsprüfungsgesellschaft 2015).

Tab. 4 Provisions of German Operators end of 2014 (Warth & Klein Grant Thornton AG Wirtschaftsprüfungsgesellschaft 2015)

Company	Provisions end of 2014 (Mio. EUR)	Interest rate for the calculations (%)
E.ON	16,567	4.7
RWE	10,367	4.6
EnBW	8,071	4.8
Vattenfall	3,014	4.0
Stadtwerke München	564	4.38
Total	38,288	Average 4.58

On behalf of the government, an expert commission reviewed the financing system and provided reform proposals to meet the actual risk related to the system of internal non-segregated funds.[21] Their recommendations and the new law published in December 2016 (BT 768/16) led to a fundamental change of the German funding system. This change was also motivated by concerns that the private utilities would not be able to cover all future liabilities with their internal non-segregated financial resources due to the experiences with high cost increases in former decommissioning and waste disposal projects. There were annually cost increases between 2.9 and 6 percent, which is much higher than the general inflation rate or the assumed

21 See KFK – Kommission zur Überprüfung der Finanzierung des Kernenergieausstiegs (2016).

nuclear-specific inflation rate (Warth & Klein Grant Thornton AG Wirtschaftsprüfungsgesellschaft 2015). Based on the reform proposals, an external segregated fund was implemented in 2016, which will have to finance all aspects related to waste disposal, i.e. interim and final storage. The fund was fed by the former provisions for these tasks totalling 23 billion EUR, including a risk premium. The utilities are still responsible for decommissioning and for the conditioning of the wastes, but all tasks as well as the operation of the interim storage facilities will be done by public companies and paid from the fund. The responsibility as well as risks, including the financial ones in the case of insufficient set-aside money, will have to borne by the public, which infringes the polluter-pays-principle (Jänsch et al. 2017).

4.3 Conclusions for Germany

Germany was able to gain some experience in the decommissioning of NPPs. The four private utilities have chosen the Immediate Dismantling strategy in nearly all cases. The public enterprise EWN chose the deferred dismantling strategy for the reactor pressure vessels. The private operators carry out themselves the dismantling process, although specialized private companies carry out some part of the work; this is especially true for the technologically challenging dismantling of the reactor pressure vessel and its internals. All NPPs currently in the post-operational stage still face several obstacles in order to be able to conclude the decommissioning process in a timely manner without escalating costs, e.g., still not a sufficient number of transport and storage casks being produced in order to defuel the reactors. The future disposal path for HLW is still highly uncertain—this also applies for the disposal of low- and medium-level wastes—and has retroactive effects on the timing, progress, and costs of the decommissioning process. Additionally, all estimated future costs are underlying many uncertainties due to cost increases and interest rates. This is especially true for all future costs related to the management for both low-and intermediate, and high-level waste. It is questionable if the financial resources set aside in the fund will cover these costs.

5 Conclusions

Overall, the three case studies show that the biggest challenges concerning the decommissioning and storing still wait for solutions. Decommissioning was in most cases neglected, only Germany has gained some experiences in decommissioning NPPs but no large-scale reactor (over 1 GW) has successfully been decommissioned. It can be stated that overall the experience is still lacking, considering the high need for decommissioning in the coming years in all of the observed countries. Until now, no scale effects could be observed, if EDF can reap scale effects due to the standardization of its fleet remains to be seen. The preferred strategy for light water reactors is Immediate Dismantling, while in some cases the radiological decay was used and the deferred dismantling strategy was applied to highly activated components. In contrast, the worldwide preferred strategy for GCRs is the Long-Term Enclosure. EDF is now also considering this strategy for its French GCR fleet due to underestimated technological challenges and missing graphite disposal routes. This postpones the end of the decommissioning of the legacy fleets in the U.K. and France well into the 22nd century. In all three cases, the decommissioning of the NPPs is critical due to the missing disposal facilities, which led to the construction of interim storage facilities.

Considering the production of the decommissioning process in the observed countries, we have two public companies EWN and Magnox Ltd. organizing the decommissioning of the legacy fleets, while the latter tenders the work to a private consortium. In Germany and France, the operators are responsible for the decommissioning of their NPPs. Some part of the work, especially the most challenging work—the dismantling of the reactor pressure vessel and its internals—has been tendered to specialized nuclear companies. In the U.K., the decommissioning of the operational NPPs has to be done by the operator, but the NDA has a "take-over" option and can decide to transfer the decommissioning responsibility to the public body. On the other hand, the high-level disposal facility is in the three countries the scope of the government. If the construction permit for Cigéo is granted, France will have the most advanced process of implementing a deep geological disposal facility while Germany and the U.K. are still in the site selection process.

The financing of decommissioning and radioactive waste management will be a long-term challenge in all three countries. All cost estimations are underlying uncertainties due to long time-scales, cost increases, and estimated interest and inflation rates. This could lead to an underestimation of future costs. Of all the observed financial systems, the old German system of internal non segregated funds seemed to be the most uncontrolled and unsecured. This led to a change in the financial system and the implementation of an external segregated waste fund. In France, the financial resources are held in internal segregated funds with administrative control and oversight by national authorities. However, this does not prevent comparatively optimistic cost estimations and due to this, likely inadequate set aside financial resources. In the U.K., the costliest aim will be the decommissioning and site remediation of the legacy fleet and Sellafield payed by the taxpayers over the next 100 years. To prevent a repetition of a shortfall of funded provisions, a system with an external segregated fund for the operational nuclear fleet was introduced. This approach seems to be the most suitable to finance the future cost of the nuclear back-end, even if it also could not overcome the problem of too low cost estimations.

References

Assemblée Nationale, 2017. "Rapport d'Information Déposé En Application de l'article 145 Du Règlement Par La Mission d'Information Relative à La Faisabilité Technique et Financière Du Démantèlement Des Installations Nucléaires de Base." Paris.

Autorité de sûreté nucléaire, 2016a. "EDF: Des Retards Sur Le Programme de Démantèlement." 2016. https://www.asn.fr/Informer/Dossiers-pedagogiques/Le-demantelement-des-installations-nucleaires/Les-strategies-de-demantelement-en-France/EDF-des-retards-sur-le-programme-de-demantelement.

Autorite de surete nucleaire, 2016b. "Le Démantèlement Immédiat." 2016. https://www.asn.fr/Informer/Dossiers-pedagogiques/Le-demantelement-des-installations-nucleaires/Les-etapes-du-demantelement/Le-demantelement-immediat.

Bredberg, Ines, Johann Hutter, Kerstin Kühn, Katarzyna Niedzwiedz, Frank Philippczyk, and Frank Thömmes, 2017. "Statusbericht Zur Kernenergienutzung in Der Bundesrepublik Deutschland 2016." Salzgitter: Bundesamt für kerntechnische Entsorgungssicherheit.

Cour des Comptes, 2014. *Le Coût de Production de l'électricité Nucléaire – Actualisation 2014*. Paris: Cour des Comptes.

DECC, 2014. "Implementing Geological Disposal." London: Department of Energy and Climate Change.

EDF, 2017. "Consolidated Financial Statements at 31 December 2016." Paris: Electricité de France.

European Commission, 2013. "EU Decommissioning Funding Data – Commission Staff Working Document." Brussels.

European Parliament, 2013. *Nuclear Decommissioning: Management of Costs and Risks*. Brussels: Policy Department on Budgetary Affairs.

Hitchin, Penny, 2010. "Excavating Chooz A." *Nuclear Engineering International Magazine*. http://www.neimagazine.com/features/featureexcavating-chooz-a/.

Hocke, Peter, and Beate Kallenbach-Herbert, 2015. "Always the Same Old Story? Nuclear Waste Governance in Germany." In *Nuclear Waste Governance – An International Comparison*, edited by Lutz Mez and Achim Brunnengräber, 177–201. Springer VS.

IAEA, 2017. *Nuclear Power Reactors in the World*. Vienna: IAEA.

Jänsch, Elisabeth, Achim Brunnengräber, Christian von Hirschhausen, and Christian Möckel, 2017. "Wer Soll Die Zeche Zahlen? Diskussion Alternativer Organisationsmodelle Zur Finanzierung von Rückbau Und Endlagerung." *GAIA*, Jahrhundertprojekt Endlagerung, 26 (2): 118–20.

KFK – Kommission zur Überprüfung der Finanzierung des Kernenergieausstiegs, 2016. *Verantwortung Und Sicherheit – Ein Neuer Entsorgungskonsens*. Berlin

Kommission Lagerung hoch radioaktiver Abfallstoffe, 2016. *Abschlussbericht Der Kommission Zur Lagerung Hochradioaktiver Abfälle*. Berlin.

Laurent, G, 2011. "EDF Nuclear Plant under Decommissioning Programme CIDEN Organization Projects Achievement." presented at the Special Seminar on the 12th meeting of the WPDDD Commemorating the 10th anniversary of the working party on the management of materials from decommissioning and dismantling (WPDD), NEA Office France.

Lehtonen, Markku, 2015. "Megaproject Underway – Governance of Nuclear Waste Management in France." In *Nuclear Waste Governance – An International Comparison*, edited by Lutz Mez and Achim Brunnengräber, 117–38. Berlin: Springer VS.

MacKerron, Gordon, 2015. "Multiple Challenges – Nuclear Waste Governance in the United Kingdom." In *Nuclear Waste Governance – An International Comparison*, edited by Luz Mez and Achim Brunnengräber, 101–16. Berlin: Springer VS.

Martelet, Bertrand, 2016. "EDF's Expertise and Position in Nuclear Decommissioning." presented at the World Nuclear Decommissioning & Waste Management Congress (Europe) 2016, London, UK.

McIntyre, P.J., 2012. "Nuclear Decommissioning – Planning, Execution and International Experience." In *Nuclear Decommissioning – Planning, Execution and International Experience*, edited by Michele Laraia, 33–48. Woodhead Publishing Series in Energy. Cambridge: Woodhead Publishing.

Ministry for Ecology, Sustainable Development and Energy, 2014. *French National Plan for the Management of Radioactive Materials and Waste 2013–2015*. La Défense Cedex: Ministry for Ecology, Sustainable Development and Energy.

NDA, 2015. "An Overview of NDA Higher Activity Waste." Cumbria: Nuclear Decommissioning Authority.

NDA, 2016a. "Explained: The New Model for Managing Sellafield – GOV.UK." Corporate Report. Cumbria: Nuclear Decommissioning Authority. https://www.gov.uk/government/publications/new-model-for-managing-sellafield/explained-the-new-model-for-managing-sellafield.

NDA, 2016b. "Integrated Waste Management – NDA Higher Activity Waste Strategy." Cumbria: Nuclear Decommissioning Authority.

NDA, 2016c. "Strategy – Effective from April 2016." Cumbria: Nuclear Decommissioning Authority.

NDA, 2017. "Annual Reports & Accounts 2016/2017." Cumbria: Nuclear Decommissioning Authority.

NLF, 2015. "Annual Report and Accounts for the Year Ended 31 March 2015." Nuclear Liabilities Fund.

OECD/NEA, 2016a. "Costs of Decommissioning Nuclear Power Plants." Paris: OECD/NEA Publishing.

NDA, 2016b. "Radioactive Waste Management Programmes in OECD/NEA Member Countries – France." Paris: OECD/NEA Publishing.

Seidel, Jan-Paul, and Ben Wealer, 2016. "Decommissioning of Nuclear Power Plants and Storage of Nuclear Waste – International Comparison of Organizational Models and Policy Perspectives." Germany: TU Berlin.

Thomauske, Bruno, 2015. "Current Status of the Final Disposal of High-Level Waste in Germany." presented at the ICOND – International Conference on Nuclear Decommissioning 2015, Bonn.

Warth & Klein Grant Thornton AG Wirtschaftsprüfungsgesellschaft, 2015. *Gutachtliche Stellungnahme Zur Bewertung Der Rückstellungen Im Kernenergiebereich.* Berlin.

Wealer, Ben, Christian von Hirschhausen, Jan Paul Seidel, and Clemens Gerbaulet, 2015. "Stand Und Perspektiven Des Rückbaus von Kernkraftwerken in Deutschland (»Rückbau-Monitoring 2015«)." DIW Berlin, Data Documentation 81. Berlin, Germany: DIW Berlin, TU Berlin.

World Nuclear Association, 2017. "Nuclear Power in France." 2017. http://www.world-nuclear.org/information-library/country-profiles/countries-a-f/france.aspx.

Future Prospects on Coping with Nuclear Waste

Gordon MacKerron[1]

Abstract

Finding safe and publicly acceptable routes for the management of long-lived nuclear wastes has been problematic in all countries that have used nuclear power. The dominant expectation on the part of Governments and the nuclear industry has been that the best option will be deep underground disposal. However even in Sweden, where political consensus has emerged over a site for a repository, disputes continue about long-term safety. Ethical issues, especially inter-generational equity, are relevant given continuing delays in implementing long-term management and where countries, like the UK, continue to build new reactors, achieving political acceptability is more problematic than where new-build is not an option. Failure to resolve nuclear waste issues is a major obstacle to public acceptance of nuclear technology.

1 Gordon MacKerron, University of Sussex, United Kingdom, gordon@mackerron.co.uk

R. Haas et al. (Eds.), *The Technological and Economic Future of Nuclear Power*, Energiepolitik und Klimaschutz. Energy Policy and Climate Protection, https://doi.org/10.1007/978-3-658-25987-7_13

1 Introduction

Radioactive waste is for many people the single most problematic issue surrounding nuclear power. For the great majority of countries using nuclear power, there has been little or no progress in resolving the waste issue. This chapter provides an overview of the main issues, with illustrations from a range of countries.

Generation of nuclear electricity produces radioactive substances, some not found in nature that are extremely hazardous. In many cases they remain potentially harmful to human health and the environment for thousands of years. Although there is now a history of nuclear generation stretching back 60 years – and a military legacy that goes back further – there is, as yet, no long-term management facility for the most problematic civilian wastes completed anywhere in the world.

This is not for want of trying. Most nuclear-using countries have official policies that endorse the option of deep geological disposal (DGD) as the favoured strategy for trying to isolate wastes from the biosphere for very long periods in the future. But few countries have made any significant progress towards building such a repository (Finland and Sweden (Swahn and Kaberger 2015) are rare, if still partial, exceptions). The failure to establish long-term management routes is a predominantly social and political question and can generally be attributed to a lack of trust on the part of affected populations, locally and nationally, in the technical solutions that have been proposed, reflecting in turn a poor history of nuclear waste governance.

2 The wastes

Radioactive wastes arise at several points in the nuclear cycle but the main problem is about so-called 'higher activity' wastes. These have the characteristics that they are highly radioactive and remain hazardous for long periods in to the future – generally measured in centuries or longer.

These higher activity wastes are of two main types. The first are 'high level' wastes (HLW), which besides being highly radioactive and long-lasting are also heat generating and so need periods of cooling before they can be managed effectively [(MacKerron 2015). These wastes are almost entirely the product of fission in reactors and consist either of spent fuel, or some of the separated components of spent fuel, including plutonium. Separation of plutonium and unburned uranium from spent fuel is known as 'reprocessing'. This is a complex chemical and mechanical process that creates new waste streams and is now undertaken by few countries (including the UK, France and Russia, with Japan expected to open its long-delayed reprocessing plant sometime soon (World Nuclear News 2017)). Plutonium is sometimes combined with uranium to make so-called mixed-oxide fuel (MOX) and then used in current commercial reactors. This however only postpones the waste issue. The reason is that it is impractical to reprocess MOX fuel, which is radioactively hotter and more difficult to manage than conventional spent fuel (von Hippel and MacKerron2015).

However, much of the world's civilian separated plutonium is currently stored (especially in the UK and France), pending decisions on whether to treat it as waste or first make it into MOX. The rationale for plutonium separation was originally found in the nuclear weapons' states desire to produce fissile material for bombs (Gowing 1974). Later a new rationale developed. This derived from the intention of several countries from the 1950s onwards to develop 'fast breeder' reactors, which would depend on large quantities of plutonium as a start-up fuel, after which reactors would generate their own fuel. Fast reactors have now been abandoned by most of their original supporting countries (the USA, the UK, France and Germany) though ambitions for fast reactors persist in Russia, India, and China (Cochran et al. 2010).

The second waste category consists of 'intermediate level' wastes (ILW) which are also radioactive and long-lasting but not heat-generating. Most countries expect to manage HLW and ILW (especially the longest-lasting ILW) together in a deep repository (OECD/NEA 2013 chapters 1 and 2). ILW arises from a number of sources, including reactor operation and – where reprocessing of spent fuel takes place – some of the waste streams that reprocessing creates. Other wastes, mostly low level (LLW) are less radioactive and are dangerous for shorter time periods. Such wastes are in most cases already managed with little controversy, often in shallow burial sites. The rest of this chapter concentrates on the higher activity wastes.

3 Proposed technological solutions

For several decades the overwhelmingly dominant expectation on the part of Governments and the nuclear industry has been that higher activity wastes will at some point be subject to buried in a deep geological disposal (DGD). This would be some distance underground – at least 500 metres or more – and would rely both on the host rock as well as engineered safety to contain the wastes safely for what is hoped to be more or less indefinite periods in to the future. While other routes for long-term management of wastes have been seriously considered – for example sub-seabed disposal, shooting wastes into space or beneath ice caps – all have fallen by the wayside in recent years in the face of perceived risks, international law obstacles or excessive cost (CoRWM 2006, Chapter 10).

Several different geologies appear in principle suitable as host environments for a DGD repository. This has included granites (Scandinavia), salt (Germany) and clay (France). (Brunnengraber et al. 2015, Part 1) While the geology is expected to form part of the barrier to the escape of radioactivity, all DGD proposals also expect that there will be multiple engineered barriers. Wastes will be sealed in canisters or drums which themselves contain internal barriers and then further materials, for example bentonite clay, may be used to provide a further barrier between the waste-containing structure and the host rock.

One variant of the DGD idea is the potential use of boreholes for the disposal of particularly difficult wastes, such as immobilized plutonium (Gibb et al. 2008). Such boreholes might be at a depth of 3 – 5 kilometers, where geological formations are in principle especially stable. If such a method became feasible – and at present there are concerns about the reliability of drilling technology to such depths – the relatively small diameter of boreholes suggests that such an option would probably only work for quite small waste volumes.

While the great bulk of opinion favours a DGD route for long-term management of higher activity wastes, there are sharp divisions between those who want a repository closed and sealed as soon as all relevant waste is emplaced, and those who favour retrievability or reversibility for some substantial future period (CoRWM 2006, Chapter 10). The argument in support of prompt closure is that the safety and security cases are stronger if a repository is closed and sealed as soon as possible. In practice such 'early' closure will probably be at least 100 years in the future for major nuclear-using countries and there are concerns that leaving a repository accessible for longer is risky, given the difficulty of knowing how societies will evolve centuries into the future.

The opposing argument – in favour of retrievability – is that maintaining access to waste, proposed in the UK case for up to 300 years (MacKerron 2015), allows for flexibility if circumstances change. There are two possible changes in circumstances envisaged in this argument. The first is that is that new technology might allow the period over which waste is hazardous to be radically reduced. Such ideas already have some currency in the notion that so-called partitioning and transmutation (P&T) might drastically reduce the half-lives of several isotopes (Schneider and Marignac 2008). However P&T would be extremely expensive and in any case could not reduce half-lives of some of the isotopes with longest lives. The second possible change is that it might be possible to recover the waste and use it as fuel for new kinds of nuclear technology. This also has current advocates but runs into the problem that further wastes will be created (OECD/NEA 2013 Chapter 4).

4 Issues of repository design and long-term safety

Beneath this widespread generic endorsement of DGD there are many technical arguments about specific repository design. For example in Sweden the confidence that was once widely shared about the so-called KBS design – which has provided a starting point for other countries' efforts, including Finland and the UK – has been eroded in recent years by concerns that the corrosion-resistance of the copper that is integral to the design might be compromised. Equally there have been long-running disputes in the US scientific community about the long-term safety of the design proposed for the (now abandoned) Yucca mountain site in Nevada (Macfarlane and Ewing 2006).

While no country has expressed the view that an alternative to DGD is a good long-term way forward, some have deliberately postponed any long-term decision. The Netherlands for example has a policy of storing wastes for around 100 years before expecting to make a long-term decision (Arentsen 2015). A few countries, for example Italy and Spain, have been influenced by this approach, but no-one seriously argues that continuing storage is an acceptable long-term way forward. This raises questions of equity between generations that are explored further below. Virtually all other nuclear-using countries are committed to a search for a way to move DGD forward as quickly – in practice generally very slowly – as possible.

In this process of trying to find a way to implement DGD the almost universal assumption has been that every country needs to take responsibility for managing its domestically generated wastes within its own borders. There have been a few exceptions to this. Russia used to take back spent fuel from the reactors it sold to its East European client states, and very small quantities of waste have been transferred between countries (for example between Georgia and the UK) (Abbott 1998). But the principle that wastes should be the responsibility of the states that generate them is still the norm.

Nevertheless proposals periodically surface to develop multi-national disposal sites – for example within Europe and more improbably in Western Australia. While these have superficial attractions, such as possible cost savings and the remoteness of some of the potential sites, thy have all floundered on mixed political and ethical grounds – the unacceptability that some countries will take on burdens that should belong elsewhere. For all practical purposes the issue of waste disposal is one that needs to be solved within each individual nuclear-waste-producing country.

Despite the near-consensus on the desirability of a DGD approach to waste management, the very limited progress to date on implementing such projects is striking. Only Finland and Sweden have made serious progress in this area. In the case of Sweden it is noteworthy that there has been serious engagement with local candidate communities – both of which were very close to pre-existing nuclear sites – and they were given the option of withdrawal if the terms offered were regarded as unacceptable (CoRWM 2006, Chapter 9). In Finland, with a tradition of consensual decision-making, there has been community consent for the construction of a repository and the process of first building an underground laboratory has been ongoing since 2011 (Auffermann et al. 2015). Construction of a full-scale repository has started and it is possible that an operating licence may be granted by 2020. Sweden has also gone through a process of canvassing communities for volunteers and two sites proved willing to act as hosts, with one now selected (Swahn and Kaberger 2015). Even in these countries, the process of emplacing wastes will take many decades, and for other countries the timescale will be substantially longer.

In all other countries progress has been much more limited. In France a candidate site – Bure, in a thinly populated part of eastern France – has been selected in a much more top-down process, but it is not clear when (or of) construction will begin (Blowers 2017, chapter 4). Elsewhere there is even less progress. In the USA, there has been a protracted process in which Yucca Mountain was selected as the preferred site for DGD, but there were a long series of legal challenges and much public and local political resistance over many years which led to the site being effectively abandoned in 2011 (Macfarlane and Ewing 2006). In Germany, where there have long been proposals for three sites for different waste categories, there have been apparently intractable delays, and a fierce anti-nuclear movement, as well as local citizen movements, that have meant that there has been no progress towards a solution for many years (Hocke and Kallenbach-Herbert 2015). In Canada a stepwise process through its NWMO, involving wide-ranging engagement has led to the potential choice of a site in Ontario but recent progress has been slow, with no agreement yet reached (NWMO 2018).

In the UK a new process was initiated in 2003 with the formation of the Committee on Radioactive Waste Management (CoRWM) charged with formulating a new approach to policy. CoRWM reported in 2006, broadly endorsing a DGD end-point and recommending that only a voluntarist approach would work (CoRWM 2006, Chapter 14)]. This was endorsed by the UK Government (Defra, BERR and the devolved administrations for Wales and Northern Ireland 2008) and after local authorities in the area around the Sellafield nuclear site provisionally volunteered there followed a three year period of intense negotiation. However this was brought to a halt in 2012, when Cumbria County Council refused permission to take negotiations any further. This means that the UK, in common with many other countries, still has no agreed site for a repository.

5 Ethics

The objective of implementing long-term management options for higher activity wastes raises important issues in ethics – specifically, equity both within and across generations. Intra-generational equity issues surround the location of waste management sites, where local communities assume the burden of responsibility while distant communities are exempted. This strongly suggests recompense for the community willing to take on a national burden.

Inter-generational equity is more problematic. The benefits of nuclear power in terms of power production have been experienced in the present and past. But in the absence, as seen above, of any satisfactory waste management solution in most countries, the costs and risks associated with nuclear waste will affect future generations, who will experience no compensating advantages (CoRWM 2006, Chapter 6) In terms of environmental principles this is a clear case of violating the 'polluter pays' principle. Attempts to avoid this violation by ensuring that funding is put aside now to cover future costs are not credible. The existence of such funds may mean that future generations have a prior claim on resources to manage waste, but the real resources needed – labour and capital equipment – fall to future generations to find and are at the expense of other potential uses of those resources in the future as well as representing risks to workers and possibly others.

On ethical grounds it is therefore important to try and find a way to manage wastes in a safe and secure way as soon as practically possible because the interests of all generations are at stake. If above-ground storage is more hazardous in the next few decades than burying waste, then the interests of this generation as well as future generations are that burial should be achieved as soon as practical. The argument for early action is therefore strong provided there is confidence that early commitment to a repository does not jeopardize the safety of future generations (CoRWM 2006, Chapter 13). But as argued above, the universal expectation is that even in countries where waste management plans are far advanced, the process of completing the burial of wastes will take up to a century or more. So even an effective waste management programme will cause some burdens to be passed to future generations and possibly expose current generations to significant risks in the meantime.

6 The politics of nuclear waste

Given the scientific consensus, established over decades, that a DGD route is potentially achievable and can in principle be made safe over many future generations, the extreme slowness in the process of implementation needs explanation. The most commonly observed reason for this universal failure is a chronic lack of public trust in the process of nuclear waste governance, and that this lack has made progress at potentially favoured DGD sites impossible – or at best exceptionally slow.

The question behind this observed lack of trust is why it should be so. Until around 2000, decision-making processes for nuclear waste management were almost universally centralized, opaque, used narrowly scientific criteria, and failed to engage with either local communities or publics. This was, especially for nuclear weapons states, partly a legacy of the secret and closed nature of all decisions about nuclear technology. Characteristically, the process was that a small elite group of government officials, nuclear industry leaders and scientists would endorse the idea of geological disposal and then announce a chosen site for a repository. This would be followed by highly vocal, local campaigns against the choice of the site, often involving coalitions of actors of quite different general political views, sometimes national and even international. Abandonment of the proposed site followed (the UK) or there would be delays lasting decades and becoming indefinite (USA and Germany)

This became known as the 'decide-announce-defend' style of decision-making (CoRWM 2006, Chapter 1), to which could then be added 'abandon' (DADA) when local communities and their wider supporters resisted. This exacerbated and to a degree helped create the lack of trust on the part of many citizens towards decision-makers in Government and the nuclear industry.

Around the turn of the century a combination of repeated failures to establish a DGD site and a new more participatory approach to decision-making in several countries – an approach which was applied to other 'difficult' areas of decision involving controversy, such as GM food – led several countries to introduce a more deliberative and open style of decision-making (Chilvers and Burgess 2008). This so-called 'deliberative turn' in decision-making processes has had some positive results. Where such processes have led to some real progress, as in Sweden, there seem to have been two conditions, one pre-existing and the other specific to waste management. The first condition seems to have been relatively high levels of initial public trust in relevant institutions including the nuclear industry. The second, strongly exemplified in Sweden, was a process of deep engagement with communities that first volunteered as repository sites, but were also given the opportunity to withdraw from the process if they found the proposals being offered were unattractive (CoRWM 2006, Chapter 9)

Countries like Canada and the UK also followed more participatory and deliberative models in trying to formulate more effective waste policies in the 2000s. In the UK, CoRWM was given a 'blank sheet of paper', with emphasis in its terms of reference to consult widely and to 'inspire public confidence' (CoRWM 2006, Annex 1), a commodity previously in short supply. CoRWM took a consciously 'analytic-deliberative' approach to its work and conducted very wide-ranging engagements with both stakeholders and non-aligned members of the public. Its 2006 report endorsed DGD as the desirable end-point but its most important recommendation[c], endorsed subsequently by Government (Defra, BERR and the devolved administrations for Wales and Northern Ireland 2008), was that communities should be invited to volunteer to become potential repository sites, to be resourced to do this effectively, and to be given a right to withdraw participation if the terms being offered were unattractive.

This new approach did improve the level of trust in the policy process but was no guarantee of success. As outlined below the ongoing revival of Government commitment to new nuclear power (HMG 2017) has made it more difficult to achieve agreement on a way forward for higher activity wastes.

While the widespread move towards more participatory forms of decision-making has improved trust and made progress more likely, it has not been a panacea. Even in its own terms it has had problems in rectifying the power imbalances that inevitably exist between local communities and the combined forces on the State and the nuclear power industry. A critique of the CoRWM process suggests that while its greater openness and engagement were helpful, it was still subject to important framing processes within which the influence of powerful incumbent actors could not be countered (Chilvers and Burgess 2008).

The politics of nuclear waste also extend well beyond the issue of local siting. Local opposition to siting proposals have often been influenced and bolstered by national and international forces. Environmental NGOs like Greenpeace have often majored on the waste (and associated reprocessing) issues and in countries like Germany a wide range of political forces – not all of them associated with environmentalism – have joined together to resist proposals for waste management solutions at national level.

This wider than purely local current of resistance to the policy of burying waste has at least two strands. One is the idea that it is impossible to demonstrate that any underground repository can guarantee that there will be no return of radioactivity to the biosphere for hundreds of thousands of years. And while most scientists in this field argue that it may be possible to make a DGD adequately safe this often falls short of endorsing a particular site, and/or the precise methods proposed to try and contain the waste within the repository. And of course no-one can guarantee the integrity of a repository thousands of years in the future (CoRWM 2006, Chapter 13)

A second strand concerns the meaning of emplacing waste deep underground. The colloquial version of this argument is 'out of sight, out of mind' involving a suspicion that the purpose of deep disposal is to make invisible the problem that the waste was created in the first place. This argument is used by those who have more wide-ranging objections to nuclear power, but is also visible among others who do not hold such views. It is often associated with scepticism about the robustness of the safety case that can be made for DGD and has been particularly strong where reprocessing of spent fuel has added complexity to the management task.

The politics (and ethics) of waste management can also vary sharply depending on the status of nuclear power within a country. Where a decision is made to abandon nuclear power the waste issue becomes purely one of managing a legacy. If there is trust that the decision to abandon nuclear power is final, then the issue becomes simply one of finding the 'least worst' way of managing the waste legacy (CoRWM 2006, Chapter 14). In such situations it has sometimes proved possible to bring together people of varying opinions about nuclear power – supporters and opponents – in the common cause of finding a solution. This was the case in 2003 in the UK when, on the formation of CoRWM in 2003, it appeared that there would be no future role for nuclear power in the UK (DTI 2003).

The problem can become more intractable when there is an ongoing programme of building new nuclear power. The politics and ethics here are different. The question is no longer how to find the least worst way to manage waste but rather whether proceeding with nuclear power is justifiable, given that more waste will ensue, and that there are alternative paths for power production that do not impose the same potential burdens on future generations. At local level this translates into uncertainty for a community in terms of how much waste they may eventually need to receive, and is likely therefore to lead to greater resistance than if it knows it will host a fixed legacy. Where such commitments to new nuclear construction are strong, as in the UK, resistance to an 'out of sight out of mind' solution to waste management has become stronger – due to a conviction that getting the waste out of the way is being used simply as a legitimation device for the advance of new nuclear construction. The stance of the UK Government on this – that it has confidence that a long-term management route for waste 'will' exist (Defra BERR and the devolved administrations for Wales and Northern Ireland 2008)– has had the tendency to reduce public trust, especially as it was as long ago as 1976 that an influential report advocated that there should be no going ahead with a substantial nuclear programme unless it could be demonstrated that a route existed (Royal Commission on Environmental Pollution1976) (not 'will exist') to manage waste effectively.

7 Conclusions

Failure across all nuclear-using countries to implement technically convincing and publicly acceptable ways of dealing with higher activity wastes has been a feature of nuclear power across the world for several decades. There have been some small advances since the Fukushima accident in a few countries, primarily Finland and Sweden, but even in those cases, no operating repository will open until 2020 at best. Further, there seem to be distinct features of Nordic political systems, especially higher degrees of public trust and consensual decision-making styles that are hard to replicate elsewhere. While issues like safety, security and proliferation risk remain problematic, it is the inability of all countries yet to find a credible long-term way forward to manage the most dangerous categories of waste that is the primary stumbling bock to acceptance of the technology.

References

Abbott, A., 1998. 'UK waives nuclear waste rule for Georgia' Nature 392: 850 (30 April 1998).

Arentsen, M.J., 2015. 'With access to the future. Nuclear waste governance in the Netherlands' in Brunnengraber et al., op. cit. Chapter in V, pp. 281-298.

Auffermann, B., Suomela, P., Kaivo-Oja, J. Vehmas, J. and Lukkanen, J. 'A final solution for a big challenge. The governance of nuclear waste disposal in Finland, in Brunnenghraber et al. op. cit. Chapter in IV, pp. 227-248

Blowers, A., 2017. The Legacy of Nuclear Power Earthscan

Brunnengraber, A., di Nucci, M.R., Loada, A.M.I., Mex, L. and Schreurs, M. (eds.), 2015. Nuclear Waste Management: an International Comparison Springer

Chilvers, J and Burgess, J., 2008. 'Power relations: the politics of risk and procedure in nuclear waste governance' Environment and Planning A: economy and space 40:8, 1881–1900.

Cochran, T., Feiveson, H., Patterson, W., Pshakin, G., Ramana, M., Schneider, M., Suzuki, T. and von Hippel, F., 2010. Fast Breeder Reactor Programs: history and status. A research report on International Panel on Fissile Materials, (February 2010)

CoRWM (Committee on Radioactive Waste Managment), 2006. Managing our Radioactive Waste Safely, CoRWM's Recommendations to Government (July 2006).

Defra, BERR and the devolved administrations for Wales and Northern Ireland, 2008. Managing Radioactive Waste Safely: a framework for implementing geological disposal, Cm 7836, (1 June 2008).
DTI (Department of Trade and Industry), 2003. Our energy future – creating a low carbon economy Cm 5761, February.
Gibb, F., Travis, K.P., McTaggart, N.A., Burley, D. and Hesketh, K.W., 2008. 'Modeling temperature distribution around very deep borehole disposals of HLW' Nuclear Technology 163: 62–73
Gowing, M., 1974. Independence and Deterrence: Britain and Atomic Energy 1945–52 (Volume 1) Macmillan.
HMG (Her Majesty's Government), 2017. The Clean Growth Strategy – leading the way to a low carbon future (October 2017).
Hocke, P. and Kallenbach-Herbert, B., 2015. 'Always the same old story? Nuclear Waste Governance in Germany in Brunnengraber et al., op. cit. Chapter in Part IV, pp. 177-202.
Macfarlane, A. and Ewing, R. (eds.), 2006. Uncertainty Underground: Yucca mountain and the Nation's High-Level Nuclear Waste MIT.
MacKerron, G., 2015. 'Multiple Challenges: Nuclear Waste Governance in the United Kingdom' in Brunnengraber et al. op. cit. Chapter in III, pp. 101-117.
NWMO (Nuclear Waste Management Organisation), 2018. Implementing Adaptive Phased Management (March 2018).
OECD/NEA, 2013. The Economics of the Back End of the Nuclear Fuel Cycle NEA no. 7061
Royal Commission on Environmental Pollution, 1976. Nuclear Power and the Environment 6th Report, Cm 6618.
Schneider, M. and Marignac, Y., 2008. Spent Fuel Reprocessing in France International Panel on Fissile Materials Research Report 4, 8 (May 2008).
Swahn, J. and Kaberger, T., 2015. Governance and management of radioactive waste in Sweden in Brunnengraber et al. (op. cit.) Chapter in IV pp. 203-225.
World Nuclear News, 2017. 'Further delay to completion of Rokkasho facilities' (December 28. 2017).
Von Hippel, F and MacKerron, G., 2015. Alternatives to MOX International Panel on Fissile Materials Research Report no. 13 (April 2015).

Riding to the Rescue?

The Changing Picture in China and the Global Future of Nuclear Power

M. V. Ramana[1]

Abstract

China has the most ambitious targets for nuclear power and some expect that it would shore up the flagging prospects for a large expansion of nuclear power around the world. But in recent years, China's nuclear program has not grown as fast as projected. This chapter explains why there are good reasons to expect nuclear power growth to slow down further. Promises that new reactor designs will be constructed in large numbers in China or exported from China to other countries seem unlikely to materialize. As a result, nuclear power's salience to future global electricity generation will continue to diminish.

1 M. V. Ramana, Liu Institute for Global Issues, University of British Columbia, Vancouver, Canada, m.v.ramana@ubc.ca

R. Haas et al. (Eds.), *The Technological and Economic Future of Nuclear Power*, Energiepolitik und Klimaschutz. Energy Policy and Climate Protection, https://doi.org/10.1007/978-3-658-25987-7_14

1 Introduction

For long, China was expected to be the engine that would propel a large-scale expansion of nuclear power in the 21st century. Starting around 2005, the country embarked on constructing a very large number of nuclear plants (see Figure 1). After a short pause following the 2011 Fukushima Daiichi accidents, China resumed new reactor construction in late 2012 and today has the most number of reactors under construction—18 reactors, nearly a third of the global total (IAEA, 2018).

The dominance of China in nuclear reactor construction testifies not just to China's emergence as an industrial powerhouse but also the decline in nuclear growth elsewhere. Indeed, in recent years, both the United States and Western Europe have seen many reactors shut down well before their license period expire in comparison with new reactor construction. Globally nuclear energy production as a share of all electrical energy generated has declined to around 10.5 percent, nearly 40 percent below the maximum of over 17 percent in 1996 (Schneider and Froggatt, 2017).

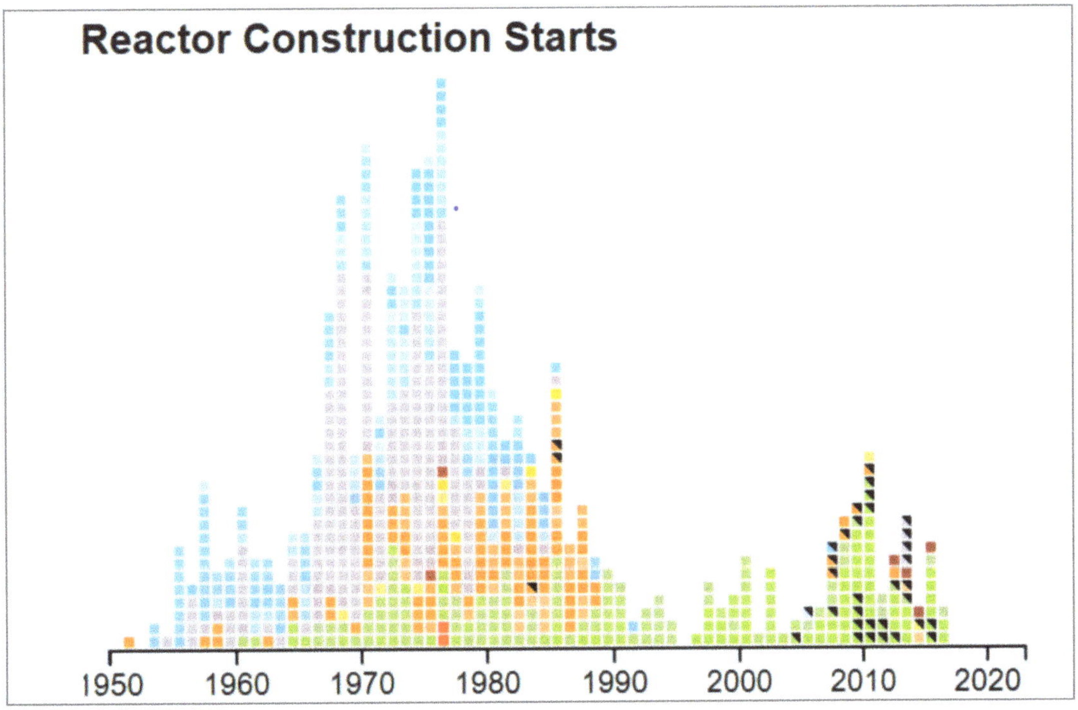

Fig. 1 Annual commencement of nuclear reactor construction

Source: "Global Nuclear Power Database" http://thebulletin.org/global-nuclear-power-database

Targets for nuclear capacity in many countries have also been declining, especially in the aftermath of the disaster that started in 2011 at the Fukushima Daichi nuclear plant (Ramana, 2016). In contrast, China has continued rolling out impressive targets for the future. More recently, the country has also started making determined attempts to export its reactors (Thomas, 2017), for example, by infusing capital into a very expensive nuclear power project in the United Kingdom, as a result of which some were even asking if China might be able to rescue Europe's nuclear energy industry (Brown, 2015; Sputnik, 2015).

Is this a realistic projection of China's role in the future of nuclear power? Will China really be able to compensate for the ongoing decline in nuclear energy? This article tries to answer these questions by examining the targets that have been set and the plans that were announced, and following these by an examination of the actual experience of reactor construction and exports. This is followed by a discussion of some drivers of reductions in nuclear plans and a brief conclusion.

2 Current status

According to the IAEA's Power Reactor Information System, as of January 2019, China had 46 operating reactors with a total net capacity of around 43 GW (gigawatts), and a further 11 reactors with a total capacity nearly 11 GW are under construction (IAEA, 2019). In 2017, nuclear power contributed 247.5 TWh, which constituted 3.9 percent of all electricity generated in China, up from 3.6 percent in 2016. The nuclear fraction has been very gradually increasing since 2010 (see Figure 2). Nevertheless, the small magnitude of that fraction implies that the large buildup of nuclear power was part of a general strategy that called for building up all kinds of electricity generation plants. In particular, China has been ramping up construction of modern renewables. In 2017, wind energy contributed 306 TWh, up by 26 percent from its contribution in 2016, while solar energy contributed 118 TWh, up by 75 percent from 2016 (China Energy Portal, 2018).

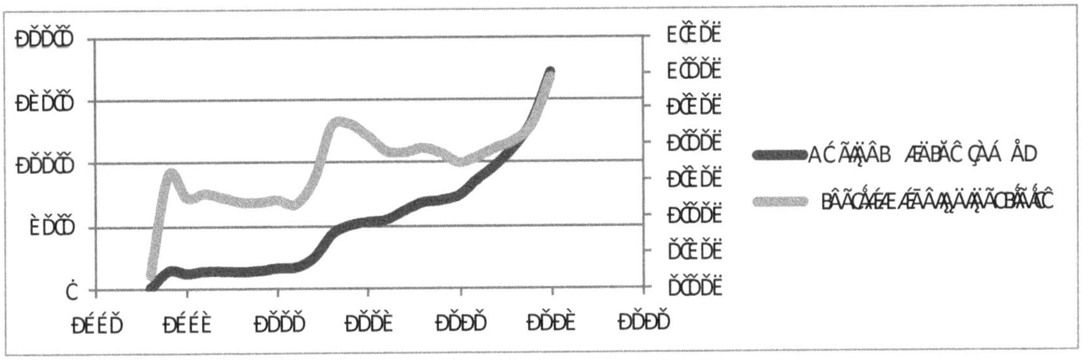

Fig. 2 Nuclear energy supplied to the grid (blue points) and the fraction of all electricity provided by nuclear power

Source: Authors calculations based on figures in BP, 2016, Statistical Review of World Energy 2016: BP, accessed October 18, 2016, at http://www.bp.com/en/global/corporate/energy-economics/statistical-review-of-world-energy.html.

3 Ambitious targets

Although China is a relatively late entrant to nuclear power, with construction of nuclear power plants starting only in the 1980s, three decades after the nuclear weapons program started, it has periodically laid out impressive targets.[2] Some of these are listed in Table 1. Perhaps the highest of these was the figure of 114 GW by 2020, released by the National Development & Reform Commission in 2010 (Power Engineering, 2010). These figures have come down and according to the latest 13th Five-Year Plan, the targets for 2020 were to have 58 GW of nuclear generating capacity in operation and a further 30 GW of capacity under construction (WNN, 2016c). Even this target will likely not be met and Chinese officials admit that it might be only 53 GW by 2020 (Stanway and Chen, 2016). But in the longer term, Chinese nuclear advocates continue to posit impressive targets, up to 150 GW by 2030, nearly thrice what is likely to be the operating capacity in 2020. For example, the China General Nuclear Power Corporation foresees a nuclear capacity of "120-150 GW by 2030" (Stanway, 2016).

2 One early projection by Professor Lu Yingzhong who was then Director of the Institute of Nuclear Energy Technology at Tsinghua University envisioned nuclear power contributing 12.7 percent of all energy (including heat and electricity) in China by 2020 with that fraction rising to 19.35 percent by 2030 (Lu, 1984).

Tab. 1 Nuclear Power Targets

Year in which Projection was Made	Year Projected For	Capacity Projected (MW)	Actual Capacity (MW)
1988	2000	6,000	2,168
1996	2010	20,000	10,282
2002	2020	40,000 – 80,000	?
2008	2020	50,000	?
2010	2020	114,000	?
2016	2020	58,000	?

Sources: (Dazhong and Yingyun, 2002; Meyers and others, 1993; Ran and Li, 1998; Sternfeld, 2009; WNN, 2016c)

When it comes to exports, too, Chinese nuclear advocates have been making extravagant claims (Thomas, 2017). This trend has become stronger after the Fukushima accidents that commenced in 2011. That accident set back nuclear programs around the world, leaving China in a comparatively better place. This has been welcomed by Chinese officials; according to Zhang Guobao, a former Administrator at China's National Energy Administration "history has given China an opportunity to overtake the world's nuclear energy and nuclear technology powers" (Stanway, 2013). In 2016, the president of China National Nuclear Corporation (CNNC) announced that "China aims to build 30 nuclear power units… by 2030"; this goal, in turn, was based on the assessment that "more than 70 countries" were "planning or are already developing their own nuclear power projects, and it is estimated 130 more nuclear power units will have been built by 2020" (Xinhua, 2016).

4 Interest in Chinese market

Because of the expectation that China will be building up its nuclear capacity rapidly, reactor vendors and national government representatives have been flocking to the country since the 1980s in the hope of selling their wares (MacDougall, 1984; Lu, 1993; Silver, 1994; Bratt, 1996; Zhou and Zhang, 2010). Foreign government officials, sometimes at very high levels, have been involved in advocating for their reactor designs (Silver, 1994; Silver, 1996b; Silver, 1996a; Zuercher, 1995; Bodgener, 2008; Ria, 2008; Cultice and Feng, 2009).

The nature of the competition is made clear by a 2008 cable from the U.S. embassy in Beijing to Washington and is worth quoting at some length:

> "Effective advocacy for U.S. nuclear suppliers is essential to ensuring access to China's rapidly growing civil nuclear power market. With the exception of the bidding process that resulted in a contract for four Westinghouse AP1000 reactors, all reactor purchases to date have been largely the result of internal high level political decisions absent any open process. Even the Westinghouse decision was arguably a political one, which was quickly followed by subsequent non-competitive purchases of the competing French and Russian plants. China is currently in the process of building as many as 50 to 60 new nuclear plants by 2020; the vast majority will be the CPR-1000, a copy of 60's era Westinghouse technology that can be built cheaply and quickly and with the majority of parts sourced from Chinese manufacturers... Pressing for open and transparent bidding processes for reactor sales (for complete plants or individual major component purchases), as well as advocating for China to pursue advanced reactor technology for future sites so that it's reactor fleet is not so reliant on aging technology through the next half century could be a more effective approach to bolster U.S. interests in China's nuclear market. Regardless of how the United States decides to advocate, it should be done continuously and from a high level in order to keep up with the French and Russians" (U.S. Embassy Beijing, 2008).

Two claims made in this cable are worth noting. First, all nuclear contracts were political decisions. In other words, they may not really be technically or economically justifiable. Neither the Chinese government nor any of the countries seeking to export reactors to the country offered a public cost-benefit analysis in support of the decision to import one or the other reactor type. Further, vendors enticed Chinese policymakers with varying degrees of technology transfer, attractive financing arrangements, and other political benefits (Ramana and Saikawa, 2011).

Second, all countries that seek to sell nuclear reactors have been engaged in high level governmental campaigning, on a continuous basis. As the *New York Times* reported in 2004: "In recent months, a procession of political leaders has pressed China to favor power plant designs and equipment from their home countries. They have included President Jacques Chirac of France; former Prime Minister Jean Chretien of Canada; Viktor Khristenko, who was named fuel and energy minister in Russia on Tuesday; and dozens of less-prominent officials. President Bush even raised the virtues of American nuclear technology with the Chinese prime minister, Wen Jiabao" (Buckley, 2004). The justification for the high level political engagement was the poor state of the nuclear reactor market elsewhere and the promises of growth in China.

5 Recent experience with reactor construction

Partly as a result of such high level intervention, China's nuclear establishment did import a number of reactors. China's entry into nuclear energy began with the construction of a series of indigenously designed reactors at the Qinshan site. However, in the late 1980s, China imported reactors with the M310 design from Framatome in France for the Daya Bay site; the majority of the reactors built in China involve the CPR-1000, a modification of the M310 design. The only exceptions were two VVER1000s and two Candu-600 reactors imported from Russia and Canada, respectively, but there was never any plan to make those the mainstay of the Chinese nuclear fleet. It was only after 2005 and the eleventh 5-year plan that imported reactors came back into consideration, with the specific aim of introducing third-generation reactors from other countries (Xu, 2008).

The relevance of the construction of the reactors was heightened in the aftermath of Fukushima because of the decision made by policy makers that China would build *only* Generation III or III+ reactors. The initial assumption was that this stipulation would lead to the adoption of AP1000 technology, in part because senior nuclear policy makers promoted the idea that this reactor design would have withstood the conditions that led to the events at Fukushima. For example, a general manager in the China Power Investment Corporation pointed out that the "reactors in the Japanese nuclear power plants, which have been affected by the massive quake, are Generation II reactors and have to rely on back-up electricity to power their cooling system in times of emergency", whereas the "AP1000 nuclear power reactors, currently under construction in China's coastal areas and set to be promoted in its vast hinterland, are Generation III reactors and have built in safety features to overcome such a problem" (Reporter, 2011). This was very much the same claim that Westinghouse itself peddled. As the Westinghouse President and Chief Executive Officer Aris Candris put it to ABC News: "Everyone has heard of what happened at the Fukushima Daiichi plant…Had an AP1000 been on that site we would have got no nuclear news post-tsunami" (Sy, 2011).

The other reason to have expected an important role for AP1000s in China's nuclear plans is that a key player in the Chinese nuclear power sector, the State Nuclear Power Technology Corporation (SNPTC), held the sole rights "to sign contracts with foreign parties to receive… 3rd generation nuclear power technology" (SNPTC, 2011). Specifically, the 3rd generation nuclear power technology that SNPTC was to contract for was the Westinghouse AP1000, which it was expecting would become "the dominant technology for China's future nuclear industry development" (Lawrence and Katz, 2007). The Fukushima accidents and the Chinese government's decision that future nuclear construction would be limited to

Generation III reactors gave SNPTC an opportunity to translate their expectation about the AP1000's dominance into reality. By May 2011, SNPTC had convinced officials at Westinghouse Electric Company that the AP1000 was going to dominate the Chinese reactor market from then on (Li and Tranum, 2011). But the actual experience of the projects involving AP1000s proved more problematic.

Before going on to the AP1000 projects, let us first look briefly at the European Pressurized Reactor (EPR) units being built at Taishan. These were originally scheduled to "be commissioned at the end of 2013 and in autumn 2014" respectively, and France's Areva had hoped "to have started work on more reactors" by then (Thibault, 2010). As of February 2017, China General Nuclear Power maintained that the two units will start commercial operations in "the second half of 2017 and the first half of 2018, respectively" (AFP, 2017). Despite the serious concerns set off in April 2015 when the French nuclear safety regulator, Autorité Sûreté Nucléaire (ASN), revealed that the reactor bases and lids Areva had supplied from its Le Creusot plant did not meet safety specifications (Chaffee, 2015), the first EPR was opened for commercial operations in December 2018 (IAEA, 2019). The delays with the EPR were not surprising, at least in retrospect, because by the time construction started in Taishan, the corresponding projects at Olkiluoto in Finland and Flamanville in France had already run into trouble and were expected to be significantly delayed (Kanter, 2009).

In contrast, the AP1000s at the Sanmen and Haiyang sites were the very first constructions of this design anywhere in the world. When construction started at Sanmen, the Shaw Group, one of the partners in the consortium building the reactor, proudly proclaimed, "As with the successful, on-time and on-schedule pour of the first nuclear concrete for the Reactor Building mat earlier this spring, we have again shown that next generation nuclear power plants can be, and are being, built in an efficient and timely manner" and looked forward "to bringing this plant on line as scheduled in 2013" (Shaw Group, 2009). Unfortunately for Westinghouse and Shaw, these promises did not come to pass.

An important source of problems, although not the only one, has been the reactor coolant pumps (RCPs) that were supplied by US manufacturer Curtiss-Wright Corporation. The RCP forces water to circulate through the reactor and transfer the heat generated by the fission reactions in the reactor core. In January 2013, Curtiss-Wright found that a piece of a "blade within the pump had separated from the... casting" and it had to recall the RCPs that had already been shipped off (NIW, 2013). The RCP had to undergo design changes and fixes that took two years to complete. The problem with the RCPs was symptomatic of a larger problem: construction of the Sanmen and Haiyang power plants had began well before the engineering of the plant's design was completed (Spegele, 2016).

Because the design has never been constructed anywhere, new problems keep surfacing.[3] One such problem was observed during tests that were conducted at the first AP1000 unit at Sanmen-1. These tests were conducted without any radioactive material being loaded into the reactor, but at high temperatures. The problem involved neutron shield blocks that are supposed to stop neutrons from the nuclear core from escaping into the rest of the reactor. During these tests, the material that was in the shield blocks had "volumetrically expanded and extruded out of the shield blocks into the nozzle gallery" and there was "internal pressurization of the shield blocks," according to a heavily redacted report on the issue presented by Westinghouse to the U.S. Nuclear Regulatory Commission in February 2017 (Cooke, 2017, p. 5). Westinghouse was forced to admit that it had "not properly considered" the possibility that the shielding material might expand in volume.

Any of these problems could result in serious safety consequences. Chinese nuclear officials have expressed concern in the past about these problems. In 2013, for example, a former vice-president of CNNC complained to *South China Morning Post*: "Our state leaders have put a high priority on [nuclear safety] but companies executing projects do not seem to have the same level of understanding" (Ng, 2013). The result has been a very long series of delays. All four AP1000s went into commercial operation in late 2018 or early 2019 (IAEA, 2019).

Cost estimates have risen too. Early estimates by China's Nuclear Energy Agency put the cost of constructing AP1000 reactors at $2300/kW (NEA, 2010, p. 48). A newspaper article from 2016 claims implicitly that the cost might be $3000/kW (Abe, 2016). In its environmental impact assessment for the Sanmen project, China's National Nuclear Safety Administration "projected a total project pricetag of 52.5 billion yuan ($8.3 billion) — more than double the original budget for the two units of 25 billion yuan"(Yu, 2018).

The problem that this higher cost of construction poses is that of economic competitiveness. In 2016, Steve Kidd, who was formerly with the World Nuclear Association, suggested that a tariff of 0.50 RMB per kWh has been mentioned for AP1000 and EPR nuclear projects, up from 0.43 RMB per kWh for projects that constructed more traditional nuclear reactor designs (Kidd, 2016). But the tariff available to the project developers was significantly lower. In April 2018, Sanmen "cleared the annual power exchange auction in Zhejiang province, making it eli-

---

3 However, it bears remembering that even with the follow on AP1000 constructions in the United States, there have been a number of technical problems with significance for safe operations, which raises fundamental questions about the soundness of that design. These problems were to cause record losses for Toshiba and drive Westinghouse to bankruptcy (Asahi Shimbun, 2017; Cardwell and Soble, 2017; Lewis, 2016).

gible to sell 766 gigawatt hours of output during 2018 at a price of 0.385 yuan per kilowatt hour ($0.061¢/kWh)...roughly 10% less than the standard nuclear tariff of 0.43 yuan/kWh" (NIW, 2018a).

The poor experience at Sanmen and Haiyang has not stopped Westinghouse from making unrealistic claims about the future of the AP1000 in China. In 2016, Jose Gutierrez, interim chief executive of Westinghouse, indulged in wishful thinking: "We expect to see a fleet of AP1000 reactors in China ... We don't know how many China wants to build, but it could be tens" (De Clercq, 2016). This does not appear to have any basis in reality and reflects an attempt to boost its plunging fortunes. Indeed, Westinghouse officials themselves may not believe in this. The Wikileaks cables reveal that nearly a decade ago, at a meeting convened by the DOE China Office, Westinghouse representative in China, Gavin Liu, "noted that because China's technical capacity is increasing, the longer it takes to start the next round of AP1000 reactors, the less scope will be available for Westinghouse" (DOE China Office, 2008).

The outlook for Westinghouse in China is bleak. As Lin Boqiang, director at the China Center for Energy Economics Research at Xiamen University told *Bloomberg News*, "The only way Westinghouse can win contracts in China is to demonstrate they can build reactors quicker and cheaper than anyone else in China's market and win hearts with actions, not words...Westinghouse so far hasn't demonstrated such abilities" (Stapczynski and others, 2015). Li Ning, from Xiamen University, told the *Wall Street Journal* that Chinese officials "are certainly very frustrated" and "they feel Westinghouse oversold the system, oversold the technology, promised more than they could really deliver" (Yap and Spegele, 2015). Of course, ever since Westinghouse filed for bankruptcy in the United States in March 2017 (Cardwell and Soble, 2017), the future of the company itself has been highly uncertain.

6 New reactor designs

China has also been at the center of efforts to rescue nuclear power by adopting new reactor designs in place of the now-standard Light Water Reactor design that has dominated nuclear power around the world. These newer reactor designs, which are mostly paper designs, are held out as solving one or more of the many problems that have plagued nuclear power.

There are at least two reasons offered for the focus on China. The first is the idea that if any country is to be capable of supporting the production of, and offering a market for, a large number of units, it would be China. Growth of nuclear power in other markets is either slow or non-existent (Schneider and Froggatt, 2016). Therefore, the idea goes, if a new reactor design is to be tried out at some scale, it would be possible only in China. As Charles Forsberg, executive director of the MIT Nuclear Fuel Cycle Project, puts it, "There have been studies that indicate that if reactors are mass-produced, they can drive down costs...The Chinese market is large enough to make that potentially possible" (Martin, 2016).

Second, there is the idea that the Chinese nuclear regulatory system might be more open to licensing non-traditional reactor designs (Eaves, 2017). One example of a non-traditional reactor design that some in China are working on, and where U.S. nuclear advocates have been pointing to China as the role model, is the molten salt reactor (Halper, 2015; Martin, 2015; Reischer, 2016). Some of the proponents are Chinese but there are also a number of foreign designers, many from the United States, who seek to have their designs first be commercialized in China.

Two other designs with a connection to China are the High Temperature Gas Cooled Reactor (HTGR) and the Travelling Wave Reactor. In an article in *Issues in Science and Technology*, for example, Richard Lester, another MIT professor, lists several features of the way the U.S. Nuclear Regulatory Commission carries out regulation of new reactor designs and offers the Travelling Wave Reactor design promoted by Bill Gates and Nathan Myhrvold as an example of a technology that was going to be developed in China because of the stricter and more rigid safety regulatory regime in the United States (Lester, 2016).

The saga of the HTGR is also relevant. This reactor design was first proposed in the 1940s and was extensively researched by German technologists, who built two reactors based on this design, but who eventually abandoned the effort to commercialize the technology. In the 1990s, South Africa set up a major program to develop a commercial HTGR called the Pebble Bed Modular Reactor (PBMR). This effort was to collapse a decade later, and the South African government abandoned the project (Thomas, 2011).

In China, work on the HTGR design started in earnest after the country signed a cooperation agreement with Germany in 1984 (MacDougall, 1984). The pilot scale HTR-10 reactor reached its criticality in 2000, achieved full power operation, and began to supply power to the grid in 2003 (Zhou and Zhang, 2010). Soon after the HTR-10 attained criticality, in 2001, the commercial scale design, called the high-temperature gas cooled reactor pebble-bed module (HTR-PM) project, that is capable of generating 250 MW of electricity, was launched (Zhang and others, 2009). The development of this reactor became a high priority under the "Chinese Science and Technology Plan" for the period 2006–2020.

In February 2008, the implementation plan and the budget for the HTR-PM project was approved by the State Council of China. The HTR-PM received final approval from China's cabinet and its national energy bureau around two weeks before the Fukushima accidents (Bradsher, 2011). However, in the aftermath of Fukushima, all nuclear construction was frozen. In December 2012, construction of HTR-PM commenced at Shidaowan in China's eastern Shandong province (IAEA, 2019). The reactor was "expected to start commercial operation in late 2017" (WNN, 2016b).

Chinese HTR proponents have painted an ambitious future for that reactor design. When construction of the plant was starting, there were plans for eventually constructing a further 18 units of the same type at the same site (NucNet, 2013). There are also plans to export these reactors to other countries, and China has entered into exploratory agreements with Saudi Arabia and Indonesia (Can, 2016; WNN, 2016a; WNN, 2016b), but has not completed construction as of January 2019 (IAEA, 2019).

China's construction of the HTR-PM and more generally its pursuit of this reactor design has been much lauded by nuclear advocates around the world, especially by those who seek to promote a focus on HTGRs or other advanced reactors as a way to rescue nuclear power from its declining fortunes. Andrew Kadak, formerly President and CEO of the Yankee Atomic Electric Company (YAEC) that operated the Yankee Atomic Nuclear Power station, said: "The industry has been focused on water-cooled reactors that require complicated safety systems. The Chinese aren't constrained by that history. They're showing that there's another way that's simpler and safer. The big question is whether the economics will pay off" (Reiss, 2009).

Unfortunately, for the proponents, the economics doesn't seem to be working out—even prior to the commencement of operations. According to the trade magazine *Nuclear Intelligence Weekly*, the high cost of generation (60 fen (¢0.9) per kilowatt hour, higher than the average 43 fen/kWh for Gen III reactors) is among the "key challenges" confronting HTGRs in China (Yu, 2016, p. 6). The high cost is no surprise; even HTGR proponents estimate that the capital cost will be about 20% higher than LWRs, although they typically will also claim that this cost will come down as more plants are built (Zhang and Sun, 2007). Another key challenge that the HTGR faces is the fact that there are a host of other small modular reactor technologies under development in China (Ramana and others, 2013; Yu, 2016). Two such designs, the ACPR50 and ACPR100 from CGN and the ACP100 from CNNC, have been in the news recently, as a result of an announcement that China was going to build maritime nuclear power platforms in the South China Sea (NEI, 2016). In part as a result of these challenges, it appears that Chinese policy makers have dropped the plan of building 18 reactors at the Shidaowan site (WNN, 2016b).

The idea of exporting HTGRs also appears to be somewhat wishful thinking. The Indonesia case provides a good illustration. News reports suggest that Wang

Shoujun, chairman of China Nuclear Engineering Group Corporation visited Batan (Badan Tenaga Nuklir Nasional, Indonesia's National Nuclear Energy Agency) in June 2016 in order to better understand the Indonesian market (Can, 2016). Despite such agreements—for example, Batan also has signed an agreement with the Japan Atomic Energy Agency on research and development of HTGRs in 2014 (WNN, 2014)—the odds of Batan being able to construct a commercial scale HTGR in Indonesia in the foreseeable future are essentially nil. Indeed, in December 2015, then Energy and Mineral Resources Minister Sudirman Said announced publicly that the government had concluded that "this is not the time to build up nuclear power capacity. We still have many alternatives and we do not need to raise any controversies" (NEI, 2015).

Other countries that have been targeted by China also have lengthy histories of ambitious announcements followed by little action; for example, over a decade ago, Argentina declared that it was embarking on "an eight year nuclear energy development program with the purpose of increasing the number of atomic plants plus resumption of uranium enrichment production" (Mercosur, 2006). Little was achieved by this program apart from the commissioning of the Atucha-II reactor in 2014, construction of which started in 1981 (IAEA, 2014b). There is little to indicate that Argentina will indeed embark on massive nuclear construction; instead it seems to be heading towards expanding its renewable energy sector (Maxwell, 2016). Indeed, in May 2018, the trade magazine *Nuclear Intelligence Weekly* reported that Argentianian officials have put plans for Chinese supplied reactors on hold (NIW, 2018b).

The pattern that may be discerned is of making tall claims about numerous reactors of one design or the other, raising hopes among nuclear power supporters for a revival of the technology based on this new design. However, these initial announcements are almost always followed by a process of slowing down, and often abandonment, with the latter steps done quietly with little fanfare. This appears to be what has happened to the HTGR design and may well be the fate of the plans to construct floating power plants, unless the real reason for the latter are to raise the stakes on the south China sea disputes. In any case, it seems hard to visualize China as a laboratory for successfully developing a reactor design that solves the problems of nuclear power.

7 Drivers for reduction

In light of all these problems, then, it is not surprising that China's current nuclear target of 58 GW by 2020 is much lower than the earlier high value of 114 GW by 2020. But problems with constructing the current generation of nuclear reactors do not constitute the only reason for this change in the outlook for nuclear power in China. There are at least three additional reasons for a lowering of targets for nuclear deployment (Ramana and King, 2017).

The first is that energy demand in China is not growing at the same fast rate it has in the past. The underlying reason for this is the deliberate shift in the nature of the Chinese economy, from one primarily focused on increasing manufacture, especially by heavy industry, to one that is actively promoting service sector and less energy-intensive sectors (Green and Stern, 2015; Green and Stern, 2016). At the same time, because of ambitious plans in the past, there is a real glut in power capacity. Most power plants, including nuclear reactors, are not being utilized at optimal levels. This trend might result in further reduction in the number of hours that nuclear reactors are operated: in March 2017, the National Energy Administration announced new rules on the priority order for different kinds of generators to supply electricity to the grid; Chinese nuclear companies are already complaining about being forced to reduce how many hours grid operators are willing to absorb the power outputs of reactors (Yu, 2017a).

The second reason is that there are very few coastal sites available for new nuclear plants to be set up. There is a limit to how many reactors can be built on existing sites. There is real and justified resistance to building nuclear power plants in inland sites, next to rivers and large lakes, water from which is already in demand for drinking, agriculture, and other higher priority uses (King and Ramana, 2015).

Finally the government seems to be paying attention local opposition to nuclear facilities; this is, again, entirely justifiable. Opposition to nuclear facilities has been growing in China since the Fukushima accidents (Buckley 2015; Lok-to 2016). One study that explored the Chinese public's willingness to pay to avoid harm found that those surveyed were "particularly concerned about the development of nuclear power in the aftermath of the Fukushima disaster and generally regard nuclear power as unsafe power generation technology" (Sun and others, 2016, p. 692). At least two nuclear facilities that were supposed to be constructed were cancelled after public protests, the most prominent being the decision in 2016 to cancel a 100 billion yuan (US$15 billion) nuclear reprocessing plant that was proposed for a location near the city of Lianyungang, Jiangsu province (Green 2016).

There is independent evidence that even industry insiders within China do expect a decline in nuclear construction going forward. For example, imports of uranium by China from other countries has been declining, from nearly 21,300 tons of uranium concentrate in 2014 down to 19,200 tons in 2015, to under 16,000 tons in 2016 (Chaffee, 2017).

8 Effect on global prospects for nuclear power

What might be the impact of a reduction in China's nuclear targets on the future of nuclear power? One handle on this is provided by critically examining the projections put out by the International Atomic Energy Agency (IAEA). The IAEA's projections have historically been well in excess of what actually materialized (IPFM, 2007, p. 85). Nevertheless, its projections are worthy of examination because they provide an indication of the nuclear industry's own outlook. Each year, the IAEA puts out two sets of projections, a low case and a high case. The first "represents expectations about the future if current market, technology and resource trends continue and there are few additional changes in explicit laws, policies and regulations affecting nuclear power" (IAEA, 2013, p. 6). In contrast, the "high case projections are much more optimistic, but still plausible and technically feasible. The high case assumes that current rates of economic and electricity demand growth, *especially in the Far East*, continue. Changes in country policies toward climate change are also included in the high case" (emphasis added). In other words, the high case projections represent something like a best case scenario for nuclear deployment.

Table 2 below lists the high case estimates for the Far East region and the world as a whole from the last seven years. Since 2010, all of the projections—for nuclear power globally and for the Far East, which is defined as China, Japan, and Republic of Korea—have been declining. Even for the IAEA, the realities of the market cannot be completely ignored. However, barring a few years, although the projected nuclear capacity in the region is declining, the fraction of global capacity constituted by the countries of the Far East has been increasing. The IAEA's low estimates also assume that similarly large fractions (35 to 40%) of the global nuclear capacity will be in the Far East. In other words, if there is to be a big revival of nuclear power, it would have to be fueled by construction in this region.

Tab. 2 Figures from the IAEA's High Case Projections

	High Case Estimate for 2050 for the Far East	High Case Estimate for the World in 2050	Fraction of Global Nuclear Capacity in the Far East	Fraction of electricity generated by nuclear power in 2050 in the Far East	Fraction of electricity generated by nuclear power in 2050 around the world
2010	450	1415	31.80%	21.60%	17.00%
2011	450	1228	36.64%	19.10%	13.50%
2012	417	1137	36.68%	16.90%	12.20%
2013	412	1113	37.02%	16.60%	12.10%
2014	398.7	1091.7	36.52%	15.10%	11.50%
2015	355	964	36.83%	14.40%	10.80%
2016	351	898	39.09%	14.50%	10.00%

Sources: (IAEA, 2010; IAEA, 2011; IAEA, 2012; IAEA, 2013; IAEA, 2014a; IAEA, 2015; IAEA, 2016)

The IAEA does not break up its estimates by country. But the much larger size of China as compared to the other two countries in the region implies that it is likely to be the dominant contributor to the IAEA's projections for the Far East. In its latest projections, the high case involves 351 GW of nuclear capacity in this region, much of which has to be in China. Even with this massive buildup, nuclear power loses market share; nuclear power contributes a slightly smaller fraction to all electricity generated in 2050 than now.

9 Conclusions

Nuclear power in China has grown dramatically in the last decade or more, in large part because of high level political decisions to promote the technology even if it was not really technically or economically justified. This rapid expansion and the ambitious targets announced by the Chinese nuclear establishment have led to the expectation that China might give the nuclear industry a new lease on life. But, as the IAEA's projections show, even if this trend is to change, and China does restart another phase of rapid expansion, nuclear power will become a smaller contributor to global electricity production than today.

This chapter has argued that because of the kinds of shifts seen in recent years, China will likely never build up the kind of nuclear power capacity that was foreseen for it even just a decade or less ago. The country is not on track to meet its current target of 58 GW of nuclear power capacity in 2020. And unless there is a substantial shift in various policies, for example, a deliberate effort to build up nuclear capacity even if it is uneconomical or otherwise undesirable, it is quite likely that the targets set in future years will, if they are to be realistic, reflect a much slower pattern of growth. There are many reasons to expect that such a policy reversal, namely for Beijing to actively promote the rapid construction of nuclear plants around the country, will not occur. In particular, there are shifts in the pattern of energy demand growth and growing public concerns about nuclear facilities that impacts the siting of reactors negatively.

The export market is not growing fast either. Despite much talk, Pakistan remains the only country to which China has exported nuclear power plants. With the rapid reductions in the costs of renewable energy technologies, especially solar photovoltaic panels, and the continued pattern of high costs and lengthy construction periods of nuclear reactors, the demand for nuclear plants is likely to decline.

One way by which nuclear enthusiasts have held on to their hope for a major revival of nuclear power is to postulate that alternate reactor designs will be introduced and constructed in large numbers. In this scenario too, China is presumed to be the main actor because of two factors: its presumed large market for nuclear reactors and the expectation that its regulatory process will approve new reactor designs more easily. But these scenarios of new reactor designs coming in to save nuclear power ignore the lengthy history of failed experiments with alternate designs and the multiple challenges faced by nuclear power, which pose conflicting priorities on reactor designers (Ramana and Mian, 2014).

Put together, these trends suggest that China is unlikely to rescue the global nuclear industry from its ongoing gradual decline.

References

Abe, T., 2016. China nuclear industry: State-owned enterprises eye overseas power projects: Nikkei Asian Review.

AFP, 2017. China delays nuclear reactor start again: Nuclear Power Daily.

Asahi Shimbun, 2017. Toshiba pulling out of overseas nuclear reactor construction: The Asahi Shimbun.

Bodgener, J., 2008. The biggest deal: Nuclear Engineering International.

Bradsher, K., 2011. China Building Nuclear Reactors With Radically Different Design: New York Times.
Bratt, D., 1996. Is Business Booming? Canada's Nuclear Reactor Export Policy: International Journal, v. 51, p. 487–505.
Brown, P., 2015. Will China rescue Europe's nuclear energy industry? Climate Home.
Buckley, C., 2004. Chance to Revive Sales Draws Nuclear Industry to China: New York Times.
Can, Y., 2016. China's nuclear giant to promote HTGR in Indonesia: People's Daily Online.
Cardwell, D., and Soble, J., 2017. Westinghouse Files for Bankruptcy, in Blow to Nuclear Power: The New York Times.
Chaffee, P., 2015. Flaws Detected in Creusot Forge's EPR Domes: Nuclear Intelligence Weekly, v. IX, p. 3.
Chaffee, P., 2017. Chinese Demand Plummets in 2016: Nuclear Intelligence Weekly, v. 11, p. 3–4.
China Energy Portal, 2018. 2017 electricity & other energy statistics, accessed April 10, 2018, at https://chinaenergyportal.org/en/2017-electricity-energy-statistics/.
Cooke, S., 2017. Sanmen Testing Raises Disturbing Design Questions: Nuclear Intelligence Weekly, v. 11, p. 4–5.
Cultice, C., and Feng, X., 2009. Doing Nuclear Business in China: Power-Gen Worldwide.
Dazhong, W., and Yingyun, L., 2002. Roles and prospect of nuclear power in China's energy supply strategy: Nuclear Engineering and Design, v. 218, p. 3–12.
De Clercq, G., 2016. Westinghouse to start first China reactor in 2017, sees tens more: Reuters.
DOE China Office, 2008. China Building Indigenous Nuclear Plant Construction Capacity, Edging Out Westinghouse: DOE China Office 08BEIJING3055_a, accessed March 5, 2017, at https://wikileaks.org/plusd/cables/08BEIJING3055_a.html.
Eaves, E., 2017. Can North America's advanced nuclear reactor companies help save the planet? Bulletin of the Atomic Scientists, v. 73, p. 27–37.
Green, F., and Stern, N., 2015. China's "new normal": structural change, better growth, and peak emissions: Centre for Climate Change Economics and Policy (CCCEP), University of Leeds.
Green, F., and Stern, N., 2016. China's changing economy: Implications for its carbon dioxide emissions: Climate Policy, p. 1–15.
Halper, M., 2015. The U.S. is helping China build a novel, superior nuclear reactor: Fortune.
IAEA, 2010. Energy, Electricity and Nuclear Power Estimates for the Period up to 2050: International Atomic Energy Agency IAEA-RDS-1/30.
IAEA, 2011. Energy, Electricity and Nuclear Power Estimates for the Period up to 2050: International Atomic Energy Agency IAEA-RDS-1/31.
IAEA, 2012. Energy, Electricity and Nuclear Power Estimates for the Period up to 2050: International Atomic Energy Agency IAEA-RDS-1/32.
IAEA, 2013. Energy, Electricity and Nuclear Power Estimates for the Period up to 2050: International Atomic Energy Agency IAEA-RDS-1/33.
IAEA, 2014a. Energy, Electricity and Nuclear Power Estimates for the Period up to 2050: International Atomic Energy Agency IAEA-RDS-1/34.
IAEA, 2014b. Power Reactor Information System (PRIS) Database, at http://www.iaea.org/programmes/a2/.
IAEA, 2015. Energy, Electricity and Nuclear Power Estimates for the Period up to 2050: International Atomic Energy Agency IAEA-RDS-1/35.
IAEA, 2016. Energy, Electricity and Nuclear Power Estimates for the Period up to 2050: International Atomic Energy Agency IAEA-RDS-1/36.

IAEA, 2019. China, People's Republic of, accessed January 27, 2019, at PRIS – Country Details at https://www.iaea.org/PRIS/CountryStatistics/CountryDetails.aspx?current=CN.
IPFM, 2007. Global Fissile Material Report 2007: International Panel on Fissile Materials.
Kanter, J., 2009. In Finland, Nuclear Renaissance Runs Into Trouble: The New York Times.
Kidd, S., 2016. China – what are today's issues influencing the reactor plans? Nuclear Engineering International.
King, A., and Ramana, M. V., 2015. The China Syndrome? Nuclear Power Growth and Safety After Fukushima: Asian Perspective, v. 39, p. 607–636.
Lawrence, D., and Katz, A., 2007. China nuclear power poised for export in `Self-Reliance' bid: Bloomberg.
Lester, R. K., 2016. A Roadmap for U.S. Nuclear Energy Innovation: Issues in Science and Technology, v. XXXII.
Lewis, L., 2016. Toshiba warns of multibillion-dollar charge on US nuclear arm: Financial Times.
Li, Z., and Tranum, S., 2011. Candris Says Fukushima Will Help AP1000 in China: Nuclear Intelligence Weekly, v. V, p. 6.
Lu, Y., 1984. The Development of Nuclear Energy in China: Energy Research Group Review Paper No. 078.
Lu, Y., 1993. Fueling one billion : an insider's story of Chinese energy policy development: Washington Institute Press, Washington, D.C.
MacDougall, C., 1984. China chooses the nuclear solution:
Martin, R., 2015. China-U.S. Nuclear Collaboration, Though Controversial, Moves Ahead: MIT Technology Review.
Martin, R., 2016. China could have a meltdown-proof nuclear reactor next year: MIT Technology Review.
Maxwell, a, 2016. Is Argentina Headed Towards a Clean Energy Revolution? NRDC.
Mercosur, 2006. Argentina re-floats nuclear energy program: MercoPress.
Meyers, S., Goldman, N., Martin, N., and Friedmann, R., 1993. Prospects for the power sector in nine developing countries: Lawrence Berkeley Lab. LBL--33741, accessed March 29, 2017, at http://inis.iaea.org/Search/search.aspx?orig_q=RN:24070433.
NEA, 2010. Projected costs of generating electricity: Nuclear Energy Agency, OECD, Paris.
NEI, 2015. Indonesia rules out nuclear as major power source: Nuclear Engineering International.
NEI, 2016. China gets onboard: Nuclear Engineering International.
Ng, E., 2013. China nuclear plant delay raises safety concern: South China Morning Post.
NIW, 2013. Weekly Roundup: Nuclear Intelligence Weekly, v. VII, p. 1.
NIW, 2018a. Weekly Roundup: Nuclear Intelligence Weekly, p. 1.
NIW, 2018b. Argentina Puts Newbuilds on Ice: Nuclear Intelligence Weekly, p. 1.
NucNet, 2013. China Begins Construction Of First Generation IV HTR-PM Unit, accessed January 10, 2013, at The Communications Network for Nuclear Energy and Ionising Radiation at http://www.nucnet.org/all-the-news/2013/01/07/china-begins-construction-of-first-generation-iv-htr-pm-unit.
Power Engineering, 2010. China raises 2020 nuclear target by 62 per cent to 114 GW: Power Engineering.
Ramana, M. V., 2016. Second life or Half-life? The Contested Future of Nuclear Power and its Potential Role in a Sustainable Energy Transition, *in* Kern, F. ed., Energy Transitions: Palgrave Macmillan, London, p. 363–396.

Ramana, M. V., Hopkins, L. B., and Glaser, A., 2013. Licensing small modular reactors: Energy, v. 61, p. 555–564.
Ramana, M. V., and King, A., 2017. A new normal? The changing future of nuclear energy in China, *in* Van Ness, P. and Gurtov, M. eds., Learning from Fukushima: Nuclear Power in East Asia: ANU Press, Canberra, Australia.
Ramana, M. V., and Mian, Z., 2014. One size doesn't fit all: Social priorities and technical conflicts for small modular reactors: Energy Research & Social Science, v. 2, p. 115–124.
Ramana, M. V., and Saikawa, E., 2011. Choosing a standard reactor: International competition and domestic politics in Chinese nuclear policy: Energy, v. 36, p. 6779–6789.
Ran, J., and Li, B., 1998. China's nuclear energy industry: rising sharply amidst a strategic shift: Xinhua.
Reischer, R., 2016. Fourth Generation Reactor Technology Development in China: ANS Nuclear Cafe.
Reiss, S., 2009. Let a Thousand Reactors Bloom: Wired.
Reporter, S., 2011. China to promote nuclear power despite explosion in Japan: WantChinaTimes.com.
Ria, 2008. Russia, China see nuclear power as a priority in economic ties:
Schneider, M., and Froggatt, A., 2016. The World Nuclear Industry Status Report 2016: Mycle Schneider Consulting, accessed August 3, 2016, at http://www.worldnuclearreport.org/The-World-Nuclear-Industry-Status-Report-2016-HTML.html.
Schneider, M., and Froggatt, A., 2017. The World Nuclear Industry Status Report 2017: Mycle Schneider Consulting, accessed May 23, 2018, at https://www.worldnuclearreport.org/-2017-.html.
Shaw Group, 2009. Shaw and Westinghouse Announce Successful Placement of Major Structural Module at Sanmen Nuclear Site in China: Business Wire.
Silver, R., 1994. With Cooperation Agreement Signed, Canada, China Talk CANDU Sales: Nucleonics Week, v. 35, p. 13.
Silver, R., 1996a. AECL, CNNC Ink Contracts For Two Candu-6s At Qinshan: Nucleonics Week, v. 37, p. 9.
Silver, R., 1996b. As Financing Deadline Looms, AECL And CNNC Renew Candu Talks: Nucleonics Week, v. 37, p. 3.
SNPTC, 2011. Introduction of State Nuclear Power Technology Corporation: State Nuclear Power Technology Corporation.
Spegele, B., 2016. Troubled Chinese Nuclear Project Illustrates Toshiba's Challenges: Wall Street Journal.
Sputnik, 2015. China Rides to the Rescue of Britain's Stalled Nuclear Program: Sputnik International.
Stanway, D., 2013. Analysis: China needs Western help for nuclear export ambitions: Reuters.
Stanway, D., 2016. China's total nuclear capacity seen at 120–150 GW by 2030 – CGN: Reuters.
Stanway, D., and Chen, K., 2016. China's debut Westinghouse reactor delayed until June 2017: exec: Reuters.
Stapczynski, S., Urabe, E., and Guo, A., 2015. Westinghouse Races China for $1 Trillion Nuclear Power Prize: Bloomberg.com.
Sternfeld, E., 2009. Development of Civil Nuclear Power Industry in China, *in* Mez, L., Schneider, M., and Thomas, S. eds., International Perspectives on Energy Policy and the Role of Nuclear Power: Multi-Science, Essex, p. 455–466.

Sun, C., Zhu, X., and Meng, X., 2016. Post-Fukushima public acceptance on resuming the nuclear power program in China: Renewable and Sustainable Energy Reviews, v. 62, p. 685–694.

Sy, S., 2011. Could New Nuclear Reactor Have Prevented Fukushima? ABC News.

Thibault, H., 2010. Construction schedule on Chinese third-generation nuclear plants races ahead of European models: Guardian.

Thomas, S., 2011. The Pebble Bed Modular Reactor: An obituary: Energy Policy, v. 39, p. 2431–2440.

Thomas, S., 2017. China's nuclear export drive: Trojan Horse or Marshall Plan? Energy Policy, v. 101, p. 683–691.

U.S. Embassy Beijing, 2008. Effective Nuclear Advocacy in China, accessed May 6, 2015, at Wikileaks at https://www.wikileaks.org/cable/2008/08/08BEIJING3362.html.

WNN, 2014. Japan, Indonesia team up on HTGR development: World Nuclear News.

WNN, 2016a. China, Saudi Arabia agree to build HTR: World Nuclear News.

WNN, 2016b. First vessel installed in China's HTR-PM unit: World Nuclear News.

WNN, 2016c. Nuclear growth revealed in China's new Five-Year Plan: World Nuclear News.

WNN, 2016d. China and Indonesia to jointly develop HTGR: World Nuclear News.

WNN, 2017. Construction milestones at new Chinese units: World Nuclear News.

WNN, 2018. Fifth Yangjiang unit connected to grid: World Nuclear News.

Xinhua, 2016. China plans 30 overseas nuclear power units by 2030: English.news.cn.

Xu, Y.-C., 2008. Nuclear energy in China: Contested regimes: Energy, v. 33, p. 1197–1205.

Yap, C.-W., and Spegele, B., 2015. China's First Advanced Nuclear Reactor Faces More Delays: Wall Street Journal.

Yu, C. F., 2016. CNEC-CFHI Deal — Boosting the HTGR Or Chinese Manufacturing? Nuclear Intelligence Weekly, v. X, p. 5–6.

Yu, C. F., 2017. NEA Codifies Nuclear Load-Following: Nuclear Intelligence Weekly, v. 11, p. 6–7.

Yu, C. F., 2018. First AP1000 Moves to Commercialization: Nuclear Intelligence Weekly, p. 3.

Zhang, Z., and Sun, Y., 2007. Economic potential of modular reactor nuclear power plants based on the Chinese HTR-PM project: Nuclear Engineering and Design, v. 237, p. 2265–2274.

Zhang, Z., Wu, Z., Wang, D., Xu, Y., Sun, Y., Li, F., and Dong, Y., 2009. Current status and technical description of Chinese 2 × 250 MWth HTR-PM demonstration plant: Nuclear Engineering and Design, v. 239, p. 1212–1219.

Zhou, S., and Zhang, X., 2010. Nuclear energy development in China: A study of opportunities and challenges: Energy, v. 35, p. 4282–4288.

Zuercher, R. R., 1995. Westinghouse Expands China Pact; Vendors Want Export Ban Lifted: Nucleonics Week, v. 36, p. 1.

Major Accidents

Three Decades after Chernobyl: Technical or Human Causes?

Nikolaus Muellner[1]

Abstract

The accident of unit 4 at the NPP Chernobyl from 1986 was arguably the worst disaster of a nuclear power plant that happened so far. It became apparent to the broader public that the vast amount of radioactive fission products that accumulate during operation of a nuclear reactor have the potential to render large areas inhabitable. The root cause of the accident was therefore of major interest for all countries who operated nuclear power plants, or who had nuclear power plants in its vicinity. Considering all information available today it is safe to draw the conclusion that the reactor design was too complex at that time, and therefore errors have been made. It is not so easy to exclude that this could happen with other designs in other countries as well.

1 Nikolaus Müllner, University of Natural Resources and Life Sciences, Vienna, Austria, nikolaus.muellner@boku.ac.at

R. Haas et al. (Eds.), *The Technological and Economic Future of Nuclear Power*, Energiepolitik und Klimaschutz. Energy Policy and Climate Protection, https://doi.org/10.1007/978-3-658-25987-7_15

1 Introduction

Thirty years after the disaster at the Chernobyl Nuclear Power Plant (NPP), all details of the accident seem to be known. The Chernobyl reactor was a pressure tube reactor of the type RBMK, which has been built several times in the Soviet Union. Experts from regulatory authorities and research institutes have analysed the existing accident data and results from accident simulations and gave their view in many published reports (GRS 1996, USNRC 1987, Snell and Howieson 1991, Sehgal 2012 – and many others). The International Atomic Energy Agency IAEA and the "Nuclear Energy Agency" of the "Organization for Economic Co-operation and Development" organized conferences and published the resulting experts' opinions (INSAG 1986, INSAG 1992, OECD/NEA 2002). In most cases, the authors conclude that human error combined with weaknesses in the reactor design of the RBMK reactor have led to the accident. In this content the term "human error" indicates violation of operating procedures and lack of knowledge of the operators of the reactor. Deficiencies in the safety culture of the power plant are typically identified as root cause for the transgressions of the operators. The long prison sentences for the main engineer Fomin of the Csernobyl NPP and his deputy Diatlov add to this picture.

This narrative of the accident and its root cause, as presented in many books, reports and papers, can be traced back to an IAEA meeting. A few months after the accident, 25th to 29th of August 1986, the IAEA organized a "Post Accident Review Meeting" were Soviet experts gave detailed information on the accident to a large number of experts from IAEA member states. The IAEA issued a report that summarized the meeting, the first report of the "International Nuclear Safety Advisory Group" (INSAG 1986). The report gave a clear statement on the reason for the accident: "… the accident was caused by a remarkable range of human errors and violations of operating rules in combination with specific reactor features which compounded and amplified the effects of the errors and led to the reactivity excursion.", where "human errors" refers exclusively to operator errors.

Five years after the accident, the report INSAG-1 was revised by IAEA and reissued as INSAG-7 (INSAG 1992). The new version is based on two Russian reports: One report (Shteynberg 1991) was issued by a commission, appointed by the Soviets' "State Committee for the Supervision of Safety in Industry and Nuclear Power". Said commission had the task to reassess the events of the Chernobyl accident. The other report (Abagyan et al., 1991) investigates in detail the cause of the accident and is authored by prestigious Soviet institutes, such as the Kurtchatov Institute or the Scientific Research and Design Institute for Power Technology (Russian abbreviation NIKIET). Both reports are annexed to INSAG-7. These reports present a completely different picture of the accident. In particular, Shteynberg (1991) contradicts the account of the accident of INSAG-1 regarding operator errors. Abagyan et al. (1991) draws a whole new picture on the RBMK designer organization. The conclusions of INSAG-7 are therefore different than INSAG-1, but retain a certain emphasis on lacking safety culture and operator errors.

The new perspective from Shteynberg 1991 and Abagyan et al., 1991 is rarely reflected in the current literature on the Chernobyl accident. Large, influential organizations such as US Nuclear Regulatory Commission (NRC) had already completed their analyses of the Chernobyl accident before the revised report INSAG-7 was published (e.g. USNRC 1987). Others, such as (OECD / NEA 2002), tend to follow the narrative of human error and violation of operating procedures notwithstanding their recent publication date. This leads to the fact that whenever the Chernobyl accident is portrayed based on literature, it is most likely that mistakes and violation of the operators are identified as main cause for the event.

References to accounts of the events given by participants in the accident night are rarely found (a counter-example can be found in Schmid 2011). The deputy chief engineer of the power station, Diatlov, who designed the test program that led to the accident of Chernobyl and who was present in the control room in the night of the accident published an article in a scientific journal after INSAG-7 was issued (Diatlov 1995). A book, which he wrote about the background and the course of the accident, was not published, but is available electronically from Internet libraries (Diatlov 2005). However, some aspects of the accident can only be understood by combining the information of the logs of the Chernobyl control system, the logbook entries, the interpretations of the various scientific institutes of the events, and the description Diatlov on the events of the night of the accident. By combining all information a new view on the root cause of the accident becomes apparent.

2 The Chernobyl reactor

The RBMK type reactor was not the first choice for the Chernobyl site (Shteynberg, 1991). In fact the RBMK design was ranked third in a feasibility study. Nevertheless it was decided to construct a RBMK reactor since the required parts and components for the RBMK design were available, while components for the other two designs would have to be manufactured and long production times were expected. The decision for RBMK reactors was taken in 1969, in 1972 it was decided to build a total of 4,000 MW electrical power (four reactor blocks). The Gidroproekt and NIKIET institutions worked together to develop the reactor design, which was subsequently examined by the Soviet State Committee for Construction and Planning. Finally, the design was approved by the Council of Ministers. The blocks Chernobyl one to four went online between 1977 and 1983.

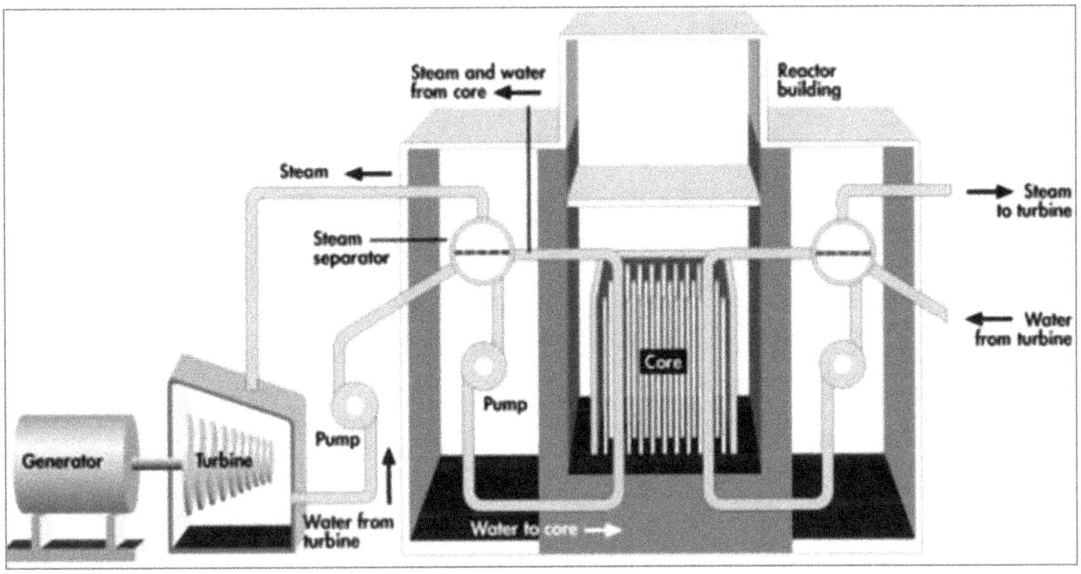

Fig. 1 Schematic of an RBMK reactor (Nuclear Energy Institute 1997)

Figures 1 to 3 show the schematic of the RBMK reactor, a section through the reactor, and the reactor core. RBMK is a Russian acronym for high-performance channel reactor and is designed as a boiling water reactor. In contrast to the western type boiling water reactors the RBMK does not feature a reactor pressure vessel. Instead it is equipped with approximately 1660 parallel vertical cooling channels (or pressure tubes). Each pressure tube can be loaded with a fuel element with an active region of about 7m, which represents the reactor core. The channels protrude vertically a graphite block (Figure 3, or (1) in Figure 2). Each single channel can be separated from the circuit by isolation valves at full load operation and a fuel element can be unloaded or loaded by the refuelling machine.

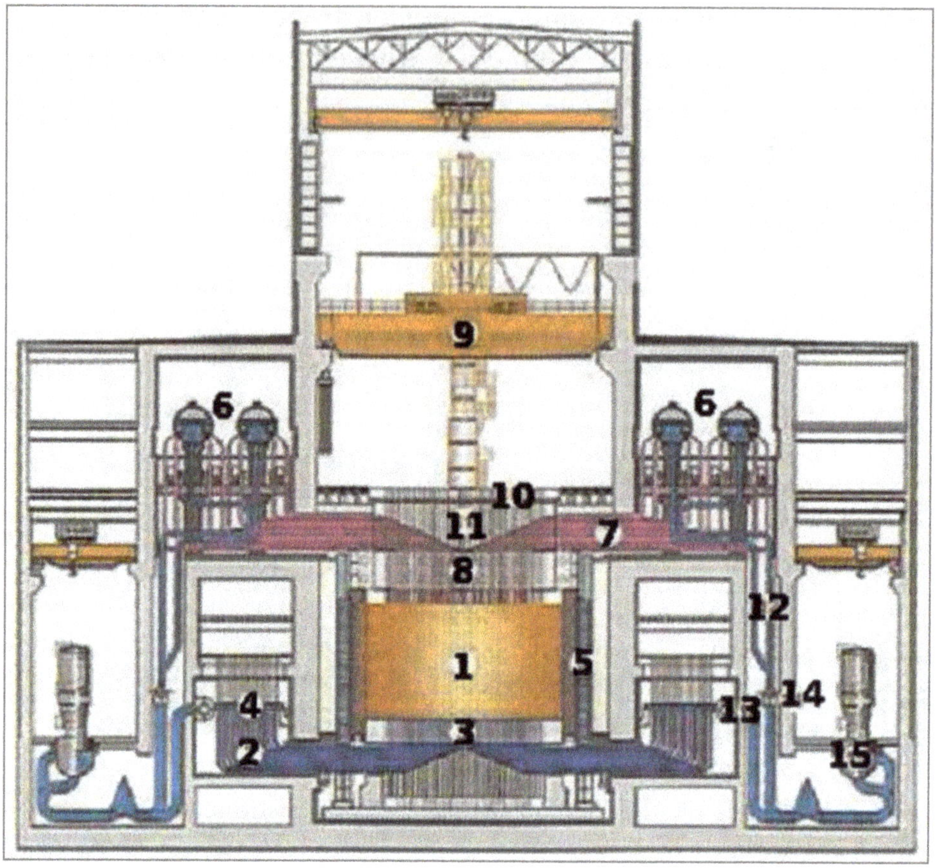

Fig. 2 Cross-section through the RBMK reactor building (D'Auria 2005). (1) core, (2) pressure tubes leading tot he core, (3) lower biological shield (4) collector, (5) lateral biological shield (6) steam drum separator (7) pressure tubes (8) upper biological shield (9) refueling machine (10) upper core plate (11) upper pressure tubes (12) recirculation lines (13) collector after pump (14) collector before pump (15) main circulation pump

The reactor cooling circuit begins and ends with the steam separators (shown in (6) in Figure 2 and as a "steam (drum) separator" in Figure 1). The steam separators separate steam from water. The separated water, together with the feed water from the condenser (see "water from turbine" in Figure 1) is pumped to the reactor core by means of a total of six main circulation pumps (with two backup main circulation pumps in total eight main circulation pumps are installed). The water is partially evaporated in the reactor core so that a two-phase mixture of water and steam is fed to the steam separator drum from the core exit. The steam separator drum separates the steam from the water, the steam is led to the turbine via the steam lines, while the water is mixed with the feed water and fed back to the core. A total of eight turbines and generators were installed at Chernobyl Nuclear Power

Plant for the four reactor blocks (two per block). It is important to note that during operation not only water, but also a certain fraction steam is present the reactor core (which is also called "void fraction").

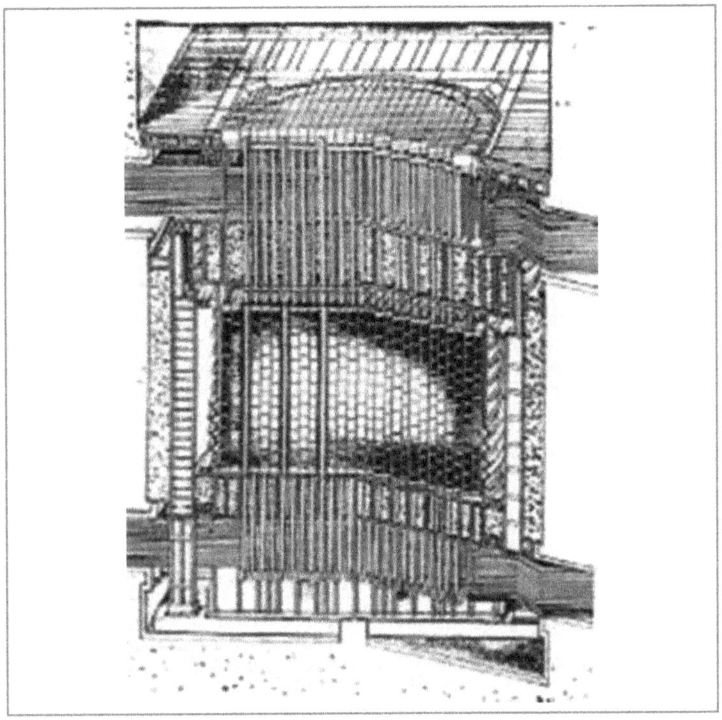

Fig. 3

RBMK reactor core (D'Auria 2005)

Reactivity initiated event

The Chernobyl accident was a so called "reactivity initiated event", which means that the control over the reactor power was lost, and that the power output of the reactor increased tremendously in a short time period. The power control of a reactor is a complex issue, and even more so in case of an RBMK reactor. There are a number of factors to consider:

The source of power in the RBMK nuclear reactor stems from the uranium-235 in the fuel pellets, which, during operation, is hit by neutrons inducing fission, a process during which thermal energy and further neutrons are generated for further fissions (chain reaction). Those neutrons however, before being able to induce further fissions, must be slowed down (moderated). Unlike light water reactors (which use water for moderation of neutrons) the RBMK reactor uses the graphite block surrounding the fuel channels as moderator (the black tiles in Figure 3). The graphite offers the advantage that a fraction of uranium 235 between 1% and 2%

in the fuel is sufficient, while water-moderated reactors need a fraction between 3% and 5%. However, light water reactors have the advantage that the coolant is also the moderator at the same time. If a light water reactor loses its coolant (i.e., a pipeline breaks, the cooling water is lost and the reactor runs dry), the chain reaction collapses and the reactor power is sharply reduced. The RBMK reactor in such a situation has to rely on its neutron absorber rods (control rods), which control the power and have to be inserted into the core.

The power of a reactor is proportional to the number of neutrons which are generated in the chain reaction. An important parameter is therefore the so called "reactivity", a measure of how the generation of neutrons is changing. Positive reactivity means more neutrons are going to be generated and the power will increase. Negative reactivity means that less neutrons are going to be generated and the power will decrease. In principle, the following factors influence the reactivity in the RMBK reactor core:

- Graphite as moderator and its temperature
- Temperature of the fuel elements
- Enrichment and burn-off of the fuel elements
- Presence of xenon in the reactor core
- The temperature and density of the cooling water when entering the reactor and the steam fraction (void fraction) in the cooling channels
- The control rods of the reactor and their positioning.

The Chernobyl accident was a so called reactivity initiated event, an event in which the reactivity (and the power) increased in an uncontrolled way. All factors above played a role in the accident, but the last two bullet points were of special importance for the Chernobyl accident.

Positive void coefficient

A key figure to characterize the behaviour of a reactor is the "power coefficient", and in case of the RBMK reactor, the "void coefficient". The power coefficient describes how the reactor is going to react to a power change: a positive power coefficient means that an increase in power will induce a further additional increase, a decrease in power a further decrease. It is desirable (although not necessarily required) to construct a reactor with a negative power coefficient: this means, an increase in power would lead to a reduction in power, a decrease to an increase. Such a reactor will by itself tend to regulate its power at a fixed power level. The other key figure, the "void coefficient", describes the feedback of the reactor to an increase of steam in the core region. A positive void coefficient means that an increase of the steam fraction in the pressure tubes in the core region leads to an increase in power, a negative void coefficient means that an increase in void leads to a decrease in power.

The light water cooled and moderated reactors have a negative power coefficient, as well as a negative void coefficient. The designers of the RBMK reactor aimed for a negative power coefficient, and accepted a positive void coefficient (during certain operational regimes). However, the designers of the RBMK were convinced that this effect was limited. Their design calculations showed that if the pressure tubes were to be filled with steam, the reactivity would first increase, but then decrease, and eventually become negative (see Figure 4, Abagyan et al., 1991). This calculation, although accepted by the regulatory authority, turned out to be wrong, as the Chernobyl accident showed.

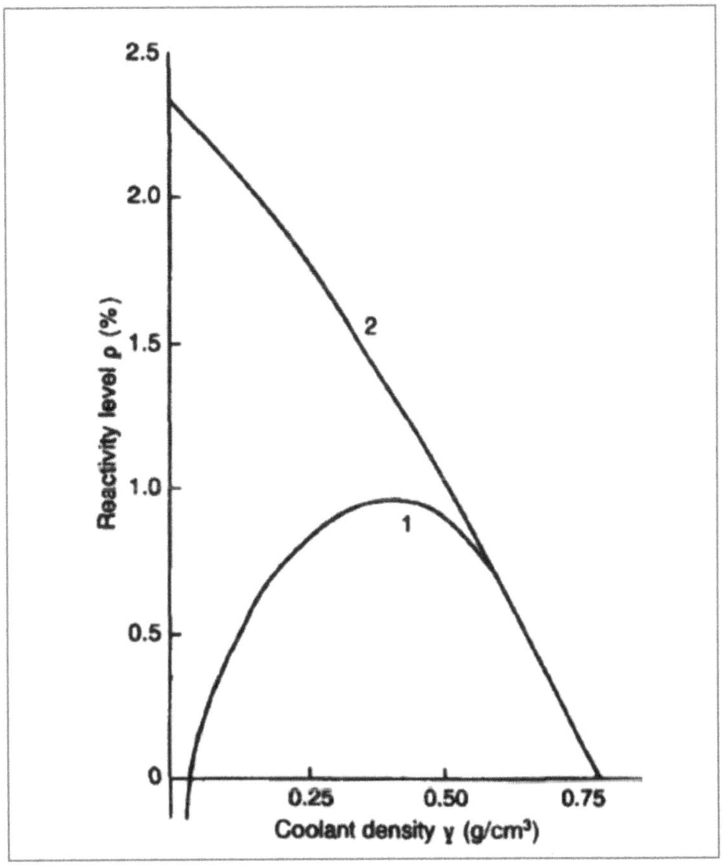

Fig. 4

Reactivity vs coolant density in a fuel channel. (1) Assumptions of the designers, reactivity negative for voided channel, (2) reality, reactivity strongly positive for voided channel (Abagyan et al. 1991)

Shteynberg (1991) points out that the impact of the steam fraction in the cooling channel on the reactivity was only evaluated for performance levels above 50% of the nominal reactor power (i.e. 1600 MWth and upwards). The accident happened at an operating power of about 200 MWth, a region which should prove to be particularly unstable. However, as Shteynberg (1991) concludes, the designers did not expect any problems with this power range (which, according to him, is shown by the lack of supporting calculations for this range) and the operators were also unaware of the risks of operating the reactor in this dangerous power region.

Control rods

As in most nuclear power plants the reactor power of the RBMK reactor is controlled by control rods. The basic principle of control rods is simple – rods with a neutron absorbing material are inserted into the reactor to "slow down" the chain reaction, or withdrawn to "accelerate" it. By inserting the control rods the power is reduced (negative reactivity is inserted), by withdrawing the control rods power is increased (positive reactivity is inserted).

According to Shteynberg (1991), as well as by Abagyan et al. (1991), the design of the control rods triggered the accident. The RBMK reactor core is large compared to other cores, therefore a large number of control rods are necessary to ensure an even power distribution. The so called "operational reactivity margin" (ORM) is a very important operational parameter for RBMK reactors which played a key role in the accident. Roughly speaking it specifies for the RBMK reactor how many control rods are fully extracted at a certain point in time. More precisely, the ORM reports the "equivalent" withdrawn control rods, which means that two control rods withdrawn to 50% count as one. Furthermore not all control rods are equally effective, and furthermore effectiveness of a rod is not equal over the whole size. The algorithm for calculating the ORM takes those effects into account. According to the operating instructions, there is a minimum value for the ORM value, which must not be exceeded. There must always be 26–30 equivalent control rods withdrawn. The main engineer may permit operation also down to an ORM value of 15 equivalent control rods. Operation below this value is not permitted.

The reason for this rule (and especially its safety relevance) it is not immediately apparent. It seems safe to assume that the operators of the reactor at the night of the accident were not aware of the safety relevance of the ORM, because Diatlov, deputy chief engineer, who was present in the control room in the accident night, wrote (Diatlov 1995) regarding the purpose of the ORM: … "a guarantee that the reactor protection operates properly. At the same time the restriction is imposed not on the maximum, which would be natural, but on the minimum". A restriction on the maximum would be, indeed, more intuitive: Inserted control rods mean that the power can be further increased as needed (by pulling them out). On the other hand, if the control rods are withdrawn from the core it means that, if needed, more negative reactivity can be introduced to shut down the reactor. So one would assume operating at power with the control rods withdrawn is safer, then with control rods inserted.

Another fact would lead the operators to neglect the ORM parameter. Usually safety relevant parameters are shown at prominent positions in the control room. The ORM value, on the other hand, was computed by the PRIZMA program (Shteynberg 1991), on a console which was usually out of view. The ORM computation would take several minutes. It was also known that the calculated ORM value was not precise, since a group of rods was not considered. In principle the ORM value could also be derived manually from the indicated penetration depth of different control rods by a hand calculation (correcting for the various effects mentioned above). Either way it would be a lengthy procedure.

But the ORM value has a critical influence on the safe shutdown of the reactor. Due to the design of the control rods, the efficiency of the reactor SCRAM (the emergency shutdown) system was linked to the ORM value. When the reactor emergency shutdown was triggered at an ORM value of 30 and more, negative reactivity could very quickly be introduced to reduce the power; at an ORM value of 15 it took six seconds to introduce the relatively small value of 1$ of negative reactivity; and at an ORM value of 7 the introduced reactivity was positive over 8 seconds (which means the reactor power increases further for eight seconds when the operator hits the emergency shutdown button). Only after that the power would decrease (Shteynberg 1991). The design of the control rod is the reason for this effect: to ensure better performance during normal operation when the control rods are fully withdrawn, displacement bodies made of graphite were fitted with telescope rods at the lower end of the control rods. This means that during normal operation the middle of the core region (approximately 7m) was filled with a graphite displacer of 5m (see Figure 5). Above and below the displacer in the reactor core was a 1 m water column. When a fully withdrawn control rod was inserted, the graphite displacer had to pass through 1m of water in the lower core region. However at this position the water acts as a neutron absorber, the graphite body acts as a moderator, which means that positive reactivity is introduced into the lower region of the reactor (power is increased). In the upper part of the reactor core, negative reactivity is inserted because the absorber material is introduced into the core (see Figure 5). If the rods are already one meter inserted, the graphite displacer will be at the bottom of the core, and no positive reactivity will be introduced. This is the reason why a minimum ORM value is enhancing safety – it means, that a certain number of rods are already inserted into the core.

According to Shteynberg (1991) the way the ORM value was displayed in the control room suggests that the designers of the RMBK reactor themselves were not aware of the critical role of the minimum ORM value. As a safety-relevant parameter, the ORM would have been automatically calculated and displayed with constantly updated values. An automatic signal to shut down the reactor in case the ORM value drops below the allowed minimum would have been implemented in the reactor protection system.

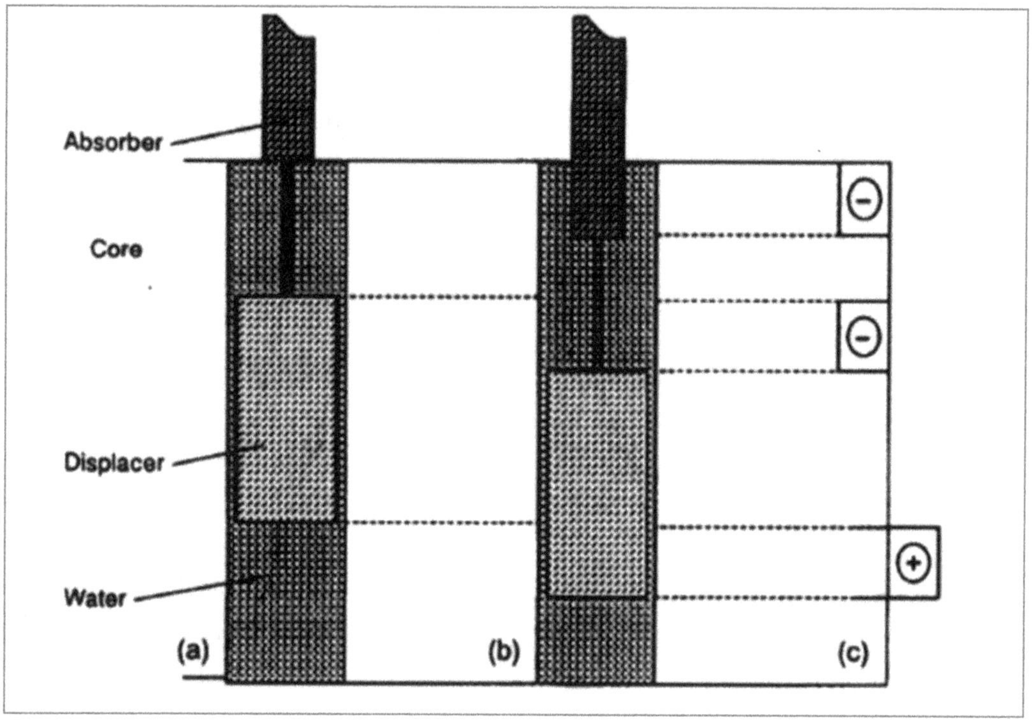

Fig. 5 Design of the control rods of the reactor protection system (Abagyan et al. 1991) (a) rod completely withdrawn with graphite displacer (b) situation while inserting the rod introducing positive reactivity over several seconds in the lower part of the core.

3 The Chernobyl accident

The course of the accident is described and commented in many places (see, for example, GRS 1996, Shteynberg 1991, Abagyan et al., 1991, USNRC 1987 and others). Here a shortened presentation is given, which focuses on the cause of the accident.

On April 25, 1986, the Chernobyl NPP Unit 4 was scheduled for a shutdown and maintenance. It was planned to carry out a test before shutting down the unit. Like all nuclear power stations, RBMK reactors need energy to cool the reactor core even after shutdown (the chain reaction is interrupted, but the decay heat, i.e. the energy from the further decay of fission products, must be dissipated). Usually the power for the residual heat removal system is drawn from the electricity grid. If, as is possible and even probable in the case of an accident, this connection is temporarily unavailable, emergency power generators (diesel generators) are put into operation. However, they need about a minute for starting up. In order to bridge the time between the loss of offsite power and connection to the emergency diesel generators, the output of the running down turbine generators should be used.

The test should confirm that this is actually possible. The experiment should start at reduced power, at between 700 and 1000 MW of thermal power (nominal power are 3200 MWth). Four of the eight main coolant pumps (six plus two reserve pumps) should remain in operation to ensure cooling of the reactor, four should be supplied by the generator connected to the running down turbine. It should be observed for how long those main coolant pumps could be supplied with power.

In attempting to reach the test conditions, the reactor power was inadvertently decreased to 30 MWth instead of the intended range. The operators decided to increase the reactor power again to be able to conduct the test. In certain accounts of the accident this is portrayed as violation of procedure, but according to (GRS 1996) an immediate re-start of the reactor in this situation would have been permissible in principle, if the ORM value were at least 30 equivalent control rods before power reduction. Shteynberg (1991) quotes the operating procedure manual, which states it is permissible to restore the power, as long as the power was not reduced below the (not clearly defined) "minimum controllable power level".

The attempt to increase the power to the planned range was not successful. The reason for this is that when the power in the reactor is reduced, more xenon is produced than is burned and there is a temporary increase in xenon. Xenon acts as a neutron poison and counteracts a power increase. Even though the ORM limit was violated again and significantly more control bars were withdrawn than planned (a breach of the operating regulations), the targeted power level could not be achieved.

The crew decided to stabilize the reactor power at 200 MWth instead and start the test from this level. In some reports, e.g. (GRS 1996, INSAG 1986), it is noted that the team in had changed the test conditions without proper consultation of the relevant departments. However, one must point out that the author of the test program, the deputy chief engineer Diatlov, was part of the team of operators at that night and present in the control room. It is, of course, a violation of procedure to modify the proposed test program ad-hoc without appropriate analysis and verification. In INSAG (1986) it was noted that a low power range for the RBMK is a particularly unstable region. However, as Shteynberg (1991) pointed out, this did not appear to be known to the designers of the reactor, nor to the operators before the accident. Furthermore INSAG (1986), and other reports referring to INSAG

(1986) states that continuous operation at that power level was forbidden. Shteynberg (1991) explicitly referred to that statement as untrue, there was no such rule. Furthermore operation at house load was an event which would be automatically triggered on some occasions. House load operation at RBMK reactors corresponds to a power level of 200–300 MWth.

The range of power where the reactor was now operated was highly unstable. In addition the power distribution showed a maximum in the upper and lower regions of the reactor, instead of dropping from a maximum in the middle upwards and downwards. A reduction in the coolant density (such as steam formation) would lead to a strong increase in output, an increase in output to an accelerated reduction of Xenon and a further decrease in the coolant density and thus to a further increase in output. The power coefficient was strongly positive.

The test program was initiated. The operators had disabled the automatic shutdown signal, which would trigger a shutdown together with start of the test, in order to be able to repeat the test if something goes wrong. INSAG (1986) states "had these trips not been disabled, the insertion of the emergency rods would have terminated the transient regardless of all the other circumstances" (as turned out later this statement is wrong). Diatlov (2005) wrote, the operators had the impression that the test started as planned and that the reactor could now be shut down. Thirty seconds after the experiment was started, the chief operator, Akimov, gave the junior operator Toptunov the order to shut down the reactor. As again Shteynberg (1991) stated, up to this event, the manual shutdown, there were no processes which could have triggered the power excursion that destroyed the reactor.

In fact it was just the actuation of the emergency shutdown system (its particular design with graphite displacers at the bottom) that destroyed the reactor, and, contrary to the statement in INSAG (1986), would the system have been actuated automatically, it would have destroyed the reactor as well. The accident occurred at 1:24 and according to Shteynberg (1991) already at 0:30 the conditions in the reactor were such that a shutdown with the shutdown system was not possible anymore. As was said earlier, due to the particular design and low value of ORM, the shutdown system increased the reactor power for eight seconds, before causing a reduction. And at 200 MWth, the reactor power coefficient was strongly positive, which means that an increase in power would lead to a further increase. Those two circumstances together led to the power surge that destroyed the reactor.

A reactor power of more than 540 MWth was recorded seconds after the actuation of the system. The increase in power together with the test conditions, which in turn reduced the throughput continuously in four of the eight circulating pumps, led to increased steam formation, which in turn led to increased power. It is assumed (Shteynberg 1991) that in the lower part of the reactor the energy input in some fuel elements exceeded 320 cal / g. This quantity of energy led to an explosion of the uranium fuel pellets, which led to a destruction of the fuel rods. The coolant came into direct contact with the hot fuel particles and evaporated spontaneously, which led to a local pressure rise. The increase in pressure resulted in the destruction of the fuel channels, which led to the formation of steam in the entire reactor core and to a further power excursion. Two loud explosions subsequently shook the entire building. The heavy upper-biologic shield (see (8) in Figure 2) was raised and turned to a vertical position. The temperatures in the reactor core rose to high levels and most likely the graphite moderator ignited, a fire that could not be extinguished for several days. The radiological releases still led to excessive limits in many hundred kilometres from the accident site.

4 Conclusions and reason for the accident

INSAG (1986) mainly blamed the operators for the accident. The crew operated the reactor in a dangerous power range, forbidden by operating procedures, violated numerous regulations, and the extremely unlikely combination of violations along with design weaknesses led to disaster. Within this reasoning it was only logical that Fomin, the chief engineer of Chernobyl NPP, and Diatlov, his deputy and author of the test program, were sentenced to prison.

The updated report INSAG (1992) and especially the expert commission reports which are annexed to INSAG (1992) draw a completely different picture. Not only the operators, but also the designers did not know about the design weaknesses. The authorities who had to examine the design did not ask the right questions.

Before the accident, however, there were several indications of the problems of the RBMK, both with regard to the steam bubble coefficient as well as the design of the control rods and their graphite displacer (Shteynberg 1991). But the response to those warnings came to slow and in some cases the warnings were ignored right away.

After re-examination of the up-to-date accounts of the accident it seems to be safe to conclude that the accident was caused mainly by design weaknesses of the reactor, together with violations of procedures by the operators.

Important technical design weaknesses can be named:

- The problematic design of the reactor control rods triggered the reactor accident in Chernobyl.
- There was an unstable power range, and the designer of the reactor were not aware of it (at least at the time of design).

- The way the central parameter ORM was displayed and the lack of automatic monitoring and response to a low ORM margin.

Essential violation of rules of the crew that can be named:

- The operational operating margin of the ORM (operational reactivity margin) was repeatedly not checked by the operating crew in the control room. If the relevant regulations had been complied with, the accident would not have occurred.
- The operating team violated procedures by changing the test conditions on the spot and by executing the test at 200 MWth, without prior analysis.
- The test program itself contained control violations which, however, would not have affected the accident sequence.

The ORM parameter, the operational reactivity margin, was not defined precisely enough by the designer. Thus, the safety relevance of the ORM rule could not be known to the operators. The "peripheral" nature of the evaluation led the operators to believe that this value was not safety-relevant. There are indications that even the designers of the RMBK reactor were not aware of the critical function of the ORM, since otherwise they would have displayed the parameter to the operators in a better way.

Coming back to the question technical or human errors a safe conclusion would be that there were components of both. Yes, there were violations of procedures by the operating crew, and yes, there were weaknesses in the technical design of the reactor. However the training and experience of the crew in the control room complied with all standards of the Soviet Union at that time (Shteynberg 1991). One important parameter (ORM value) was ignored, but by a whole team in the control room. There is no evidence that another team would have behaved differently.

The design weaknesses were certainly not known at the time the RBMK design was developed. Although there were new calculations in the years 1980 and 1985, which suggested that a completely steam-filled channel would not lead to a fading of the chain reaction, but would lead to a power excursion. However, no proper attention was paid to these calculations. There were also references to the "positive", i.e. reactivity-increasing effect of the control rods (Shteynberg 1991). It is therefore also important that there was no open technical debate on the technical weaknesses of the RBMK reactor in the Soviet Union, and that no consequences had been drawn by supervisory authorities and political authorities.

The lessons learned from the Chernobyl accident nowadays points toward the importance of operator training. While this is certainly a valuable lesson, there is another lesson to be learned. There were warnings on the design errors that did not receive proper attention. There are enough examples in Western countries where a warning of a supposed design weakness led to a lengthy review, while affected reactors remained in operation. One example is the possible deboration of the pump seal of a pressurized water reactor in the case of a small leak. There was the fear that a power excursion might occur if a deborated coolant plug could be transported into the reactor core. The investigations have finally shown that the concern was unfounded – but the reactors were still operating until clarification. A further example of a safety – critical issue that has long been discussed in the "Western" world (and is still discussed): If a severe accident occurs, are the containment sump filters clogged because of their narrow mesh width, or can the cooling water be recirculated? Here, too, the reactors were operated without waiting for the complete clarification of the question.

The general perception of the Chernobyl catastrophe is still very much oriented along the lines of INSAG (1986). Human failure together with design weaknesses is determined as the root cause. Lessons for other nuclear power plants are generally confined to underlining the important role of safety culture. Chernobyl is perceived as a problem that is limited to the RBMK reactor in the Soviet Union. In Western nuclear power plants, an accident such as in Chernobyl is unthinkable. However, it is questionable whether this distinction is fully justified. Of course, an accident will not take place in exactly the same way – but undetected faults in the technical design can lead to accidents or severe accidents in other reactor designs as well. Warnings of design weaknesses in general do not lead to a shutdown of the affected units, even in the Western industries, but are examined while the plants continue operation or are sometimes even ignored. With all the differences between RBMK reactors and Western pressurized water reactors, there are similarities that are often overlooked. Reports such as D'Auria (2005) comparing the safety level (and related interpretations) of RBMK reactors to the safety level of Western reactor concepts without classifying the western designs as significantly superior are of little concern.

Designers of reactors are humans and humans make mistakes. The computer systems that the designers of the RBMK reactors had available led them in the wrong direction regarding the behavior of the reactor in the formation of steam in the pressure tubes. The supervisory bodies were not in a position to identify these errors. There was too little or no reaction to hints that the design calculations could be faulty. The operators, after all, did not follow instructions whose meaning they did not understand. All these are reasons for a catastrophic accidents against which also Western reactors are not immune. The RBMK reactor was designed and operated by humans. In view of the complexity of this machine, mistakes have been made – and the consequences of mistakes can be severe when dealing with nuclear power plants.

References

A. Abagyan, E. Adamov, L. Bol'shov, E. Chukardin, V. Petrov, E. Velikhov, 1991. A Report By A Working Group Of Ussr Experts, Causes and Circumstances of the Accident at Unit 4, of the Chernobyl Nuclear Power Plant and ‚Measures to Improve the Safety of Plants with RBMK Reactors, Moscow. In: INSAG International Nuclear Safety Advisory Group (1992): The Chernobyl Accident: Updating of INSAG-1: A Report By the International Nuclear Safety Advisory Group, Safety Series No. 75-INSAG-7, Vienna: IAEA

F. D'Auria, 2005. Deterministic Safety Technology in 'RBMK, TACIS Project 30303 Final Report, Pisa, ISBN 88–902189-0-8

A. Diatlov, 1995. Why INSAG has still got it wrong, Nuclear Engineering International, Vol 40, No 494, September 1995, In: Feature on Chernobyl Accident Nuclear Engineering International, 8 April 2006, http://www.neimagazine.com/features/featurewhy-insag-has-still-got-it-wrong

A. Diatlov, 2005. Chernobyl: Kak eto bylo. Moscow: Nauchtekhlitizdat. http://www.lib.ru/MEMUARY/CHERNOBYL/dyatlow.txt

GRS Gesellschaft für Reaktorsicherheit, 1996. Tschernobyl – Zehn Jahre danach. Der Unfall und die Sicherheit der RBMK-Anlagen, Bericht Nr 121, Köln

INSAG International Nuclear Safety Advisory Group, 1986. Summary Report on the Post-accident Review Meeting on the Chernobyl Accident A Report by the International Nuclear Safety Group INSAG Series No. 1 Subject Classification: 0603-Nuclear power plants STI/PUB/740, Vienna: IAEA

INSAG International Nuclear Safety Advisory Group, 1992. The Chernobyl Accident: Updating of INSAG-1: A Report by the International Nuclear Safety Advisory Group, Safety Series No. 75-INSAG-7, Vienna: IAEA

NEI Nuclear Energy Institute, 1997. Source Book "Soviet-Designed Nuclear Power Plants in Russia, Ukraine, Lithuania, Armenia, the Czech Republic, the Slovak Republic, Hungary and Bulgaria, Fifth Edition, NEI, Washington, DC

OECD/NEA Nuclear Energy Agency Organisation for Economic Co-Operation and Development, 2002. Chernobyl – Assessment of Radiological and Health Impacts 2002 Update of Chernobyl: Ten Years On

S. Schmid, 2011. When safe enough is not good enough: Organizing safety at Chernobyl, Bulletin of the Atomic Scientists 67(2) 19–29, DOI: 10.1177/0096340211399404

B. Sehgal, 2012. Nuclear Safety in Light Water Reactors, Severe Accident Phenomenology, Academic Press, Amsterdam

N. Shteynberg, Chairman, 1991. Report By A Commission To The USSR State Committee, For The Supervision Of Safety In Industry And Nuclear Power 1 Causes and Circumstances of the Accident at Unit 4 of the Chernobyl Nuclear Power Plant on 26 April 1986. Moscow. In: INSAG International Nuclear Safety Advisory Group, 1992. The Chernobyl Accident: Updating of INSAG-1: A Report By the International Nuclear Safety Advisory Group, Safety Series No. 75-INSAG-7, Vienna: IAEA

G. Snell, J. Howieson, 1991. Chernobyl – A Canadian Perspective, AECL Atomic Energy Canada Limited

USNRC United States Nuclear Regulatory Commission, 1987. Report on the accident at the Chernobyl nuclear power station, NUREG-1250. Washington, DC

The Reality after Fukushima in Japan

Actual Damage to Local People

Tadahiro Katsuta[1]

Abstract

This study analyses the government's efforts and the actual situation of the victims of Fukushima Daiichi nuclear power plant accident five years after the accident. As of September 5, 2015, about 99 thousand Fukushima prefecture residents had been forced to evacuate from their homes. Currently, the government is seeking to lift evacuation orders aggressively. However, evacuees have mixed feelings. The amount of legally required compensation for damages continues to increase; it reached 7.65 trillion yen (US$76.5 billion) in the latest review as of the end of March 2016. TEPCO is practically bankrupt and has been collecting funds from all Japanese citizens. As of the end of December 2015, 51 people were diagnosed with malignant or suspected malignant thyroid cancer in the second examination conducted by Fukushima Prefecture. Government measures, i.e., disaster recovery plans, compensation for damages, and scientific approaches, have been used as means to avoid taking responsibility through the use of power, the use of money to keep victims silent, and the use of science as an excuse; these measures are driving the victims into a corner instead of supporting them. Ultimately, two common causes of these problems are related to the nuclear energy policy of the past and the nuclear energy policy for the future.

---

1 Tadahiro Katsuta, School of Law, Meiji University, Tokyo, Japan, tkatsuta@meiji.ac.jp

R. Haas et al. (Eds.), *The Technological and Economic Future of Nuclear Power*, Energiepolitik und Klimaschutz. Energy Policy and Climate Protection, https://doi.org/10.1007/978-3-658-25987-7_16

1 Introduction

Five years have passed since the Tokyo Electric Power Company (TEPCO)'s Fukushima Dai-ichi Nuclear Power Plant Accident (hereinafter referred to as the Fukushima accident) which occurred due to the magnitude 9.0 Great East Japan Earthquake of March 2011. Now, have the victims of Fukushima accident been able to get back to normal life without any problem? Unfortunately, they are still facing many problems. Recovery has been slow in Fukushima due to the existence of areas designated as evacuation zones, their dissatisfactions are suppressed by the compensation for damages, and they suffer from the fear of developing cancer through the participation in the thyroid examination. This study examined these three challenges as the current main issues faced by the victims to clarify the current status of the victims who are suffering not only from the effects of the Fukushima accident but also because of government measures. The complex current conditions are described in Chapter 2 and the essential issues obscured by the current situation are clarified in Chapter 3.

2 Aspirations and the reality five years after the Fukushima accident

2.1 Evacuation plan[2]

The residents of Fukushima prefecture, in which Fukushima Dai-ichi nuclear power plant is located, are exhausted from the prolonged evacuation and hope to be freed from the life as an evacuee. On the other hand, they have doubts about the hasty lifting of evacuation orders by the government due to their anxiety about radiation-related issues.

The Reconstruction Agency, which was established after the Great East Japan Earthquake, set the five years following the earthquake of 2011 as the intensive reconstruction period, and the term from April 2016 to March 2021 as the reconstruction and creation period[3]. They emphasized that the number of earthquake evacuees

2 Reconstruction Agency, Current Status of Reconstruction and Challenges, March 2016. http://www.reconstruction.go.jp/english/topics/Progress_to_date/image/20160307_Current_Status_of_Reconstruction_and_Challenges_rev1.pdf

3 Reconstruction Agency, "The Process and Prospects for Reconstruction", March 2016, http://www.reconstruction.go.jp/english/topics/Progress_to_date/image/20160307_process_and_prspects.pdf

from Iwate Prefecture, Miyagi Prefecture and Fukushima prefecture has dropped to 174 thousand people as of February 2015 from its peak at about 470 thousand.

But if we look at the figures closely, it becomes clear that Fukushima prefecture alone has been left behind. As of September 5, 2015, about 99 thousand Fukushima prefecture residents—the number is greater than half of the total number of evacuees—had been forced to evacuate from their homes. Specifically, about 55 thousand people had evacuated to other areas within Fukushima prefecture; among these evacuees, approx. 51 thousand people had been compelled to live in temporary housing. The remaining approx. 43 thousand people had evacuated to other prefectures across Japan. According to the questionnaire survey involving 1,000 earthquake victims conducted in March 2016 by Japan Broadcasting Corporation (NHK), 310 people had to evacuate more than 5 times; among these people, 250 people were Fukushima prefecture residents. Besides, the percentage of people who responded that they do not feel that disaster recovery has been achieved was 17.6 % in Iwate Prefecture, 18.2% in Miyagi Prefecture, and 49.9 % in Fukushima Prefecture[4].

About 70 thousand people have evacuated from the designated evacuation zones due to the Fukushima accident: specifically, about 24 thousand people evacuated from the difficult to return zone, about 23 thousand people from the restricted residence zone, and 24 thousand people from the zone in preparation for the lifting of the evacuation order. The prolonged evacuation period has exhausted the Fukushima prefecture evacuees. As of the end of September 2015, the total number of disaster-related deaths—i.e. deaths that were not caused directly by the earthquake and tsunami but were due to indirect causes such as deterioration of physical conditions as a result of evacuation—was 3,407 people. These people had been living in 9 prefectures and Tokyo. Of these, Fukushima prefecture had the highest number at 1,979 deaths[5]. According to the latest information released by Fukushima prefecture, the number of deaths has risen to 2,038 (as of 1 April 2016)[6].

Since the sum of deaths including deaths directly caused by the earthquake and tsunami is 3,866, the number of deaths caused by indirect reasons has exceeded that of the deaths caused by direct reasons. As the government hasn't provided a

4 NHK (Japan Broadcasting Corporation), Great East Japan Earthquake, A Survey of 1000 Survivors, (in Japanese), http://www.nhk.or.jp/d-navi/link/shinsai5/shinsai5.pdf, accessed 26 April 2016.

5 Reconstruction Agency, "The number of disaster-related deaths due to the Great East Japan Earthquake" 25 December 2015. (in Japanese) see http://www.reconstruction.go.jp/topics/main-cat2/sub-cat2-6/20151225_kanrenshi.pdf Accessed 12 April 2016.

6 Deaths and injuries due to the Great East Japan Earthquake (as of 1 April 2016), https://www.pref.fukushima.lg.jp/uploaded/life/198319_445221_misc.xlsx, accessed 26 April 2016.

definition of the term "disaster-related death," bereaved family members must prove with great difficulty that the death of the deceased family member is related to the disaster. Therefore, the potential number of disaster-related deaths may be higher.

According to the statistics collected by the Cabinet Office, the number of suicide related to the Great East Japan Earthquake has decreased everywhere else but Fukushima prefecture. The number of suicides committed in Iwate prefecture and Miyagi prefecture in 2011 following the earthquake was 17 and 22, respectively; this number in 2015 dropped to 3 and 1, respectively. On the contrary, the number of suicides increased from 10 in 2011 to 19 in 2015 in Fukushima prefecture[7].

Currently, the government is seeking to lift evacuation orders aggressively. Among the zones in preparation for the lifting of the evacuation order, orders covering a part of Tamura city and a part of Kawauchi city were lifted in 2014, and an order covering a part of Naraha town was lifted in September 2015. In June 2015, the government announced that they will enable the lifting of evacuation orders for all restricted residence zones and zones in preparation for the lifting of the evacuation order by March 2017[8]. If this plan materializes, 47 thousand people will be able to return to their homes.

However, evacuees have mixed feelings. According to the results of the NHK survey, 45.8% of Fukushima evacuees responded that it is too early. For example, in February 2016, the government held a briefing in Minami-souma city and stated that they hope to lift the evacuation order in April. In response to this, numerous residents commented that it is too soon to lift the order since progress has been slow in implementing decontamination activities[9]. In March 2016, Fukushima prefecture released the results of its questionnaire survey. Among the people who had evacuated to other prefectures and had no home to return to in Fukushima prefecture after April 2017 which is when the program for offering rental houses free of charge will be terminated, about 70% of them did not wish to return to Fukushima while about 10% wanted to return to the prefecture and about 20% responded that

7 Cabinet Office, "Number of suicides related to the Great East Japan Earthquake" 13 March 2016. (in Japanese) see http://www8.cao.go.jp/jisatsutaisaku/toukei/pdf/h27joukyou/jishin.pdf Accessed 12 April 2016.

8 Nuclear Countermeasures Headquarters, "Accelerating post-nuclear disaster Fukushima recovery efforts" (Revised version), 12 June 2015. (in Japanese) see http://www.meti.go.jp/earthquake/nuclear/kinkyu/pdf/2015/0612_02.pdf Accessed 12 April 2016.

9 Tokyo Shimbun, "Residents oppose plan to lift evacuation order in April at an explanatory meeting in Minami-souma city", 21 February 2016. (in Japanese) see http://www.tokyo-np.co.jp/article/national/list/201602/CK2016022102000126.html Accessed 12 April 2016.

they are still debating on whether or not to return[10]. These response results may be due to the following reasons: the fact that their lives at the evacuation destination have taken root, concerns over changing the children's living conditions, and fear of radiation-related issues.

Decontamination work in the designated areas to be decontaminated under the direct control of the government has finished in 6 municipalities among the 11 municipalities within Fukushima prefecture and the plan is to finish decontamination in the remaining municipalities by the end of FY2016[11]. But anxiety in Fukushima prefecture is strong. According to the NHK survey, 38.7% of evacuees responded that their fear of exposure had not changed even five years after the accident. In December 2015, the Ministry of the Environment announced that they will not decontaminate areas more than 20 km away from daily activities area in Fukushima prefecture[12]. However, as a result of local opposition, the ministry changed the policy to carry out decontamination in satoyama areas—border zones of agricultural land and forested land traditionally regarded as one area—where people may enter easily[13].

2.2 Compensation for damages

TEPCO continues to pay compensation for nuclear damages to the people who suffered damages such as individuals, sole proprietors, and corporations.

Legally required compensation costs have continued to increase and the total reached 7.65 trillion yen (US$ 76.5 billion) in the latest review as of the end of March 2016. Out of that total, the amount of the agreed-upon compensation was 5.92 trillion yen (US$ 59.2 billion). Compensation costs include medical examination

10 Fukushima Prefecture, "Interim report on the residence intentions survey", 25 March 2015. (in Japanese) see https://www.pref.fukushima.lg.jp/uploaded/attachment/158116.pdf Accessed 12 April 2016.

11 Ministry of the Environment "Progress map of decontamination activities implemented under the direct control of the government" 4 March 2016. (in Japanese) see http://josen.env.go.jp/material/pdf/josen_gareki_progress_201603.pdf Accessed 12 April 2016.

12 Environmental recovery review meeting, "Direction of radioactive materials management measures for forests (draft)" 21 December 2015. (in Japanese) see http://www.env.go.jp/jishin/rmp/conf/16/mat05.pdf Accessed 12 April 2016.

13 Project team of relevant ministries and agencies for recovering forests and the forest industry in Fukushima, "Comprehensive approach for recovering forests and the forest industry in Fukushima", 9 March 2016. (in Japanese) see http://www.reconstruction.go.jp/topics/main-cat1/sub-cat1-4/forest/160309_3_siryou1.pdf Accessed 12 April 2016.

costs, compensation for psychological damages, voluntary evacuation expenses, and business loss expenses[14]. In terms of the number of claims, approx. 899 thousand cases by individuals, approx. 1.3 million cases by individuals (losses due to voluntary evacuation), and approx. 4.02 million cases by corporations and sole proprietors had been filed as of April 2016[15].

TEPCO has been showing consideration for the circumstances and feelings of the victims[16]. However, according to the Nuclear Damage Compensation Dispute Resolution Center, which was established as means to provide Alternative Dispute Resolution (ADR) by mediating disputes between victims and TEPCO to enable them to reach agreements without having to go to court, 4,239 claims were made in 2015 and the mediation process is still ongoing indeed for 2,746 of those claims[17].

Reparation does not cover the only TEPCO. The company has received the financial assistance from the nuclear power operators and government. That is, TEPCO has been bankrupt substantially and has attracted funds not only the consumer of electric companies without TEPCO but also the entire Japanese people.

According to the Nuclear Damage Compensation and Decommissioning Facilitation Cooperation, which was established newly to manage compensation funds, the cooperation received 508.3 billion yen (US$ 5.1 billion) from nuclear power operators including TEPCO, an additional 110.0 billion yen (US$ 1.1 billion) from TEPCO alone, and 9 trillion yen (US$ 90 billion) of government bonds from the government as of FY2014[18].

According to the estimation released in March 2015 by the Board of Audit of Japan, the government will need 30 years at the maximum to collect the debt owed

14 TEPCO, "New Comprehensive Special Business Plan" 31 March 2016. (in Japanese) see http://www.meti.go.jp/press/2015/03/20160331005/20160331005-1.pdf Accessed 12 April 2016.

15 TEPCO, Records of Applications and Payouts for Indemnification of Nuclear Damage (as of 22 April 2016). http://www.tepco.co.jp.cache.yimg.jp/en/comp/images/jisseki-e.pdf, accessed 26 April 2016.

16 TEPCO, Five Promises to the Afflicted, https://www4.tepco.co.jp/en/images/5promises.pdf, accessed 26 April 2016.

17 Nuclear Damage Compensation Dispute Resolution Center, Activities Report, March 2016. http://www.mext.go.jp/component/a_menu/science/detail/__icsFiles/afieldfile/2016/04/20/1347876_009.pdf

18 Nuclear Damage Compensation and Decommissioning Facilitation Corporation, FY2014, Business Report (in Japanese), http://www.ndf.go.jp/soshiki/zai_h26jigyo.pdf, accessed 26 April 2016.

by TEPCO when it is assumed that the government provides financial assistance at the maximum government bond amount of 9 trillion yen (US$ 90 billion)[19].

2.3 Thyroid cancer diagnosis

In contrast to Fukushima prefecture's responses to evacuation plans and compensation, the prefecture has continued to deny the possibility of children's thyroid cancer together with the government. This may be because of the involvement of the government in the diagnosis process.

Fukushima prefecture is continuing its health survey which includes surveys of external and internal doses and thyroid examinations[20] . In regard to the thyroid examination, the preceding survey—ultrasonic wave examination for residents who were under 18 years old and lived in Fukushima prefecture at the time of the accident—was conducted from FY2011 to FY2013. Of the about 370 thousand subjects, 300 thousand people were examined (participation rate: about 82%)[21].

As of the end of June 2015, 113 people were diagnosed with malignant or suspected malignant thyroid cancer[22]. Of these, 99 people underwent surgery. Although this result is higher than the Japan's thyroid cancer statistics, the Fukushima Prefectural Citizens Health Survey Committee has not recognized these thyroid cancer cases as the result of the Fukushima accident; the Committee's reasoning is that these people were exposed to less radiation when compared with the case of Chernobyl accident and that some of the subjects may have been over-diagnosed.

A full-scale survey has been started involving the subjects of the preceding study and children who were born after the accident. If nodules or cysts that are

---

19 Board of Audit of Japan, "Report on the results of the accounting audit regarding the implementation status of government's assistance provided to TEPCO for compensation for nuclear damage" March 2015. (in Japanese) see http://www.jbaudit.go.jp/pr/kensa/result/27/pdf/270323_zenbun_01.pdf Accessed 12 April 2016.

20 According to an estimation of external exposure dose rate based on a questionnaire survey, 93.8% of the respondents were exposed to doses between 0 to 2mSv as of December 2015. However, only 560 thousand people responded out of the 2.05 million subjects (27.4%). As for internal exposure measurements using a whole body counter, 281,228 people were exposed to less than 1mSv while 26 people were exposed to doses between 1mSv to 3mSv. Source: "Overview of the residents health survey" https://www.pref.fukushima.lg.jp/site/portal/43-7.html (in Japanese) Accessed 12 April 2016.

21 ibid.

22 Fukushima Prefectural Citizens Health Survey Committee, "Interim report on the prefectural citizens health survey", March 2016. (in Japanese) see http://www.pref.fukushima.lg.jp/uploaded/attachment/158522.pdf Accessed 12 April 2016.

larger than a predetermined size are found in the first examination, those people undergo a second examination. As of the end of December 2015, 51 people were diagnosed with malignant or suspected malignant thyroid cancer in the second examination. Unfortunately, only 29 of them submitted a basic survey questionnaire that provides data on their exposure dose at the time of the accident. Among these values, the highest dose was 2.1 mSv[23].

In May 2015, a research group of Okayama University published a paper of epidemiological studies related to frequent occurrence of childhood thyroid cancer[24]. According to the group, based on the results of the screening tests of Fukushima Prefecture, at the maximum, the number of thyroid cancer incidences in a certain area of Fukushima prefecture was 50 times higher than Japan's average annual number of thyroid cancer incidences. Accordingly, the group concluded that excessive occurrence of thyroid cancer has already been detected. However, this paper has received criticism and the academic debate on this issue has been continuing[25].

Diagnosis results are reviewed by the Fukushima Residents Health Survey Committee for the purpose of obtaining professional advice from experts belonging to research institutes and universities across Japan.

In October 2012, it was revealed that this committee had held secret preparatory meetings prior to the open review meeting to pre-arrange the discussions of the committee members; it was also discovered that they had created a scenario to lead the discussion at the time of the meeting[26]. Furthermore, it was revealed that the former Chair of the committee had sent out a document to thyroid specialists across Japan in January 2012 to urge them not to respond to requests from the parents of

---

23 Prefectural Citizens Health Survey Committee, "Thyroid examination (full-scale examination) implementation status" 15 February 2016. (in Japanese) see http://www.pref.fukushima.lg.jp/uploaded/attachment/151272.pdf Accessed 12 April 2016

24 Tsuda, Toshihide et al., "Thyroid Cancer Detection by Ultrasound Among Residents Ages 18 Years and Younger in Fukushima, Japan: 2011 to 2014", Epidemiology: May 2016 – Volume 27 – Issue 3 – p 316–322., see http://journals.lww.com/epidem/Citation/2016/05000/Thyroid_Cancer_Detection_by_Ultrasound_Among.3.aspx Accessed 12 April 2016.

25 Takahashi, Hideto et al., "Re: Thyroid Cancer Among Young People in Fukushima", Epidemiology • Volume 27, Number 3, May 2016, see http://journals.lww.com/epidem/Fulltext/2016/05000/Re___Thyroid_Cancer_Among_Young_People_in.36.aspx Accessed 12 April 2016.

26 Management Investigation Committee, Correction of survey results concerning the management of the meeting of Fukushima Residents Health Survey Committee. November 15, 2012, (in Japanese) http://www.pref.fukushima.lg.jp/uploaded/attachment/45898.pdf

the examination participants for a second opinion – an approach in which patients/ guardians choose the treatment by obtaining the opinion of several experts[27].

3 Challenges indicated by the reality

3.1 Problems of accident response measures

1. Evacuation issues: The government is giving higher priority to the external foreign policy over the lives of the evacuees and is trying to take advantage of the Fukushima accident. For example, the government announced in 2015 that it would communicate to the whole world at the Tokyo Olympic Games in 2020 that Japan has recovered from the 2011 disaster by regarding the event as a symbol of Japan's recovery[28]. Plans for lifting evacuation orders and decommissioning activities are scheduled around the 2020 Tokyo Olympic Games in the government's disaster recovery plan[29].
2. Compensation issues: The future of compensation payments by TEPCO is uncertain. From now on, TEPCO will have to become competitive and operate its business more efficiently due to the liberalization of electricity retail sales in April 2016. Although the government has not announced the total cost of Fukushima accident yet, it will reach at least about 13.3 trillion yen including decommissioning and decontamination cost according to a calculation using data released by TEPCO[30].

---

27 Michiyuki Matsuzaki, Opinion, What is happening to the children in Fukushima?, May 2015 (in Japanese). http://1am.sakura.ne.jp/Nuclear/kou131Matsuzaki-opinion.pdf

28 Press Conference by Prime Minister Shinzo Abe on the Upcoming Fourth Anniversary of the Great East Japan Earthquake, March 10, 2015. accessed 26 April 2016. http://japan.kantei.go.jp/97_abe/statement/201503/1210209_9916.html

29 See Ref.2

30 (1) Decommissioning and contaminated water treatment costs of 2 trillion yen: Although TEPCO has already set aside a reserve of 1 trillion yen (US$ 10 billion), the government has asked the utility to secure another 1 trillion yen (US$ 10 billion) within 10 years. (2) Compensation costs of about 7.1 trillion yen (US$ 71 billion): The total of legally required compensation costs according to the latest data is about 7.7 trillion yen (US$ 77 billion, see Table 3). (3) Decontamination costs of 3.6 trillion yen (US$ 36 billion): The Ministry of the Environment has estimated the decontamination cost at about 2.5 trillion yen (US$ 25 billion) and the interim storage facilities cost at about 1.1 trillion yen (US$ 11 billion). See Ref. 13.

3. Thyroid cancer diagnosis: Due to lack of clear information about the relationship between radiation exposure and cancer, the anxiety of people about the effects of radiation has increased more by the responses of the government and Fukushima prefecture. Although the health investigation committee of Fukushima prefecture is operating with the Fukushima prefectural health fund, since this fund received grants of 78.2 billion yen from the Ministry of the Environment[31] and 25 billion yen from TEPCO[32], the neutrality of this committee is unclear.

Traditionally, the Japanese government has tended to avoid dealing with radiation-related problems. For example, on October 20, 2015, Fukushima Bureau of Ministry of Health, Labour and Welfare (MHLW) recognized the leukemia developed by a worker who worked on decommissioning tasks after the Fukushima accident as an occupational disease[33]. However, MHLW stated that "this recognition does not prove scientifically the causal relationship of radiation exposure and its health effects." The government's responses imply that it is trying to avoid an increase in workers' compensation due to recognition of occupational diseases.

After the Fukushima accident, the government created and released a quick reference table of radiation exposure in order to eliminate the people's radiation-related concerns. However, it was discovered that they had secretly corrected the figures without providing sufficient explanation. In the table, the level of natural background radiation in Japan was changed from 1.5 mSv/year of the April 2011 version to 2.1 mSv/year in the May 2013 version. Furthermore, the comment "No observable increase in cancer incidence" for exposure levels under 100mSv was deleted[34].

3.2 Common factors

The results obtained are shown in Table 1. Government measures, i.e., disaster recovery plans, compensation for damages, and scientific approaches, have been used as means to avoid taking responsibility through the use of power, the use of

---

31 Ministry of the Environment, Support of Fukushima prefecture health research business, accessed 26 April 2016. http://www.env.go.jp/chemi/rhm/support.html

32 See Ref. 13.

33 Ministry of Health, Labour and Welfare, "Result of review at the 'review meeting on occupational/non-occupational ionizing radiation disease' and approval as occupational disease/injury" 20 October 2015. (in Japanese) see http://www.mhlw.go.jp/file/05-Shingikai-11201000-Roudoukijunkyoku-Soumuka/kouhyousiryou.pdf

34 National Institute of Radiological Science, Dose scale, accessed 26 April 2016, http://www.nirs.go.jp/data/pdf/hayamizu/e/20130502.pdf

money to keep victims silent, and the use of science as an excuse; these measures are driving the victims into a corner instead of supporting them. Furthermore, it seems that these efforts are being made to obscure the responsibility rather than to resolve the problems, and in hopes that the victims will give up on seeking solutions.

Ultimately, two common causes of these problems are related to the below described past and future nuclear energy policies.

Common factor 1: Promotion of the aggressive nuclear energy policy of the past

The cause of the current confusion concerning Fukushima accident responses is the claim aggressively made by the government and power companies in the past that a nuclear accident will not occur. As a result, the responses by the government and TEPCO were slow. The victim's and general citizens' distrust in the government and TEPCO still remains.

Common factor 2: Promotion of an aggressive nuclear energy policy for the future

The government is trying to forcefully settle all problems related to the Fukushima accident at an early stage because it is trying to maintain the already set out nuclear energy policy for the future. From that standpoint, evacuation, compensation and exposure problems are all inconvenient facts and the government is afraid that these facts will have a negative effect on its efforts to maintain the nuclear energy policy. On the other hand, victims and the general public continue to have anxiety about the future.

Tab. 1 Measures and purposes of the government and TEPCO

Issue	Responsible party	Victim	Solution	Reality	The real purpose
Evacuation plan	Government	Residents	Disaster recovery plan, Lifting of evacuation orders	Use of power	Diplomatic message
Compensation	TEPCO	Residents (Japanese citizens)	Compensation system	Keep victims silent by the money	Revival of the company
Thyroid cancer diagnosis	Government, Fukushima prefecture	Residents (Children)	Scientific investigation	Use of science as an excuse for reaching definitive conclusions	Elimination of social anxiety

4 Conclusions

At present, five years after the Fukushima accident, the government's responses so far to the evacuation problems, compensation issues and the risk of thyroid cancer have been insufficient. It is obvious that the government's intention behind these insufficient measures is to maintain the nuclear energy policy.

Therefore, the victims have been hurt not only by the impact of the Fukushima accident but also by the government's responses. People affected by the nuclear disaster caused by the nuclear promotion policy of the past are now suffering from the current promotion of the nuclear energy policy for the future.

Distributing the Costs of Nuclear Core Melts

Japan's Experience after 7 Years

Eri Kanamori, and Tomas Kåberger[1]

Abstract

The costs of managing the consequences of the Fukushima-Daiichi nuclear have been significant already, and the estimated total future costs have increased over time. The immediate payments have been possible by direct payments from the Japanese government. However, most these payments are not acknowledged as government spending. Instead, a complicated system of envisioned re-payments have been created.

Based on the three Special Business Plans published by TEPCO since the nuclear disaster, this evolving perception of the economic consequences and the increasingly complicated repayment schemes are described.

The conclusion of the authors are that the repayment schemes are not compatible with a future efficient, competitive electricity market.

It is suggested that other governments who implicitly or explicitly accepting economic liabilities for nuclear accidents prepare themselves in order to avoid un-necessary indirect cost after future reactors accidents.

1 Eri Kanamori, Ritsumeikan University, Osaka, Japan, kanamori@ba.ritsumei.ac.jp; Tomas Kåberger, Chalmers University of Technology, Göteborg, Sweden, tomas.kaberger@chalmers.se

R. Haas et al. (Eds.), *The Technological and Economic Future of Nuclear Power*, Energiepolitik und Klimaschutz. Energy Policy and Climate Protection, https://doi.org/10.1007/978-3-658-25987-7_17

1 Background

In this paper, we attempt to describe the way the Government of Japan (GOJ) and Tokyo Electric Power Company (TEPCO) have managed to cope with the successively increasing acknowledged costs of the Fukushima core melts-downs, and following radioactivity leaks in 2011. The evolving scheme of dealing with the cost are found in TEPCO's special business plans.

But first, in this paper, a short description of the consequences of the core melts and of Japan's national energy policy. This, as a background in order to understand the challenges of TEPCO and the design of their recent special business plans:

It was on March 11, 2011 that the three operating TEPCO reactors at Fukushima-Daiichi nuclear power plants proved unable to cope with the effects of an earthquake. The nuclear reactor core melts in Fukushima and consecutive explosions resulted is emission of radioactive substances into the air and water. Despite winds bringing most of the air emissions out into the Pacific, there was also some contamination on land, and over a hundred thousand people were instructed to evacuate their homes, while many in addition relocated at their own initiative, without evacuation orders.

Systematic, comprehensive studies of health effects are not published. Only increased thyroid cancer among children in the affected areas are documented [Tsuda et al. 2015], but sometimes denied to have any relation to the contamination. Thus economic measures of health costs are uncertain and not included in the official cost estimates.

The efforts to control the emissions are still engaging in the order of 5 000 people at Fukushima-Daiichi Nuclear Power Plant. This is many more than the number of people employed when the plant produced electricity.

Removing spent fuel from damaged fuel pools, pumping contaminated water into new-built storage tanks by 2018 containing a million cubic meters, building treatment plants extracting as many radioactive isotopes as possible, and constructing an ice wall around the reactors in order to reduce the amount of water flowing into the most contaminated parts of the plant are some of the short-term efforts. For the longer term, attempts are made to find out where the melted reactor fuel is, and then to develop technologies and strategies for the long term decommissioning of the plant. The work can be followed at the web-site of TEPCO. [TEPCO 2011–2018]

Estimating the total cost from the accident until all the remains of the reactors are brought to a condition where further spreading of radioactivity will be avoided in the very long term is difficult.

The purpose of this paper is not to describe the long-term technical solutions, nor the long term financial solutions. This introduction was intended as a sketchy background to the financial challenges of managing the consequences. The intention with this paper is to provide a description of how the short-term costs have been managed in Japan since the accident.

Similarly, a brief background of the energy policy follows to understand the way the consequences have been managed:

Japan's energy policy since the 1970s has focused on the development of nuclear power. Despite the absence of domestic uranium reserves, the import of this fuel has been seen as less problematic than dependence on fossil fuel imports. Uranium is easier to store, and a reprocessing ambition, making breeder reactors or production of mixed oxide fuels possible, made the envisioned future import volumes small compared to oil dependence.

Despite strong political backing, the nuclear strategy encountered problems. New reactors were expensive, slowing down the expansion. Even operation of existing reactors have faced problems and the peak in nuclear electricity generation was as early as 1998, with 327 terawatt-hours (TWh) delivered. Still, the last full year before the 2011 Fukushima failure, more than 290 TWh were produced by 54 nuclear reactors in Japan. [BP, 2017]

The Fukushima-Daiichi failure was a serious disaster. Still the "4th Strategic Energy Plan" [GOJ, 2014], a cabinet decision made in April 2014, which is the basic and comprehensive Japanese energy policy today, aims at the "Re-establishment of nuclear energy policy" (p.47). The plan says that "Nuclear power is an important base-load power source as a low carbon and quasi-domestic energy source, contributing to stability of energy supply-demand structure, on the major premise of ensuring its′ safety" (p.24). It also says that "Even after the TEPCO's Fukushima nuclear accident, use of nuclear energy is expected to expand in the world. The scale of the expansion is particularly remarkable in Asian nations where energy demand is rapidly increasing. Japan, with its experience of the accident, is expected to make contributions in the fields of safety, nuclear non-proliferation and nuclear security as an advanced nuclear nation" (p.50). The latest "5th Strategic Energy Plan" [GOJ, 2018] holds the same attitude.

From these statements, it is obvious that Japan's energy policy is still supporting nuclear power. Under this national policy, TEPCO's special business plans have been made and, as we shall describe below, approved by the Government. In the national policy, allocation of the Fukushima costs is not explicitly dealt with. Thus the costs are not visible in the national budget, and not described as a cost of the government. Instead, the matter has been included in TEPCO's special business plans. It is assumed in Japan from provisions of the "Act on Compensation for Nuclear Damage" that only TEPCO is responsible for the accident while the Government is not.

While the government is prescribing how to manage the costs, and in fact providing the necessary funds, the roles of the current government and future tax- and ratepayers are not clearly presented and rarely debated in public.

This paper is an analysis of how the three consecutive special business plans for TEPCO, published after the accident, have evolved; How they present the solution of the challenging task of finding the money necessary to manage the immediate costs of the failure in reactor control after the earthquake.

2 TEPCO's special business plans

In the 2012 special business plan of TEPCO, the acknowledged costs for Fukushima were estimated as 25 billion euros.[2] (Figure 1) This is to pay compensation to victims who were ordered to evacuate from defined areas in the Fukushima prefecture.

In the 2014 plan, the estimated costs became 106 billion euros and it included not only compensation but also some decontamination, interim storage, and decommissioning. And in the 2017 plan, the total costs were assessed as 215 billion euros. The scheme of dealing with these costs will be explained chronologically.

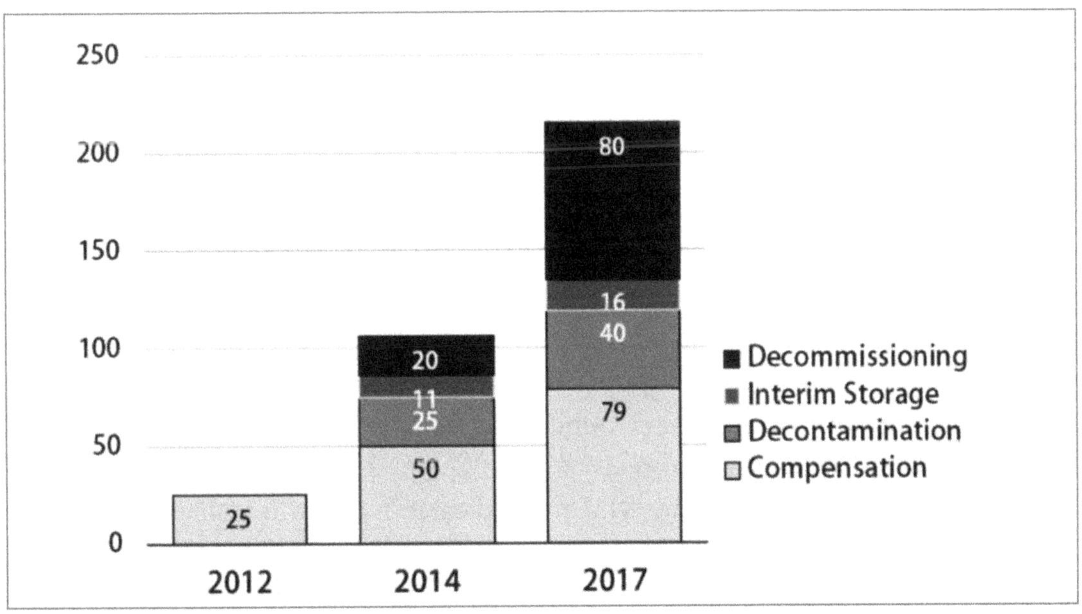

Fig. 1 The estimated cost of Fukushima accident

2 In this paper, 1 trillion yen is assumed equal to 10 billion euros. In 2018, this is only the right order of magnitude as one Euro is close to 130 JPY.

2.1 The 2012 TEPCO Special Business Plan

In the 2012 plan (TEPCO, 2012a), called "Comprehensive Special Business Plan", Fukushima costs were seen as only compensation and estimated at 25 billion euros.

In the financial year ending 31 March 2011, TEPCO recorded a 12 billion euros net loss. This net loss was mainly from asset impairment and efforts to cool and control reactors. Because of the proven loss, TEPCO's equity was eroded from 21 billion euros in 2009/2010 to 12 billion euros in 2010/2011 (non-consolidated), despite raising a few billions from issuing new shares in October 2010. So, the 25 billion euros of compensation was already bigger than TEPCO's equity. Thus, this early estimated expense was unaffordable for the company. Under normal economic conditions of limited liability companies in the world, this should have resulted in a bankruptcy where shareholders lost their assets in the company.

However, the government of Japan, instead, decided to rescue the shareholders. The government provided 25 billion euros to TEPCO, for the company to be able to pay compensation, and an additional 10 billion euros in exchange for TEPCO's new, emitted shares giving the government ownership of 54.69%. This prevented TEPCO from going bankrupt and gave incumbent shareholders an opportunity to stay as owner. At the same time the government avoided making unlimited commitments.

The 10 billion euros payment was mainly to make TEPCO able to pay for its ongoing business activity, such as buying additional fuels and paying interests to banks. Strictly speaking, the government did not do it directly, but through the "Nuclear Damage Liability Facilitation Corporation" (NDF) which they created after Fukushima. As a result, it is the NDF that holds 54.69% of TEPCO's shares and TEPCO is now controlled by the Government via the NDF.

The scheme of 2012 plan is shown in Figure 2, above. The two government aids are shown as arrows 1 and 2. Arrow 1 shows that the NDF provided 25 billion euros for TEPCO to pay compensation to victims. In order to do this, the Government provided its bonds of 50 billion euros to the NDF. These government bonds were issued for special purpose of supporting compensation and had three features which are different from ordinary bonds: yielding no interests, prohibited to be transferred to a third party, and ready to be redeemed at any time on demand for the purpose. Therefore, when the NDF needed cash in order to assist TEPCO paying compensation, and the NDF demanded it officially to the GOJ, the NDF could acquire cash at any time. From 2012 onwards, the redemption has been repeated monthly. The Government derived the cash from its annual energy budget. It seems that there was already a concern that 25 billion euros might not be sufficient for paying compensation, since the government special purpose bond was up to 50 billion euros already at this stage. At the same time, as arrows 2 show, the NDF invested

10 billion euros in 54.69% of TEPCO's share. To finance this additional aid, the NDF borrowed the money from banks based on a guarantee from the government.

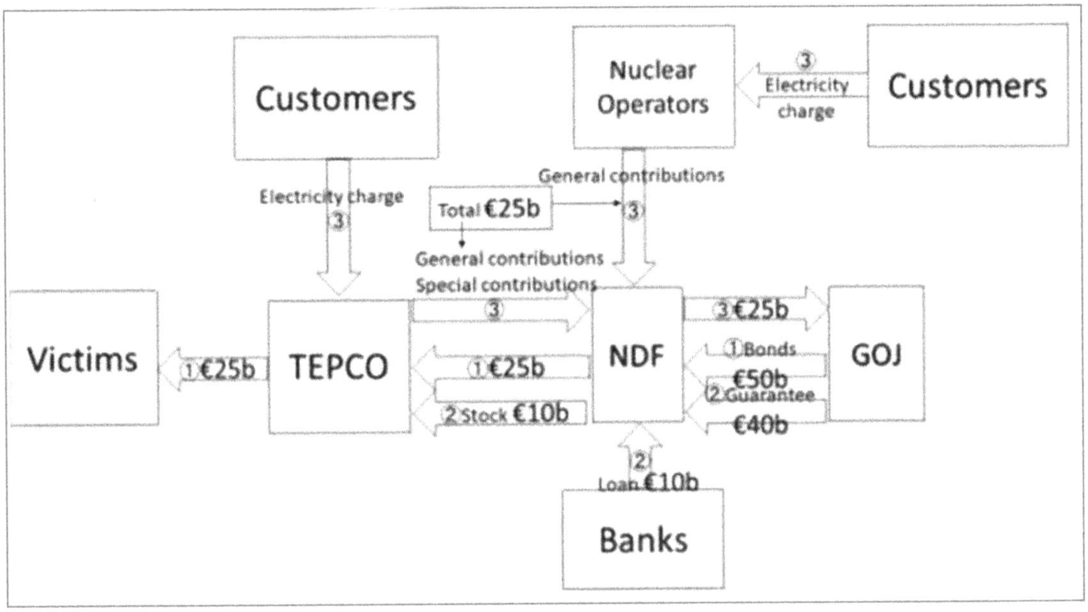

Fig. 2 The scheme of the 2012 plan

The money for compensation to victims, it is said, will be paid back to the NDF in a complicated way. A new law was imposed whereby the responsibility to contribute to repayment was shared between all nuclear operators.

In other countries, the liable party would be a matter for courts under existing laws. Here, the Japanese government and diet retroactively distributed the economic liability to all nuclear operators. It may be seen as a retroactive insurance fee that could have been imposed on them beforehand.

As the electricity system, at the time, was operated by regional, vertically integrated, monopolies, costs were easily passed on to electricity customers. While an insurance fee, in a competitive market economy, would have given consumers the opportunity to opt for lower cost sources of electricity, the retroactive charge and the monopoly in the market simply forced customers to pay.

In Figure 2, arrows with the number 3 represent the planned process of repayment. The money is called "contributions." There are "general contributions" from TEPCO and other nuclear operators and "special contributions" only from TEPCO. In the year of 2016, TEPCO and other operators paid about 1.6 billion euros as general contributions and TEPCO paid an additional 1.1 billion euros. It was considered that the payment of contributions would last for ten years or so. Of course, this money is expected to continue to come from electricity consumers.

2.2 The 2014 TEPCO Special Business Plan

In 2014, the "New Comprehensive Special Business Plan" was established which disclosed that not only compensation payments would be needed, but also paying for decontamination, interim storage and decommissioning (TEPCO, 2014). The estimated compensation cost was doubled from 25 to 50 billion euros. Decontamination costs of 25 billion euros, interim storage of 11 billion euros and decommissioning of 20 billion euros were now taken into consideration.

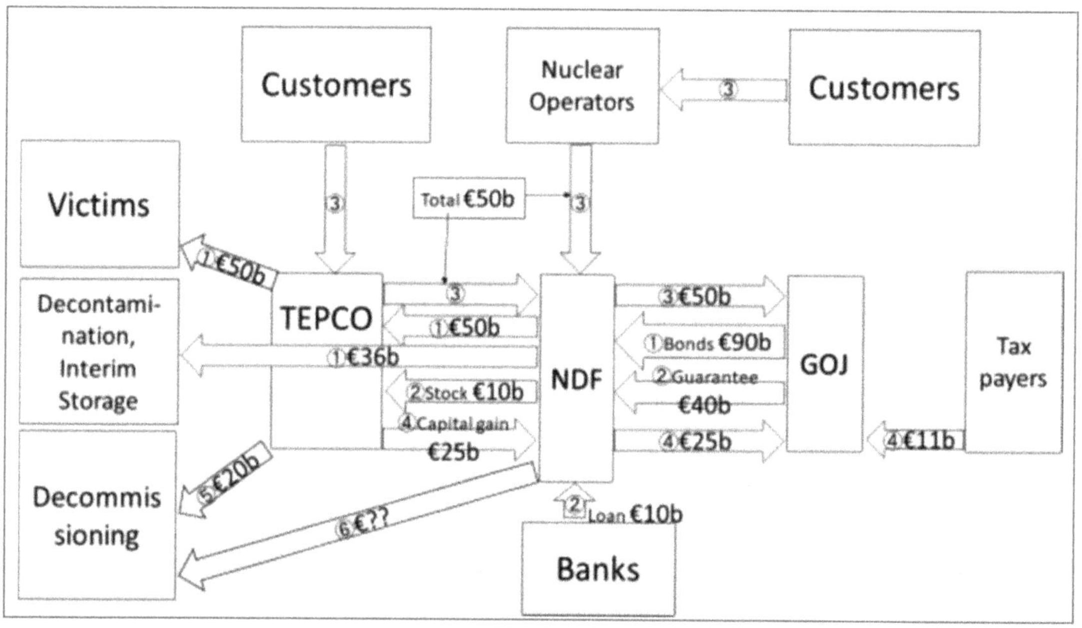

Fig. 3 The scheme of the 2014 plan

Arrows with the number 1, in Figure 3, show how TEPCO was expected to pay 50 billion euros in compensation to victims, the NDF provides the same amount in order for TEPCO to do this. The system for planned repayments stayed the same as in the 2012 plan. The obvious change being that total "contributions" increased to 50 billion euros (arrows numbered with 3).

As for decontamination and interim storage, the two costs are shown together in Figure 3 for practical reasons, and the total amount was 36 billion euros. The NDF directly makes payments for decontamination and interim storage, not through TEPCO's financial statements. To cover this 36 billion euros, GOJ issued additional special purpose bonds of 40 billion euro. Therefore the total bonds provided to NDF increased from 50 to 90 billion euros (arrow numbered 1 from GOJ to the NDF).

The next question was how to arrange the pay back of this latter amount. As seen before, the contribution from TEPCO and other nuclear operators remained 50 billion euros, which is the same amount of compensation (arrows numbered 3).

Nuclear power companies would not want to, and could not, increase electricity charges further, as a re-regulation process introducing some competition in the electricity sector had started and such obligations would weaken the competitiveness of the companies with nuclear power.

Instead new ways of repayment were invented; capital gain and a special tax. It was planned that 25 billion euros for decontamination is to be covered by imagined future capital gain of TEPCO's share which the NDF holds, termed capital gain. The 11 billion euros of interim storage costs shall come from the Government energy budget (arrows 4).

As for the last category of Fukushima cost, 20 billion euros of decommissioning of Fukushima-Daiichi nuclear power plant is planned to be financed by TEPCO's management effort through cost reduction and sales of assets (arrow 5). This is a very ambitious task for TEPCO management, considering the plans to increase competition in the Japanese markets. And all this, while electricity consumption is falling, and decentralised solar electricity is increasing its' market share. To be able to collect the benefits of efficiency as profit, the achieved efficiency improvements must not only cover this cost compared to today operations, the improvements must be so much better than improvements among competitors, in order to generate the profit necessary to cover the costs.

In addition, there was one more important feature in the decommissioning scheme of 2014 plan. GOJ began to support research and development (R&D) of decommissioning methods applicable to the crippled reactors in Fukushima (arrow 6). It is unknown how much the NDF will pay for R&D of Fukushima decommissioning. The law was changed and NDF before abbreviation was, as a consequence, renamed as "Nuclear Damage Compensation and Decommissioning Facilitation Corporation", and the function of decommissioning facilitation was added to NDF. (GOJ, 2014, p.48)

2.3 The 2017 TEPCO Special Business Plan

The 2017 plan revealed that Fukushima estimated cost increased to 215 billion euros (TEPCO, 2017). Compensation, decontamination, interim storage and decommissioning were given as 79 billion, 40 billion, 16 billion, and 80 billion euros respectively. The scheme to cover costs became even more complex.

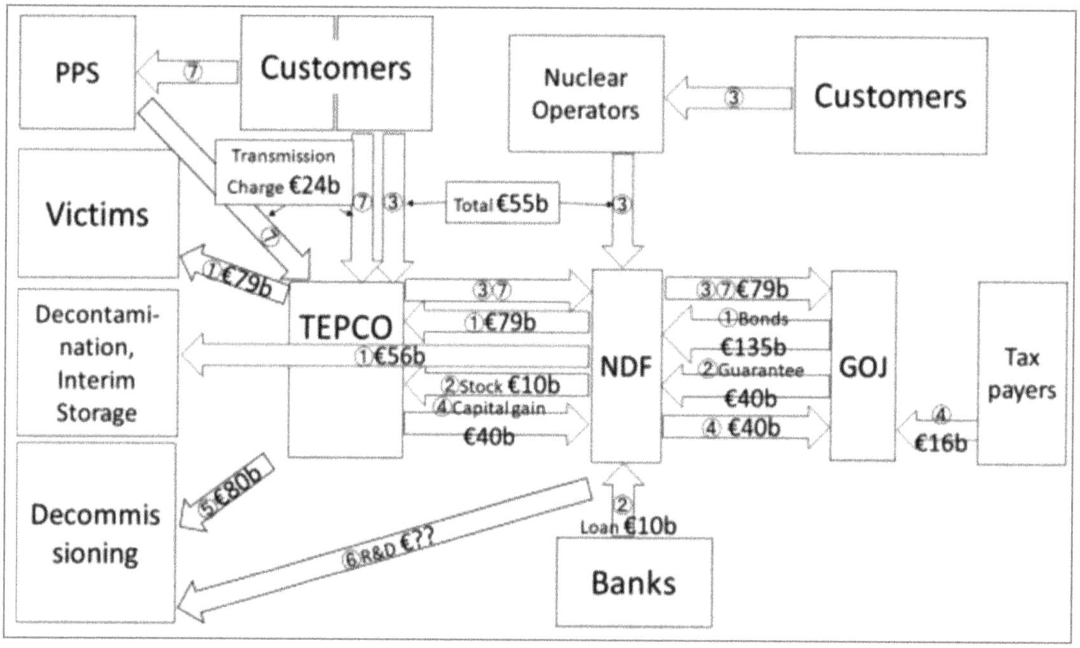

Fig. 4 The scheme of the 2017 plan

Firstly, compensation of 79 billion euros is provided by the NDF as in the 2014 plan (arrows 1). But changes were made in the repayment process. The contribution did not go up to 79 billion euros. Instead, a part of the compensation was to be repaid through a new channel: from a transmission charge (arrows 7). This, in practice, means that the economic liability for the nuclear accident consequences will be paid also by customers who are buying no nuclear electricity at all. The interpretation that the contribution was a retroactive insurance payment by all nuclear operators is now no longer possible. Instead all electricity consumers are paying.

TEPCO's consumers split into two groups after April 2016 when some competition was introduced in the Japanese electricity retail market. Some consumers remained TEPCO's customers and continued to pay electricity charges to TEPCO. Others switched retailers and are now paying electricity charges to other suppliers. TEPCO could not collect money from the latter consumers anymore.

So, the new way of collecting money from these consumers by using transmission charge was invented. Now, TEPCO's customers pay additional charges to TEPCO. And PPS who use transmission services of TEPCO also pay an extra charge. This charge is set to yield the 24 billion euros needed to pay back to the NDF and GOJ together with the contributions. Accordingly, the contribution becomes 55 billion euros in total. In other words, 55 billion euros (arrows 3) and 24 billion euros (arrows 7), both charged on consumers, will be used to pay back to the money of compensation.

Next is decontamination and interim storage. In the 2017 plan, the total cost of the two was 56 billion euros. But this amount cannot be financed by the 90 billion which the NDF provided to TEPCO in the 2014 plan, because from that sum, they already pay compensation of 79 billion euros. So, in the 2017 plan, the Government decided to provide additional bonds of 45 billion euros, resulting in total amount of 135 billion euros (arrows 1). The repayment scheme of the two costs is the same as 2014 plan: Decontamination of 40 billion euros is to be paid back from capital gain, while interim storage of 16 billion is from Government's energy budget which means from other tax income or further national borrowing.

It might be necessary to note here that the anticipated capital gain of 40 billion euros is not a credible expectation in the absence of any privilege given to TEPCO in the future competitive electricity market. The NDF holds 54.69% of TEPCO's share, after paying about 10 billion euros. In order to get 40 billion euros capital gain, the share price must increase to 50 billion euros. 50 billion euro for half of TEPCO means around 100 billion euro for 100% of the company. Therefore, TEPCO's market value in the future must become more than 100 billion euros. This is hard to imagine if customers in Japan are to enjoy electricity at prices that can make industrial customers competitive, and give households a decent standard. The reality among large peers in Europe is that Edf, with roughly the dubble installed generation capacity to TEPCO had a market value of some 20 billion euros at the end of 2016.

Finally, decommissioning costs, that are estimated to be 80 billion euros in the 2017 plan, will be financed by TEPCO's management efforts. There is no change about this scheme. Then, what is TEPCO's management efforts in 2017 plan? In Fuel/Thermal Power business, TEPCO plans to reduce maintenance cost by 30% (TEPCO, 2017, p.28). This, again, is very ambitious. It is so, not only considering the Fukushima events but in relation to the track record of TEPCOs management before 2011. Pursuing cost cuts must not affect the safety of its business and the efforts to avoid more accidents. This is particularly relevant in the nuclear business, where they are planning to restart of Kashiwazaki-Kariwa nuclear power station

(TEPCO, 2017, p.37), a plant that has experienced severe difficulties in operation even before the Fukushima failure.

In Transmission and Distribution business, the plan says that TEPCO will raise 1.2 billion euro/year on average through rationalization (TEPCO, 2017, p.30). Rationalization, again means cost reductions. If this is possible, one may ask why it has not been done already.

3 Discussion

The full costs of the Fukushima nuclear reactor core melts are not yet possible to quantify. Still, the costs already identified and acknowledged by the Japanese government are so large they are beyond the paying capacity of any reactor owner in the world, and clearly beyond the paying capacity of TEPCO.

In a stepwise process, the unavoidable costs of the Fukushima disaster have been identified and distributed by the Japanese government. Worth noting, is that in this process the government has protected TEPCO owners from bankruptcy. This could be seen as introducing a moral hazard: While the report by National Diet of Japan Accident Investigation Committee has concluded TEPCO is responsible for the consequences by neglecting warning of earthquakes and tsunamis, pointing out that "Risk of the arrival of a tsunami was known by both Nuclear and Industrial Safety Agency and TEPCO in 2006, but TEPCO neglected this risk and no measures were taken" (1.2 of PART 1), TEPCO has been protected from the economic consequences by the government. The government has placed the economic cost and responsibility on the customers and tax payers, but not on the owners of the plant. The moral hazard appear as other industries may come to expect similar protection and abstain from taking costly precautions when learning about risks in their activities.

Now, the Japanese people are compelled to pay for most of the costs. Still, the schemes set up for this purpose. The complexity of the plans, and the lack of transparency in reporting the use of money provided by the Governments or regulated contributions collected from ratepayers, makes public debate and democratic control difficult or impossible.

Some part of the plan deserves public scrutiny as the current government plans may appear unrealistic, thus creating problems and deficits for future Governments. It is for example hard to believe that a 40 billion euros capital gain will be secured by TEPCO in the competitive electricity markets that exists in most countries and that is now also said to be developed in Japan. Similarly, it is hard to believe that significant profits can be made by cost reducing management efforts if the markets are really competitive. These plans seem to rely on assumptions about electric power companies being able to in some way tax consumers rather that supplying in a competitive market.

Thus one may see these plans as only postponing the understanding of what the government is going to have to provide through increasing taxes on the Japanese people in an already weak government budget situation.

4 Conclusions

Seven years after the nuclear core melts the experience shows a lack of readiness, and absence of any plan for how to manage the economic consequences of an accident of this magnitude.

Still, the improvised solutions, have for the seven years both kept the government's borrowing capacity intact, and allowed TEPCO to avoid going bankrupt.

We dare predict that the 2017 TEPCO Special Business Plan will be impossible to fulfil. Further improvised and complicated solutions may follow. However, it is increasingly unlikely that the current idea of nuclear power companies in different ways paying will be compatible with the global development of low cost renewable electricity generation technology and competitive electricity markets. Instead the development appears likely to make it impossible to generate profits by nuclear or other large scale thermal power plants. Even without the Fukushima related liabilities, Japan's electric power companies would have economic challenges.

A less complex solution, worth investigating further, would be that the government directly takes over all the nuclear liabilities from the power companies and have the power companies paying by transferring the transmission network to a government controlled national transmission system operator. That solution would support, rather than conflict with, the ongoing electricity market reform, and transition to low cost electricity. It would, in one move, unbundle transmission from generation and create the conditions necessary for an efficient electricity market reducing costs for Japan's future electricity supply.

The experience is also relevant when considering restarting reactors, re-introducing the risks of further reactor accidents in Japan. Had Japan seen winds bringing more contamination over land – or even worse, in the direction of Tokyo – the financial consequences may not have been manageable by the government or the Japanese society by itself.

This observation leads to the need to find market based solutions that will distribute costs globally. Here the idea of a compulsory insurance via catastrophe bonds appears as an option deserving further consideration [Radetzki & Radetzki 2000, Kåberger 2018].

The experience of Japan may prove valuable for consideration also in other nuclear countries. Though many of these are operating under different legislation and international conventions regarding liability for reactor accidents, the real magnitude of the economic consequences are often not understood by people within the democratic decision making processes in Europe or North America.

References

BP, 2017. Statistical Review of World Energy. http://www.bp.com/statisticalreview (accessed 2017-07-01).

Government of Japan (GOJ), 2014. 4th Strategic Energy Plan, April 2014, http://www.enecho.meti.go.jp/en/category/others/basic_plan/pdf/4th_strategic_energy_plan.pdf (accessed 2017-10-05).

Government of Japan (GOJ), 2018. 5th Strategic Energy Plan, July 2018, http://www.meti.go.jp/english/press/2018/pdf/0703_002c.pdf (accessed 2018-09-04).

Kåberger, T., 2018. Economic Management of Future Nuclear Accidents. In Haas, R., Ajanovic, A., Mez, L. (ed.): The Technological and Economic Future of Nuclear Power. Springer VS 2019.

National Diet of Japan Accident Investigation Committee, 2012. *The Report by National Diet of Japan Accident Investigation Committee (Kokkai Jiko Cho)*, Tokuma Shoten (in Japanese).

Radetzki M. & Radetzki M., 2000. Private Arrangements to Cover Large- Scale Liabilities Caused by Nuclear and Other Industrial Catastrophes. *Geneva Papers on Risk and Insurance*, Vol. 25 No 2, April 2000.

Tsuda, T., Tokinobu, A., Yamamoto E. Suzuki E., 2015. Thyroid Cancer Detection by Ultrasound Among Residents Ages 18 Years and Younger in Fukushima, Japan: 2011 to 2014. *Epidemiology* • Volume 27, Issue 3, p 316–322, doi: 10.1097/EDE.0000000000000385

Tokyo Electric Power Company (TEPCO), 2011–2017. http://www.tepco.co.jp/en/index-e.html

Tokyo Electric Power Company (TEPCO), 2011. Financial Report for the year ending 31 March 2011 (in Japanese), http://www.tepco.co.jp/ir/tool/yuho/pdf/201106-j.pdf (accessed 2017-10-05).

Tokyo Electric Power Company (TEPCO), 2012a. Comprehensive Special Business Plan, May 2012, http://www.tepco.co.jp/en/press/corp-com/release/betu12_e/images/120509e0104.pdf (accessed 2017-10-05).

Tokyo Electric Power Company (TEPCO), 2012b. Financial Report for the year ending 31 March 2012 (in Japanese), http://www.tepco.co.jp/ir/tool/yuho/pdf/201206-j.pdf (accessed 2017-10-06).

Tokyo Electric Power Company (TEPCO), 2013. Financial Report for the year ending 31 March 2013 (in Japanese) http://www.tepco.co.jp/ir/tool/yuho/pdf/201306-j.pdf (accessed 2017–10-06).

Tokyo Electric Power Company (TEPCO), 2014. New Comprehensive Special Business Plan, 15 January 2014, http://www.tepco.co.jp/en/press/corpcom/release/betu14_e/images/140115e0206.pdf (accessed 2017–10-05).

Tokyo Electric Power Company Holdings (TEPCO), 2017. Outline of the "Revised Comprehensive Special Business Plan (The Third Plan), http://www.tepco.co.jp/en/press/corpcom/release/betu17_e/images/170518e0101.pdf (accessed 2017–10-05).

Alternatives

On New Thinking and Designs of Electricity Markets

Heading towards Democratic and Sustainable Electricity Systems

Reinhard Haas, and Hans Auer[1]

Abstract

In recent years increasing shares of variable renewable energy sources (RES) have changed the structure of electricity markets in several countries. The core objective of this paper is to provide insights into the conditions necessary to bring about a more democratic and sustainable electricity system by integrating even larger quantities of variable RES. Our major finding is that a market-based approach would ensure that competitive forces rather than governmental interferences – such as capacity mechanisms – shape the future of the electricity markets. This transition towards a competitive and sustainable future electricity system will be based on an approach of "new thinking" which requires a paradigm shift in the whole electricity system. This includes switching to a more flexible and smarter concept allowing a greater scope for demand participation, storage options and other flexibility measures.

1 Reinhard Haas, Technische Universität Wien, Austria, haas@eeg.tuwien.ac.at; Hans Auer, Technische Universität Wien, Austria, auer@eeg.tuwien.ac.at

R. Haas et al. (Eds.), *The Technological and Economic Future of Nuclear Power*, Energiepolitik und Klimaschutz. Energy Policy and Climate Protection, https://doi.org/10.1007/978-3-658-25987-7_18

1 Introduction

For a long time the electricity system has been determined by the generators. Until the mid-1990s, and in many countries even longer, large generation companies, which were often highly vertically integrated, dominated the electricity system. This was supported by the assumption of existing economies-of-scale. Huge power plants, mainly nuclear and hard coal, were constructed along the lines "the bigger, the cheaper". This strategy was accompanied by high growth rates in electricity consumption.

Over the course of time these patterns have changed. At first, growth rates fell from 10% per year in the 1960s to about 5% in the 1980s and 90s and to almost zero (in some OECD countries) in recent years. That is to say, today, there is less cake to be shared between generators, especially, given the preference for renewable generation.

The first signs of criticism of such a generation-focused paradigm emerged already in the 1970s. Lovins (1978) was one of the first to predict three major developments: (i) that future electricity consumption rates would decrease; (ii) that decentralized generation mainly from PV systems would increase, and (iii) that the importance of demand-side issues would grow. In addition to this, with the liberalization of electricity markets the picture began to change. The core objective of liberalization was the introduction of competition in generation in order to harvest the full benefits of electricity supply for both citizens and industry. Due to huge excess capacities after the first phase of liberalisation, the principle of "prices equal short-term marginal costs" in spot markets emerged.

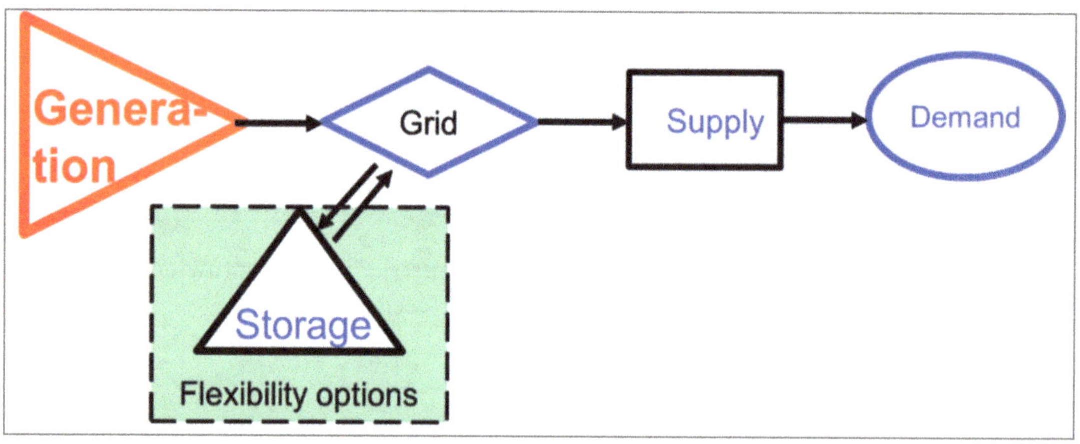

Fig. 1 "Old" thinking in electricity markets

It is important to note that in this first phase of liberalized markets the old "one-way-thinking" still prevailed,. which was characterized by the fact that the generators were at the core of the system and of the thinking of the policy makers, see Fig. 1.

In recent years, mainly due to the increase of variable renewables the capacity (factors) of the conventional plants has decreased leading to losses in revenues for their owners. This has resulted in growing concerns that most of these plants will be shut down thereby leading to decreases in supply reliability. Consequently, energy-only markets* have been questioned and calls for capacity payments have been launched.

The core intention of this paper is to serve as a primer for introducing truly competitive, democratic and sustainable electricity systems in every country world-wide. It is motivated by the current discussion on how to integrate large shares of variable RES but the basic intention goes beyond that. The aim is to show how to attain real competition in electricity markets, including all dimensions such as generation and storage as well as demand-side options.

In addition, the European Commission has set ambitious targets for increasing the share of electricity from renewable energy sources (RES), e.g. EC (2009). Indeed, in the EU-28, in recent years electricity generation from variable sources such as wind and solar has increased dramatically, with Germany, Spain, Italy leading. In the EU-28, between 1997 and 2016 "new" renewables (excluding hydro) mainly from wind grew from less than 1% to about 16%. For 2030, the EU has set further goals of a share of 27% (compared to about 14% in 2016) energy from RES. This target is for all uses – heating, electricity and transport. Consequently, also electricity generation from RES will grow, as documented in the National Renewable Energy Action Plans (NREAPs), however, it is not clear to which absolute level. Another major motivation for this paper is to show what is needed in order to integrate these higher quantities into the electricity system.

These increasing shares of variable RES have especially in Germany changed the usual pattern of electricity markets in Western Europe. Yet, variable RES-E do not provide electricity simultanously with demand. It is important to note, that almost all other generation technologies do not either. The fact that these must run capacities are offered at Zero costs over a large time per year have led to the argumentation that fossil plants like Combined-cycled gas turbines (CCGT) or coal power stations become economically less attractive because of the lower fullload hours per year. This argument has led to the call for capacity mechanisms (CM) in addition to the current "energy-only" markets. The idea is that specific owners of a flexible power plant should be paid for holding the plant ready for operation.

Due to these developments, currently, the whole electricity system is at a crucial crossing. On the one hand, the way to a sustainable electricity system based mainly on RES could be paved in the next years. In this context we emphasize especially the considerable price decreases of PV which has brought this technology close to cost-effectiveness on household level, see Haas et al (2013). On the other hand, there are forces which try to retain the old centralized fossil and nuclear-based generation planned economies. Capacity mechanisms (e.g. in France and England) should help to freeze this anachronistic pattern.

The core objectives of this paper are: (i) to explain how a truly competitive market-based electricity system can be brought about in the future without continous governmental interferences; (ii) to explain why capacity payments do not contribute to such a system but rather preserve the present system and (iii) to show that generators will no longer be the heart of the system but rather balancing groups and the suppliers.

A specific intention of this paper is to bring together all important aspects for heading towards a sustainable as well as competitive future electricity system. It considers technical options and aspects of market design and applies it to a further increase of RES in the electricity system. Moreover, it links the concept of residual load to price signals from the wholesale markets, the relevance of flexibility measures on the demand-side as well as demand response due to these price signals.

2 How prices in electricity spot markets come about

To analyze the impact of variable RES on the prices in wholesale electricity markets it is first important to understand the current market rules and market structures, see Auer/Haas (2016). Of key relevance is to understand how prices in European electricity markets currently come about. In this context it is important to look at the historical dynamics. The liberalization process in Europe started in the late 1980s in the UK and gradually migrated to continental Europe with the implementation of the EU-directive on Common Rules for the Internal Electricity Market (EC, 1997). One of the major features of the liberalized electricity markets was that the pricing regimes changed. In former regulated markets, prices were established by setting a regulated tariff, which was calculated by dividing the total costs of electricity provided by the number of kWhs sold – with some differences between different groups of customers. The major change that took place after liberalization was that prices on the wholesale electricity markets were now expected to reflect the marginal costs of electricity generation. Since then the price formation is mainly based on a fundamental approach where the intersection of a merit order curve on the supply-side and the demand curves results in the corresponding market price at every point-of-time, see Haas et al (2013b) and Fig. 2.

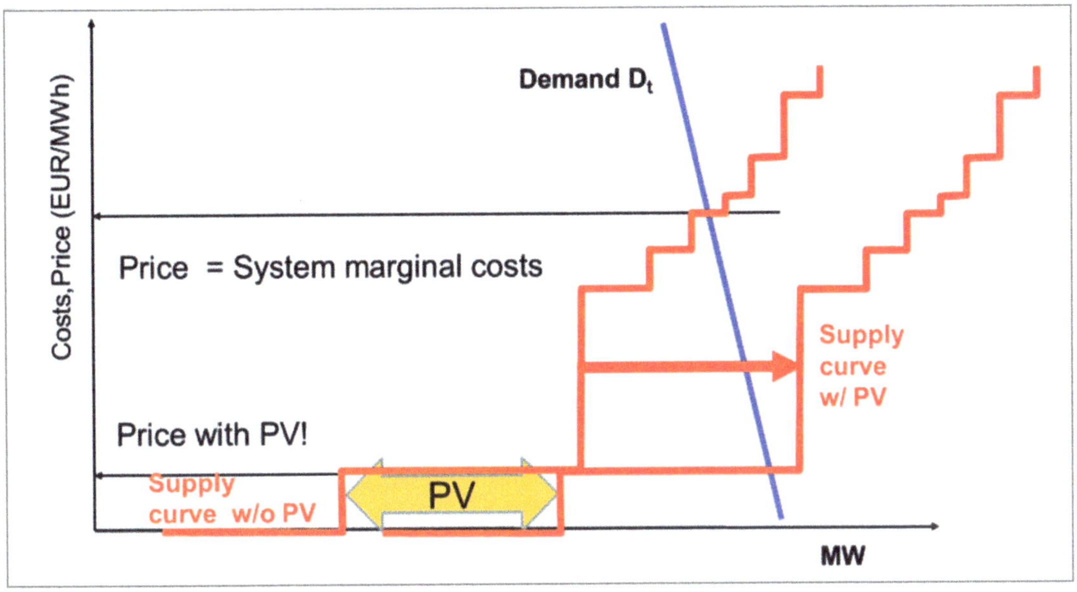

Fig. 2 Merit order supply curve with and without additional PV capacities at on-peak time of a bright summer day with short term marginal costs for conventional capacities

The typical historical pattern of electricity generation in the Western Central European electricity market consisted since decades of conventional fossil, nuclear and hydro capacities. Since the late 1990s in western central Europe, most of the time nuclear contributed the largest share, followed by fossil and hydro. Non-hydro renewables were not a significant factor until recently. However, since 2013 renewable electricity contributes the largest quantity in the EU-28. At the time when liberalization started huge already depreciated excess capacities existed in Europe. This led to the expectation that prices will (always) reflect the short-term marginal costs (STMC) as illustrated in Fig. 2.

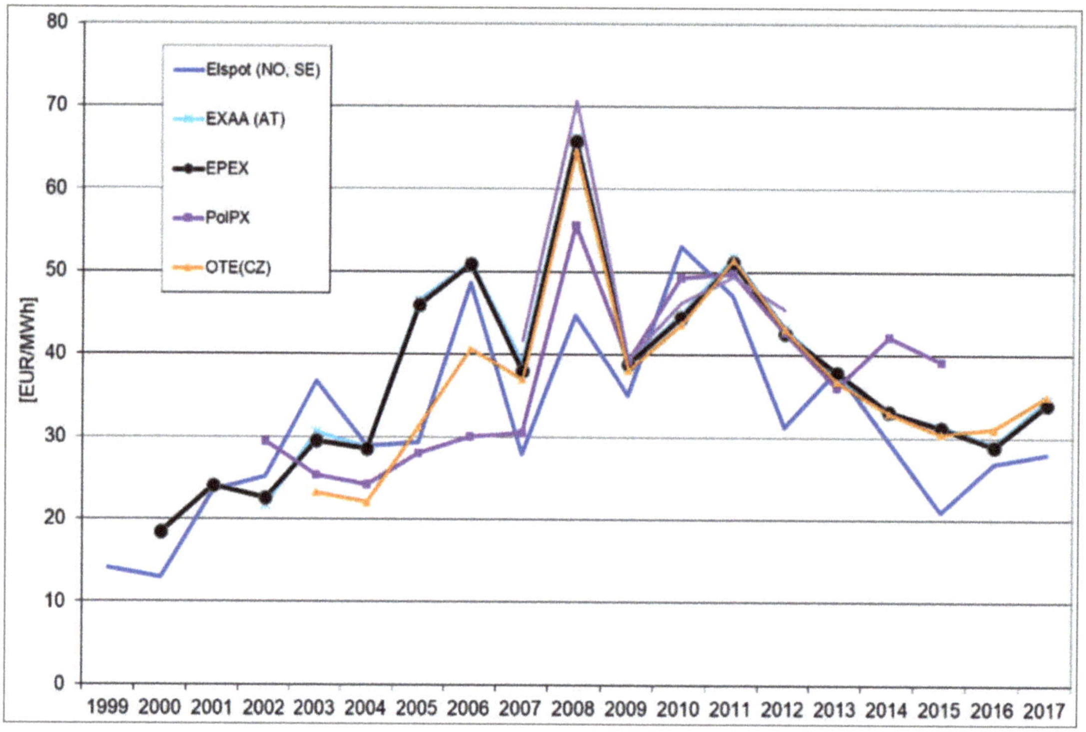

Fig. 3 Development of spot market (day-ahead) electricity prices in several European wholesale markets (1999-2017, 2017 preliminary)

3 The impact of variable RES on prices in electricity spot markets

Between 2011 and 2016 remarkable decreases in day-ahead prices at the Western European power exchanges were observed see Fig. 3. The major reason for this decline in day-ahead prices was the increase of variable RES with zero short-term marginal costs.

It is the remarkable rise This increase of renewables has started to impact spot prices, trading patterns and the dispatch of conventional generation by about 2011. The explanation is simple. Assume e.g. a sunny day with ample solar generation. Then the supply curve is shifted to the right as schematically shown in Fig 2, which essentially pushes nuclear and fossil fueled generation "out of the market", Haas et al 2013b.

This impact of variable RES on electricity prices is already known since volatile hydro power was used for electricity generation. The best example is the Nordic market, mainly Norway and Sweden, where since decades almost only technologies with Zero short-term marginal costs meet the whole supply. Since about 2007–2010

– in Denmark already earlier – there was experience with temporarily high wind in the systems, see e.g. Nicolosi (2010). In recent years increasing generation from photovoltaic systems was added to the production portfolios, mainly in Germany, Italy and Spain, and has contributed to temporarily very low – sometimes even negative – prices.

4 The end of the myth of base load

The core question is, what the impact of the aggregate of various variable RES on the wholesale electricity markets is. Aside from the above-described effects, variable RES will also influence the costs at which fossil generation – especially natural gas – is offered. The reason is that they would lead to much lower fullloadhours, e.g. only 1000 instead of 6000 h/yr before. Yet, the revenues earned from these hours must cover both the fixed and variable costs, see also Haas et al (2013a). Hence, in a market with large shares of renewable energy sources the role of conventional capacities will change see e.g. Nielsen et al (2011).

This leads to the following categories of presumed "problems": (i) Prices decrease to Zero or become even negative at a number of days; (ii) a lack in contribution margin to fixed costs for conventional flexible power plants. However, it is not yet clear, on how many days very high and on how many days very low (or negative) prices will prevail and how high or how low these prices will be.

Of further relevance in this context is how the price spread in European markets will evolve in the future as larger amounts of PV, solar thermal and wind generation are added to the network. The consequence for electricity prices are shown in Fig. 4 where a hypothetical scenario with high levels of generation from wind, PV and run-of-river hydro plants over a week in summer are depicted using synthetic hourly data for an average year in Austria. The figure leads to significant volatilities in electricity market prices with total costs charged for conventional capacities – black solid line –within very short-term time intervals.

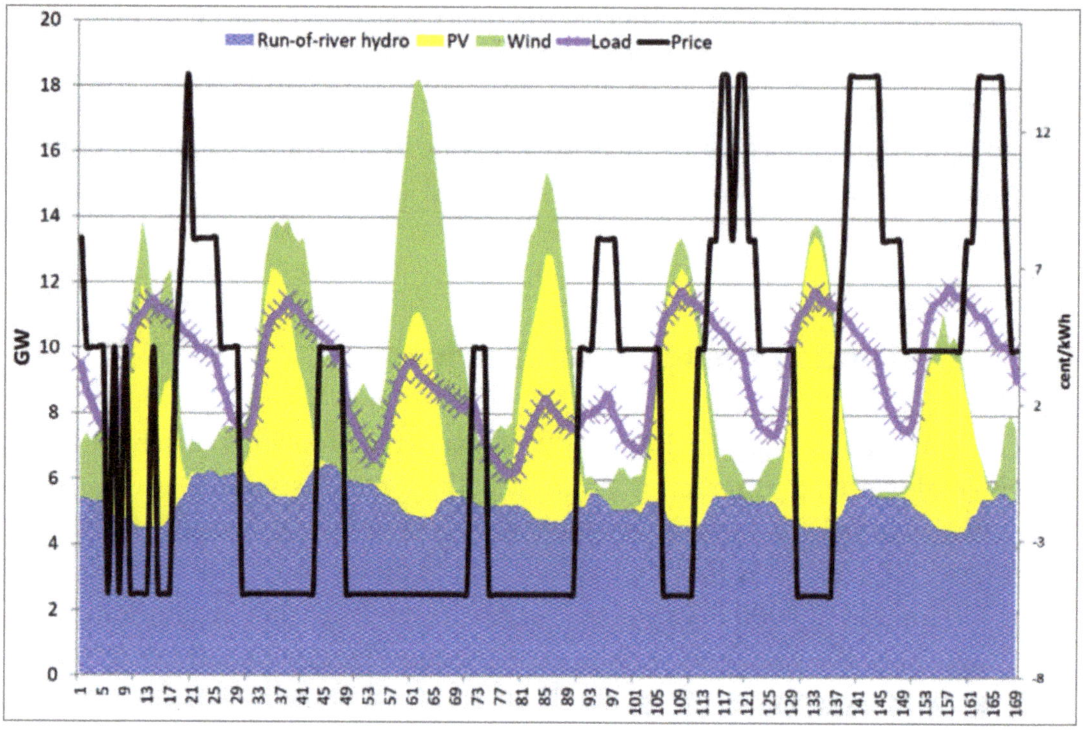

Fig. 4 Development of variable RES from wind, PV and run-of-river hydro plants over a week in summer on an hourly base in comparison to demand and resulting electricity market prices with total costs charged for conventional capacities (Source: own analysis, adapted from Auer/Haas 2016)

Our method of approach is based on the following principles: (i) Most relevant is the coverage of residual load which is the difference between final electricity demand and generation provided by non-flexible electricity generation from variable RES as well as coal and nuclear power plants, see Fig. 5; this is modeled on an hourly base over a calendar year based on historical RES electricity generation; (ii) Deduction of available conventional and backup capacities including must-run; (iii) flexibility on the demand side based on consumer behavior incl. flexibility instrument such as storage etc.; (iv) hourly electricity prices equal to short-term marginal costs and scarcity as well as excess pricing.

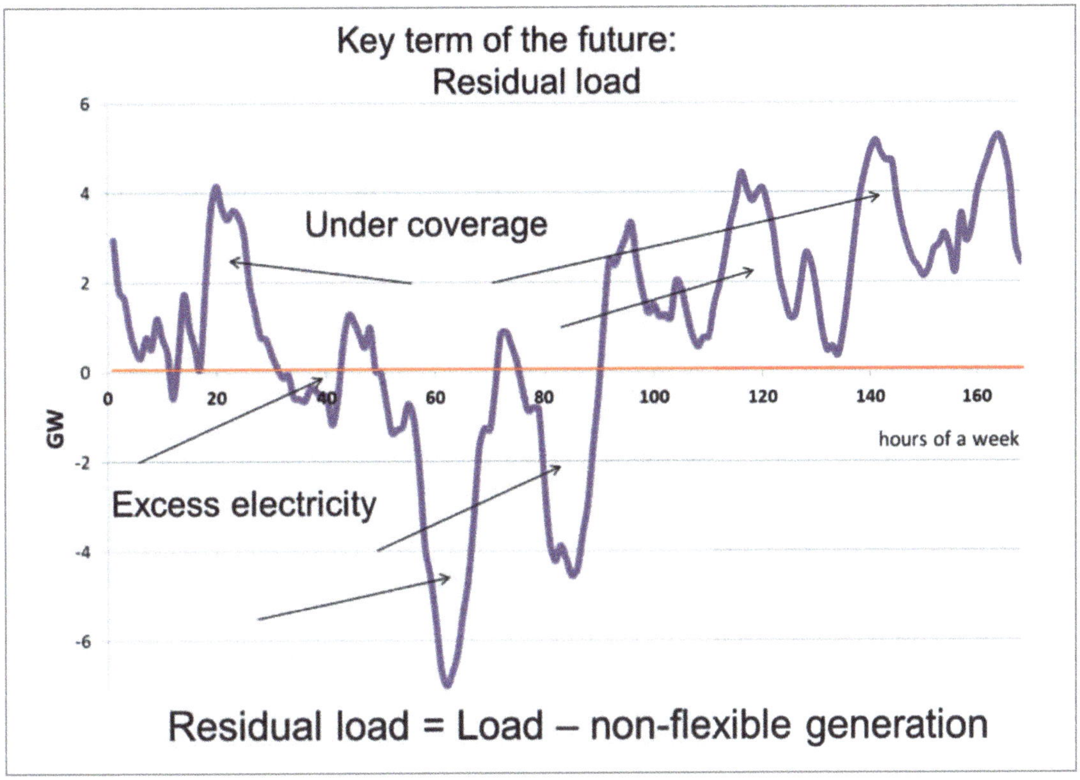

Fig. 5 New thinking: The concept of residual load referring to Fig. 4

For the residual load shown in Fig. 5 a price pattern as described in Fig. 4 may emerge. Hence, in the long run the impact of variable RES on the price spread is that it will increase. The intuitive explanation is that when renewables are plentiful, say during windy or sunny periods, the prices will be extremely low, approaching zero or possibly going negative, while at other times – when demand is high and renewables are scarce – prices can be much higher due to strategic bidding by fossil generators exercising market power.

While Fig. 5 shows the concept of residual load over a week, Fig, 6 shows the corresponding graph over a year classified by magnitude in decreasing order. In Fig. 6 the classified residual load curve over a year in the case of high shares of variable renewables is described including the relevant areas for the discussion. The crucial areas in this load duration curve are on the top left and on the bottom right. In the circle on the top left the question is how to cover under shortage on these hours, in the circle on the bottom right the question is how to use this excess generation of electricity.

For both areas there are in principle two options:

- By regulated capacity payments ?

or

- By competition between supply-side and demand-side technologies and behaviour (incl. Storages, grid and other flexibility options)?

Important remarks: Flexibility measures will contribute in a competitive way to reduce these price spikes and consequently the price spreads and lead to new equilibria between supply and demand!

As an example in Fig. 6 the profile of residual load in Austria 2013 and the development in a scenario up to 2030 with a much higher share of variable renewables is described. The major finding of Fig. 6 is that the duration curve of the residual load profile will become steeper and that the number of hours with excess generation will become higher. This effect will lead straightforward to higher price spreads and will also increase the attractiveness of storage, flexible peaking units and other flexibility options.

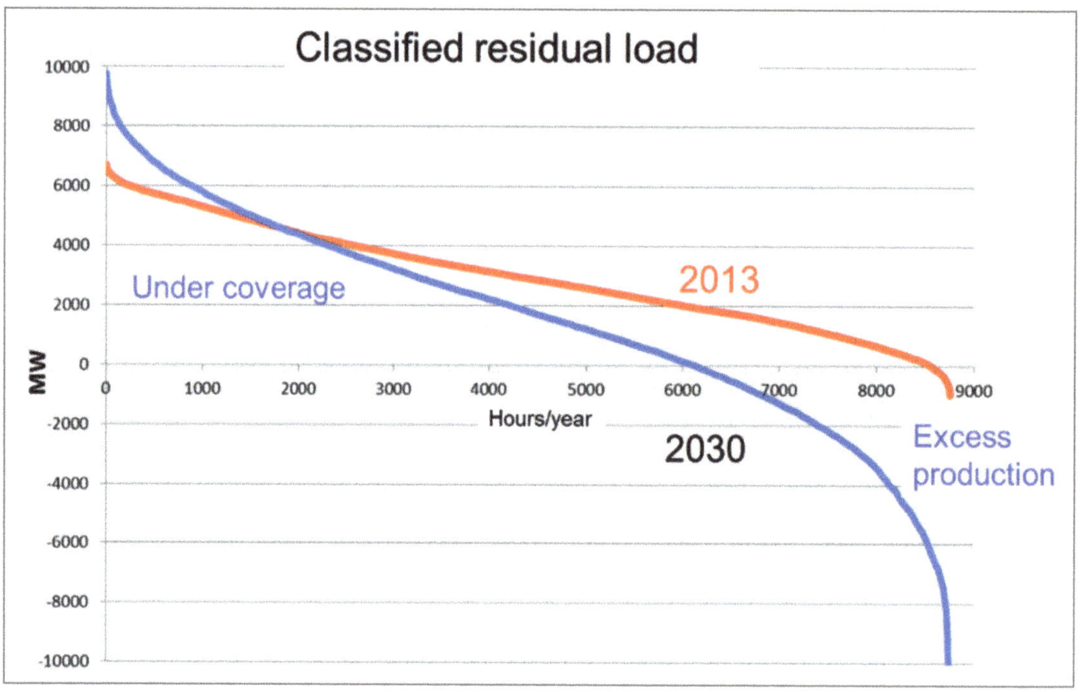

Fig. 6 Development of residual load in Austria 2013 and in a scenario up to 2030 with high share of variable RES

5 The core problems of regulated capacity payments

If the price pattern described above in Fig. 4 is not accepted by politicians another option are capacity payments. The fact that the renewable must run capacities are offered at Zero costs over a large time per year have led to the argumentation that fossile plants like CCGT or coal power stations become economically less attractive because of the lower fullload hours per year. This argument has led to the call for "capacity" payments in addition to the current "energy-only" markets. The idea is that specific owners of a flexible power plant should be paid for holding the plant ready for operation.

The major reason, why at least currently there is no need for centralized CP in Europe that there are still many other options in the market, which we think are by far not yet exhausted. However, to exhaust these options some dogmas has to be changed. Especially the historically prevailing and still existing definition of supply security – that every demand has to be met at every point-of-time regardless of what are the costs – has to be revised in a way that compares the costs of (all) supply-side and demand -side options as well as customers WTP for capacity depending on time.

The major open questions regarding centralized capacity payments are, see Haas (2014):

- Which quantity of capacity should get payments and where?
- How to split in existing and new capacity?
- How to tune with grid extention? Every grid extention has undoubtedly an impact of necessary capacities in a specific area
- Who would plan? On national or international level?

There are three core problems regarding CP:

1. All regulatory capacity payments for power plants destort the EOM and lead to wrong price signals for all other options
2. Price peaks at times of scarce resource should revive the markets and lead to effective competition
3. We should strive to retain system resource adequacy by ensuring correct price signals and without capacity payment
4. Every capacity payment reduces the shares of the variable renewables

6 A market design approach for supply security

One major argument for the call for centralized CM is to retain supply security in the electricity system. The historical (anachronistic) definition of supply security is: At every point-of-time every demand has to be met regardless of the costs. In this context it is important to note that supply security is an energy economic term. It is different from technical system reliability.

The core problem is that so far world-wide the demand-side has been neglected widely with respect to contributing to an equilibrium of demand and supply in electricity markets. Major exceptions are: (i) in the 1980s and 1990s in the U.S., Sweden, Denmark and other advanced countries DSM-measures have attracted attention. After the liberalization of the electricity markets most of these programmes disappeared. (ii) In Denmark – the leading country for integrating variable renewables especially wind – has integrated a lot of power-to-heat technologies, that now play an important role in energy markets.

The major reason for this ignorance of the demand-side is that in times of regulated monopolies every demand could be met due to significant excess capacities. And still in the liberalized markets huge excess capacities remained. This aspect – to develop the impact of demand-side and customers WTP – is essentialy for a real electricity market and it is actually regardless of the aspect of an integration of larger shares of RES.

Such a market-based approach would take into account customers willingness-to-pay (WTP). The equilibrium between demand and supply would come about at lower capacities. It is also important to note that at points-of-time where WTP is lowest, e.g. in the evening, the marginal costs (MC) of providing capacity could be highest, see also Auer/Haas (2016).

7 Flexibility: The key term of the future

Our major findings for integrating large quantities of variable RES-E into the electricity system by using market-based principles and how, straightforward, a sustainable electricity system could work, are that the following conditions have to be fulfilled, see also EC (2015):

- Of core relevance for integrating larger shares of RES-E in a competitive way is a pricing system in revised energy-only markets where the prices signal provide information on scarcity or excess capacities at every point-of-time (at least at quarters of an hour);

- Another important issue is that the demand-side market is developed, see above. So far consumers have never been asked what the value of capacity is for them and what they are willing to pay for specific quantities of capacity. An important analysis in this context has been conducted by Praktiknjo (2013). He clearly identifies two findings: (i) there is a quite different WTP between different groups of customers; (ii) it is very unlikely that generating electricity is always cheaper than saving capacity.

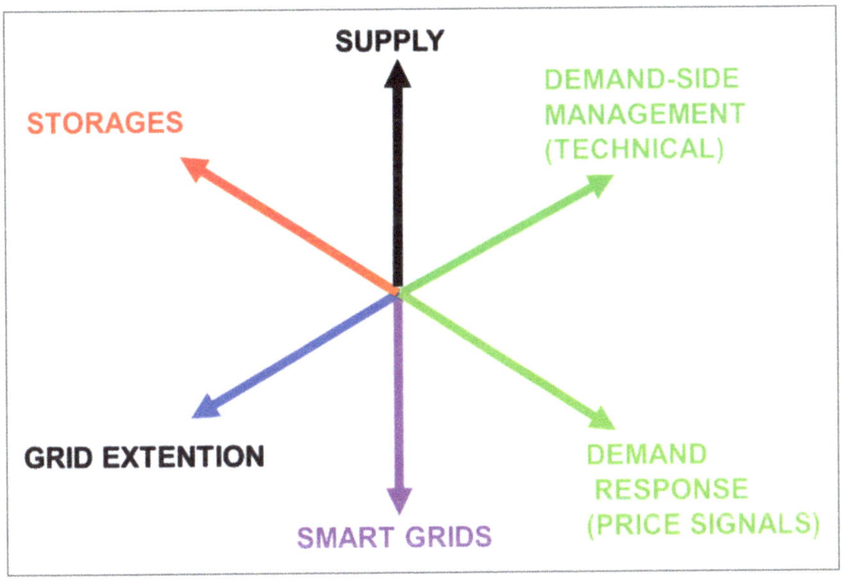

Fig. 7 Dimensions of electricity markets

- More flexibility in the organization of the market is required: To better integrate electricity from RES in the market the time intervals in markets should be reduced (more emphasis on intraday markets, shorter trading intervals (from hours to ¼ hours); shorter ahead leading times for market clearing and forecasting of electricity generation from variable RES);
- Most important to balance variations in residual load is to implement an optimal portfolio of flexibility options which already exists today. A very comprehensive review of energy system flexibility measures to enable high levels of variable renewable electricity is provided in Lund (2015). Currently these potentials are not fully harvested due to low economic incentives, see next chapter. The most important flexibility options to balance variations in residual load are, see Fig. 7:

- short-term and long-term storages such as batteries, hydro storages, or chemical storages like hydrogen or methane;
  - Technical demand-side management measures conducted by utilities like cycling, load management, e.g. of cooling systems;
  - Demand response due to price signals mainly from large customers to price changes, time-of-use pricing;
  - Transmission grid extention leads in principle to flatter load and flatter generation profiles;
- Smart grids: They allow variations in frequency (upwards and downwards regulation) and switch of voltage levels and contribute in this context to load balancing
- Balancing groups will play a key role in this new concept. These are the entities which finally have to balance generation, flexibilities and demand options.

8 New vs old thinking: Further development of the wholesale electricity market design

Regardless of the issue of increasing quantities of variable RES in the electricity system there are some measures to be introduced which would improve the wholesale market structure and competitiveness basically. In addition to a revised EOM these are:

- more flexibility in the organization of the market is required;
- shorter ahead leading times for market clearing and forecasting of RES-E production;
- long term contracts (futures, forwards) should be made available even for longer time periods than 6 years if the market needs it.

Finally we state that the transition towards a competitive and sustainable future electricity system will be based on the following principle of "new thinking", which is to accept a paradigm shift of the whole electricity system – including switching from an inflexible and one-way system where variable load is met with changes in generation to a more flexible and smarter system allowing two-way electricity flows – to our understanding – a greater scope for demand participation by consumers needs to be included. In addition, suppliers (or balancing groups) are the most important part of the whole energy service providing chain, see also Fig. 8.

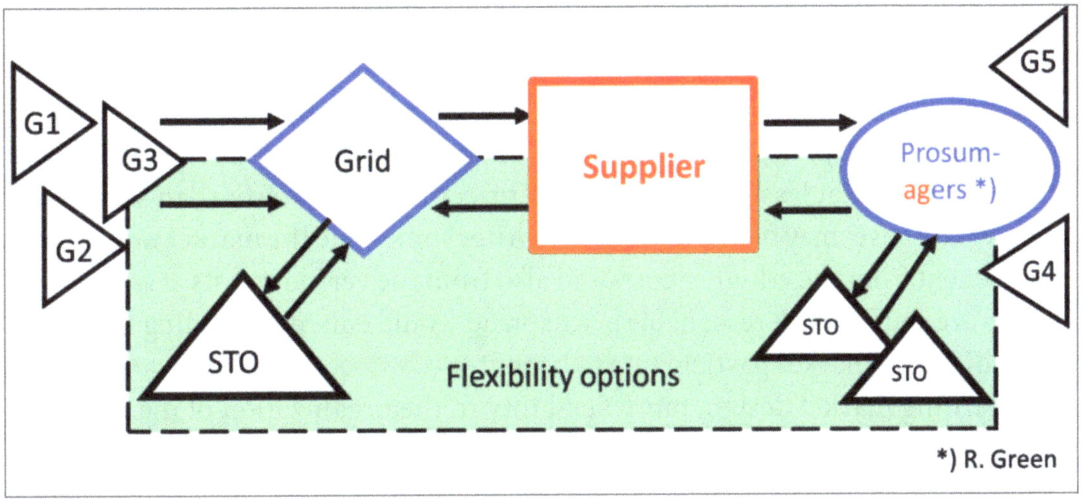

Fig. 8 New thinking in electricity markets: a supply-oriented bidirectional system with very high flexibility

As indicated in Fig. 8 in future decentralized PV systems along with decentral battery storages may play an important role. The astonishing changes in the solar industry epitomize the over-all way PV is heading to. (WNISR 2015): "There seems to be a general recognition that the fall in production costs of RE technologies, particularly of PV, coupled with the expected falling costs of electricity storage will accelerate the transformation of the power sector."

And the IEA, which has been tradionally skeptical with respect to RES states in the WEO (2017): "PV is on track to become the cheapest source of new electricity in most countries world-wide".

9 Conclusions

The major conclusions of this analysis are:

- The key to a sustainable competitive electricity system is the full exhaustion of flexibility options based on correct price signals in the wholesale as well as in the retail market. Currently on both levels the market does not yet provide proper price signals to trigger flexibility options (e.g. technical demand-side management, economic demand-response due to price signals as well as short-term and long-term storage options) which would balance the residual load profile more effectively.

- It is not possible to force variable RES into the system by means of technical planning. Proper financial incentives are necessary. Correct price signals are crucial in a revised energy-only market along with scarcity and excess pricing signals; the only "negative" aspect of a market without a capacity component would be that, at least in the short run, prices higher or lower than short-term marginal costs may occur temporarily. After some time the market would learn to benefit from these higher costs and also from the very low costs at times when RES are abundant. A reasonable price spread would emerge providing incentives for different market participants to benefit.
- Regarding market design, more flexibility in the organization of the market is required: To better integrate RES-E in the market, time intervals in markets should be reduced (i.e. more emphasis on intraday markets, shorter trading intervals – from hours to quarters hours; faster market clearing and shorter forecasting times regarding wind and solar).

In conclusion,

This transition towards a competitive and sustainable future electricity system will be based on an approach of "new thinking" which requires a paradigm shift in the whole electricity system where no longer the generators are the centre but the balancing groups respectively the supply companies. This includes switching to a more flexible and smarter concept allowing a greater scope for demand participation, storage options and other flexibility measures.

Finally we state is that the evolution of such a creative system integrating variable RES in Western Europe may also serve as a model for RES-based electricity supply systems world-wide.

References

Auer H., R. Haas: On integrating large shares of variable renewables into the electricity system, Energy, 1–10 (2016).

EC: Directive on the promotion of the use of energy from renewable sources, Brussels, 2009.

EC: Directive 96/92EC of the European Parliament and of the Council Concerning the Common Rules for the Internal Electricity Market. Official Journal L27 of 1/30/1997, Luxemburg. 1997.

EU: EU Energy in Figures, Brussels 2012.

European Parliament and Council: Directive of the European Parliament and of the Council on the promotion of electricity produced from renewable energy sources in the internal electricity market, Directive 2001/77/EC – 27 September 2001, Brussels, 2001.

European Economic and Social comittee: Launching the public consultation process on a new energy market design (COM(2015) 340 final). Brussels 2016.
Haas R, Lettner G, Auer J, Duic N. The looming revolution: How Photovoltaics will change electricity markets in Europe fundamentally", Energy 57, 2013, 38–53.
Haas R., On the future design of electricity markets: Capacity payments or smart solutions? In: C. Brebbia et al "Energy Quest" Wessex, 2014.
Haas R, Auer H, Resch G., Lettner G. The growing impact of renewable energy in European electricity markets in: Evolution of Global Electricity Markets—New Paradigms, New Challenges, New Approaches, edited by Fereidoon P. Sioshansi. (Amsterdam: Academic Press, Elsevier, 2013). ISBN: 978-0-12-397891-
Lovins A., Soft Energy, 1978.
Lund H. Renewable energy systems. 2nd edition, ISBN: 9780124104235, Academic Press (2014)
Lund PD, Lindgren J, Mikkola J, Salpakari J: Review of energy system flexibility measures to enable high levels of variable renewable electricity. Renewable and sustainable Energy Reviews, 45, 785–807, (2015),Nicolosi M. Wind power integration and power system flexibility - an empirical analysis of extreme events in Germany under the new negative price regime. Energy Policy 2010;38:725-768.
Nicolosi M. Wind power integration and power system flexibility - an empirical analysis of extreme events in Germany under the new negative price regime. Energy Policy 2010;38:725-768.
Nielsen S, Sorknæs P, Østergaard PA. Electricity market auction settings in a future Danish electricity system with a high penetration of renewable energy sources e a comparison of marginal pricing and pay-as-bid. Energy 2011; 36: 4434–44.
Praktiknjo A. Sicherheit der Elektrizitätsversorgung im Spannungsfeld der energiepolitischen Ziele Wirtschaftlichkeit und Umweltverträglichkeit, PhD thesis, Berlin, 2013.

www.ingramcontent.com/pod-product-compliance
Lightning Source LLC
LaVergne TN
LVHW082113200926
848484LV00008B/1945